THE
INTERNATIONAL SERIES
OF
MONOGRAPHS ON PHYSICS

GENERAL EDITORS

R.K. ADAIR
H. EHRENREICH
R.J. ELLIOT
D.H. WILKINSON

INTRODUCTION TO
PHASE TRANSITIONS
AND CRITICAL
PHENOMENA

BY
H. EUGENE STANLEY
Boston University

OXFORD UNIVERSITY PRESS
New York   Oxford

Oxford University Press

Oxford New York Toronto
Delhi Bombay Calcutta Madras Karachi
Petaling Jaya Singapore Hong Kong Tokyo
Nairobi Dar es Salaam Cape Town
Melbourne Auckland

and associated companies in
Beirut Berlin Ibadan Nicosia

First published in 1971 by Oxford University Press, Inc.,
200 Madison Avenue, New York, New York 10016

First issued as an Oxford University Press paperback, 1987

Oxford is a registered trademark of Oxford University Press

Library of Congress Cataloging-in-Publication Data

Stanley, H. Eugene (Harry Eugene), 1941–
Introduction to phase transitions and critical
phenomena.

Reprint. Originally published: Oxford: Clarendon Press,
1971. (International series of monographs on physics)
Bibliography: p.
1. Phase transformations (Statistical physics)
2. Critical phenomena (Physics) I. Title. II. Series:
International series of monographs on physics (Oxford,
Oxfordshire)
[QC175.16.P5S72 1987]   530.4   87-12357
ISBN 0-19-505316-8 (pbk.)

2 4 6 8 10 9 7 5 3 1

Printed in the United States of America
on acid-free paper

TO IDAHLIA

PREFACE

THIS monograph is intended to serve as an introduction to the inter-disciplinary field of phase transitions and critical phenomena. It is a short book, and is not designed to review all of the recent developments in this rapidly-developing area. I have attempted, however, to provide an introduction that is sufficiently thorough that much of the current research literature can profitably be read.

The subject matter concentrates almost exclusively upon phenomena near the *liquid–gas* and *ferromagnetic* critical points, and the analogies between fluid and magnetic transitions are emphasized throughout the book. The decision not to treat in detail the phase transitions that occur in a wide variety of other systems was made in order that the novice reader (not familiar with, say, superfluidity) should appreciate the concepts underlying current research in critical phenomena.

I assume as background some familiarity with elementary thermo-dynamics and statistical mechanics, and I also assume that the reader has some notion of what a phase transition is. Therefore Chapters 1 and 2 are designed to provide a brief review and to establish the notation to be used; they may certainly be omitted by a number of readers. Likewise the chapters in Part III concerning the 'classical' van der Waals, mean field, and Ornstein–Zernike theories may be read quickly or else skipped altogether. Conversely, the material presented in certain of the appendices, and in Chapters 12–15, is more compact and is in-tended for the advanced student.

The material treated in this monograph proved to be adequate for a one-term course on phase transitions and critical phenomena given at MIT. I have also used the manuscript to supplement an introductory course in thermodynamics and statistical mechanics, and I found the material in Chapters 2–6 and 10–11 particularly useful. For those who may consider using portions of this book as supplementary material for courses in solid state physics or applied mathematics, I would recommend Chapters 6–9, 11, and 13–15. In so far as it has proved feasible, I have attempted to keep the chapters reasonably independent of one another so that the reader may skip about in the text if he wishes.

Since phase transitions and critical phenomena form an interdisciplinary field involving the work of chemists, mathematicians, physicists, and engineers, it has not been possible to find a notation that all classes of readers will find natural. To help in this respect, I have provided a list of symbols and their definitions in the 'notation guide' which follows the list of contents.

In writing this monograph I have continually regretted the fact that my aim of writing a short book has not permitted me to treat as many topics as I would otherwise have desired. Therefore, in the lists of *Suggested further reading* appearing at the ends of each chapter I have provided references to particularly readable works which should serve to extend the text along the lines I would have liked to. The bibliography at the end of the book includes only the articles referred to in the text; a considerably more extensive bibliography will appear in my companion volume, *Readings in Phase Transitions and Critical Phenomena*.

A large number of people have generously assisted with the preparation of the manuscript. First and foremost of these is my wife, Idahlia, to whom this volume is dedicated. I am deeply moved by the prodigious efforts of my research students, who have spent many hours of work on the manuscript. Particular sections were either written or re-written by Gerald Paul, Sava Milošević, Richard Krasnow, Charles Gordon, Frederic Harbus, Harvey Botman, and Jørgen Randers. Alexander Hankey solved all the exercises in the original lecture notes, and many of his solutions are incorporated into the present manuscript. David Njus prepared the drawings, and by his alertness and ingenuity in many cases improved their pedagogical value. The entire manuscript was also read carefully by Kenneth Ogan, Arthur Cook, Jill Punsky, Koichiro Matsuno, Jeffrey Golden, Nihat Berker, Douglas Karo, Stephen Schwartz, Judith Herzfeld, David Lambeth, Howard Lee, and Richard Lucash. The subject index, author index, and notation guide were kindly prepared by J. Punsky, S. Milošević, and K. Matsuno respectively.

I wish to express my appreciation to many of my professional colleagues who have supplied me with information and help in the preparation of this book. I have the particular pleasure of thanking Drs. A. J. Guttmann, P. C. Hohenberg, and J. B. Lastovka and Professors G. B. Benedek, H. Z. Cummins, M. H. Edwards, R. B. Griffiths, B. Jancovici, E. H. Lieb, J. D. Litster, J. E. Mayer, G. Sposito, G. Stell, K. Stierstadt, and L. Tisza for their thoughtful criticisms of the lecture notes

from which this monograph has developed. I am also greatly indebted
to Professor Robert B. Griffiths and to four former teachers, Professors
Max Delbrück, Thomas A. Kaplan, Charles Kittel, and J. H. Van
Vleck, for setting extremely high examples of intellectual honesty and
clarity.

It is a pleasure to thank Professor Charles Kittel and the Physics
Department of the University of California, Berkeley, for their hospi-
tality during the academic year 1968–1969, and to thank the Miller
Institute for Basic Research in Science for financial support in the form
of a postdoctoral fellowship. This monograph developed from a set of
lecture notes prepared for 'Physics 290g', spring quarter, 1969. To the
students and faculty who attended these lectures, and to those MIT
students who attended various sets of lectures given at this institution, I
wish to express my gratitude for many stimulating discussions which
contributed significantly to my own understanding of the subject.

I want to thank Mrs. Vera Sarantakis, Mrs. Janet Pollock, and Miss
Susan J. Leonard for their diligence and patience in typing the final
manuscript, and to thank a countless number of secretaries at Berkeley
for having typed portions of the original lecture notes.

I am greatly indebted to the staff of the Oxford University Press for
their gracious assistance in so many ways. I am also deeply appreciative
of the warm advice and thorough criticism of the early manuscript by
one of the series editors, Dr. Walter Marshall. Were it not for his
encouragement, the original lecture notes would never have developed
into the present monograph.

Severe self-criticism, even as I study the final page proof, means that
I must resist an urge to re-write the treatment of several topics that I
now realize can be explained more clearly and precisely. I do hope that
readers who notice these and other imperfections will communicate
their thoughts to me.

Cambridge, Massachusetts   H. E. S.

January 1971

CONTENTS

NOTATION GUIDE

FOR those cases in which a symbol has more than one meaning, we list the chapter(s) in which the symbol is used. Symbols are not listed here if they occur only in the immediate context of a statement defining them, and conventional mathematical symbols are not listed. The abbreviation c.p.e. denotes 'critical-point exponent'. The reader should notice that all inequalities are written in the form $x \geq y$; that we have denoted the four thermodynamic potentials by U, E, G, and A; that, while in most cases we use the same symbol for analogous fluid and magnetic quantities, the correlation functions are denoted respectively by $G(\mathbf{r})$ and $\Gamma(\mathbf{r})$; that different symbols are used for time-dependent quantities (e.g., $G(\mathbf{r})$ and $\mathscr{G}(\mathbf{r}, t)$); that for magnetic systems we adhere to the relation $G(T, H) = -kT \ln Z(T, H)$ (Wannier 1966); that the energy of parallel spins is $-J$ except in Chapter 6 (where it is $-2J$), and that small arrows (►) denote equations referred to frequently.

Symbol	Meaning	Page	Used only in chapter
$a\,[a']$	c.p.e. for thermal conductivity $T > T_c\,[T < T_c]$	203	
a, ℓ	parameters in van der Waals equation	68	
$A(T, V)\,[A(T, M)]$	Helmholtz potential for fluid [magnet]	22	
$\mathscr{A}\,[\mathscr{A}']$	coefficient for specific heat, $T > T_c\,[T < T_c]$	44	
a_ϵ, a_H	scaling parameters	181	
a_M, a_S	scaling parameters	274	
a_ℓ, b_ℓ, c_ℓ	coefficients of series expansions	137, 148	9
$b\,[b']$	c.p.e. for shear viscosity, $T > T_c\,[T < T_c]$	203	
B	coefficient of Botch–Fixman correction	225	
$\mathscr{B}$	coefficient for order parameter	10, 42	
$B_S(y)$	Brillouin function	81	
$c\,[c']$	c.p.e. for bulk viscosity $T > T_c\,[T < T_c]$	203	
$\mathscr{C}\,[\mathscr{C}']$	coefficient for K_T, χ_T for $T > T_c\,[T < T_c]$	43	
$\mathbf{C}$	Curie constant	3, 83	
$C(\mathbf{r})$	'direct' correlation function	102	7
$\hat{C}(\mathbf{q})$	Fourier transform of $C(\mathbf{r})$	102	7
C_H	specific heat for constant magnetic field	32, 35	
C_M	specific heat for constant magnetization	35	
C_P	specific heat for constant pressure	25	
C_V	specific heat for constant volume	25	
C_V°	specific heat for non-interacting limit	75	

Symbol	Meaning	Page	Used only in chapter
d	dimensionality of lattice	46	
D	dimensionality of spin	111	
$\mathscr{D}$	coefficient for critical isotherm	43	
D_S	sound wave damping constant	211	
D_T	thermal diffusivity	211	
D_ℓ	$(mn)^{-1}(\frac{4}{3}\eta + \zeta)$	211	
e	energy density	275	
e_1	energy density measured from its equilibrium value	276	
$E(S, P)\,[E(S, H)]$	enthalpy for fluid [magnet]	22	
$\mathscr{E}$	electric field	206, 288	
$f(x),\ F(x)$	arbitrary functions	—	
$\mathscr{F}$	shape function	234	
g	Landé factor	80	
$\mathbf{g}$	momentum density	274	
$g(\ell)$	number of graphs with ℓ lines	142	
$G(T, P)\,[G(T, H)]$	Gibbs potential for fluid [magnet]	22	
$\bar{G}$	Gibbs potential per particle	92	
$G(\mathbf{r})$	static pair correlation function, fluid	46, 95	
$\mathscr{G}_{nn}(\mathbf{r}, t) = \mathscr{G}(\mathbf{r}, t)$	density–density correlation function	204	
$\mathscr{G}_{hn}(\mathbf{r}, t)$	heat density–number density correlation function	278	
h	Planck's constant	94	7
h	tanh $(\bar{\mu}H/2kT)$	84	6
h	dimensionless magnetic field, site model	131, 191	8, 12
h	heat energy density	276	App. D
$\hbar$	scaled magnetic field	186	
H	magnetic field	8	
$\tilde{h}$	effective magnetic field, cell model	192	12
$\mathscr{H}$	Hamiltonian	80	
$\mathscr{H}^{(D)}$	Hamiltonian for isotropically-interacting D-dimensional spins	111	
H_c°	kT_c/m_0	43	
H_{eff}	effective magnetic field	82	
$\tilde{H}$	$\bar{\mu}SH/kT$, normalized magnetic field	88	
i, j, k, m, n	lattice sites	—	
$I(\mathbf{q})$	intensity of scattered radiation, static case	99	
$\mathscr{I}(\mathbf{q}, \omega)$	intensity of scattered radiation, dynamic case	209	
$I^\circ(\mathbf{q})$	intensity of scattered radiation in non-interacting limit	99	
I_B	intensity of Brillouin component	213	
I_R	intensity of Rayleigh component	213	
J or J_{ij}	exchange energy	90, 111	
$\mathscr{J}$	J/kT	92	
$\mathscr{J}_c$	J/kT_c	152	

Symbol	Meaning	Page	Used only in chapter
$\mathbf{j}_e$	energy current density	275	
k	Boltzmann constant	3	
$\mathbf{k}_0$	wave vector of incident radiation	99	
$\mathbf{k}_S$	wave vector of scattered radiation	99	
K_S	adiabatic compressibility	26	
K_T	isothermal compressibility	3, 25	
K_T°	isothermal compressibility for an ideal gas	43	
L	length of cell in Kadanoff construction	191	12
$L_j(T)$	Landau expansion coefficient of the Helmholtz free energy	168	
ℓ_{jk}	expansion coefficient of $L_j(T)$	169	
ℓ	dummy index of summation; integer	—	
m	mass of a molecule	3	
$M = M(T, H)$	magnetization	8	
$M_H(T)$	constant-field magnetization	42	3, 4
$M_T(H)$	constant-temperature magnetization	62	4
M_0	$M(T = 0, H = 0)$	81	6
m	scaled magnetization	186	
$\mathcal{M}$	magnetization operator	36	
m_0	magnetic moment per spin	43	
n	N/V or $\langle n(\mathbf{r}) \rangle$	26, 95	
$n(\mathbf{r})$	number density at point $\mathbf{r}$	94	
n_1	number density measured from its equilibrium value	276	
n	N/N_A, number of moles	67	5
n	index of refraction	206	13
N	total number of particles (or spins)	26	
N_A	Avogadro's number	67	
p	$(P - P_c)/P_c$, dimensionless pressure	74	5
P	pressure	1	
$\check{P}$	P/P_c, normalized pressure	72	
P_c	critical pressure	2	
P_c°	pressure of ideal gas at $\rho = \rho_c$, $T = T_c$	43	
$\mathcal{P}$	number of nearest neighbour pairs in a lattice	139	
$\mathcal{P}_D^N$	Padé approximant of order $[D, N]$	162	
q	coordination number ($\equiv$ number of nearest neighbours)	90	
$\mathbf{q}$	momentum transfer vector	99	
r	distance	46	
r_c	radius of convergence	153	
R	Debye persistence length	104	
$\mathcal{R}$	kN_A, ideal gas constant	67	
s_i	Ising spin on site i	91	
$\tilde{s}_\alpha$	magnetic moment of cell α	192	

Symbol	Meaning	Page	Used only in chapter
S	entropy	22	
S	spin quantum number	80	
$\mathbf{S}_i$	vector spin on site i	80	
$S_{nn}(\mathbf{q}) = S(\mathbf{q})$	static structure factor; Fourier transform of $G(\mathbf{r})$	100	
$\mathscr{S}_{nn}(\mathbf{q}, \omega) = \mathscr{S}(\mathbf{q}, \omega)$	dynamic structure factor; Fourier transform of $\mathscr{G}_{nn}(\mathbf{r}, t)$	205	
$\mathscr{S}_{nn}^{L}(\mathbf{q}, \omega)$	Fourier–Laplace transform of $G_{nn}(\mathbf{r}, t)$	278	
t	time	204	
T	temperature	1	
$\tilde{T}$	T/T_c, normalized temperature	72	
T_c	critical temperature	2	
T_{c1}	critical temperature for divergence of susceptibility	119	
T_{c2}	critical temperature for spontaneous magnetization	119	
T_N	critical temperature for antiferromagnet	11	
T_λ	λ temperature for ^{4}He	19	
$\mathbf{T}$	stress tensor	275	App. D
$\mathbf{T}$	transfer matrix	132	8
u	$\exp(-4J/kT)$	158	9
$U(\mathbf{r})$	interparticle potential	76, 96	
$U(S, V) [U(S, M)]$	internal energy for fluid [magnet]	22	
v	$(V - V_c)/V_c = \tilde{V} - 1$	74	5
v	$\tanh(J/kT)$	139	8, 9
v	c/n, velocity of light in medium	206	13, 14
v_s	sound velocity	211	
V	volume	22	
V_c	critical volume	71	
$\bar{V}$	V/N	68	
$\tilde{V}$	V/V_c, normalized volume	72	
$\mathbf{v}_1$	velocity measured from its mean value	276	
w	transition probability per unit time	279	
x	$\bar{\mu}H/kT$	80	6
z	homogeneity parameter	235	15
z	$1 - \tanh(J/kT)$	164	9
Z	partition function	77	
Z_c	$P_c V_c/\mathscr{R}T_c$	72	
$\mathscr{Z}$	grand partition function	94	
$\alpha [\alpha']$	c.p.e. for C_H and C_V, $T > T_c [T < T_c]$	44	
α	characteristic inverse time	283	App. E
α_P	coefficient of thermal expansion, $V^{-1}(\partial V/\partial T)_P$	26	
α_H	$(\partial M/\partial T)_H$	35	
α_M	$(\partial H/\partial T)_M$	36	

Symbol	Meaning	Page	Used only in chapter
β	$1/kT$	77	
β	c.p.e. for $\rho_L - \rho_G$ and $M(T, H = 0)$	10, 42	
β_G	c.p.e. for $\rho_c - \rho_G$	4	
β_L	c.p.e. for $\rho_L - \rho_c$	4	
$\tilde{\beta}$	β/β_c, reduced inverse temperature	153	9
$\gamma\,[\gamma']$	c.p.e. for K_T and χ_T, $T > T_c\,[T < T_c]$	3, 43	
$\Gamma(\mathbf{r})$	pair correlation function, magnet	45	
$\Gamma(\mathbf{r})$	dimensionless correlation function, fluid	102	7
Γ_B	half-width of Brillouin component	211	
Γ_R	half-width of Rayleigh component	211	
Γ_L	relaxation rate for longitudinal spin fluctuations	244	15
δ	c.p.e. for critical isotherm	3, 43	
δ_s	$H \sim M^{\delta_s}, S = S_c$	62	
$\Delta_{2\ell}[\Delta_\ell']$	gap exponent, $T > T_c\,[T < T_c]$	50	
ϵ	$(T - T_c)/T_c$, normalized temperature	4, 10	
$\epsilon_T[\epsilon_\eta;\,\epsilon_s]$	crossover temperature between regions 1, 2 [2, 3; 3, 4]	254	
ζ	$S \sim -M^{\zeta+1}, T = T_c$	55	
ζ	bulk viscosity	203	
ζ_0	bulk viscosity far from critical point	247	
η	$G(r) \sim r^{-(d-2+\eta)}, T = T_c$	46	
η_E	$\Gamma_{EE}(r) \sim r^{-(d-2+\eta_E)}, T = T_c$	62	
η	shear viscosity	203	
η_0	shear viscosity far from critical point	247	
η^*	$\eta(q = \kappa, \Omega = \Omega_\eta^*)$	251	
θ	vapour pressure curvature exponent	51	
κ	$1/\xi$, inverse correlation length	105	
κ_0	coefficient of inverse correlation length	214	
λ	arbitrary critical-point exponent	39	3
λ	molecular field parameter	82	6
λ	wavelength	214	
λ_s	c.p.e. for singular part of function	41	
$\lambda_\pm$	eigenvalues of transfer matrix	132	
Λ	thermal conductivity	203	
Λ_0	thermal conductivity far from critical point	247	
Λ^*	$\Lambda(q = \kappa, \Omega = \Omega_T^*)$	251	
μ	chemical potential	33	
μ_B	Bohr magneton	80	
$\bar{\mu}$	$g\mu_B$	80	
$\nu\,[\nu']$	c.p.e. for correlation length for $T > T_c\,[T < T_c]$	46	
ξ	correlation length	5, 46	

Symbol	Meaning	Page	Used only in chapter
$\xi_0\,[\xi'_0]$	coefficient of correlation length for $T > T_c\,[T < T_c]$	46	
ρ	mN/V, mass density	1	
ρ_c	critical density	2	
ρ_G	density at gas side of coexistence curve	3	
ρ_L	density at liquid side of coexistence curve	3	
σ	$T - T_c \sim M^\sigma$, $S = S_c$	62	4
σ	$M(T, H)/M(0, 0)$, reduced magnetization	84	
τ	relaxation time	285	
φ	$C_H \sim H^{-\varphi}$, $T = T_c$	55	
χ_T	isothermal susceptibility	8	
χ_S	adiabatic susceptibility	35	
$\bar{\chi}_T$	χ_T/χ_T^0, dimensionless susceptibility	147	9
χ_T^0, χ_{Curie}	isothermal susceptibility in non-interacting limit	43, 147	
ψ	$S \sim -H^\psi$, $T = T_c$	55	
ω_B	Brillouin frequency	211	
ω_c	characteristic frequency	234	15
Ω	relaxation rate	249	
Ω^*	$\Omega(\epsilon, q = \kappa)$	251	

1

WHAT ARE THE CRITICAL PHENOMENA?
A SURVEY OF SOME BASIC RESULTS

O u r purpose in this first chapter is to provide the reader with a brief qualitative description of what actually happens near the critical point, and with an appreciation of why the study of these critical phenomena has burgeoned in recent years.

1.1. Classical era of critical phenomena

Many of the basic facts of critical phenomena were observed fifty or even a hundred years ago, while other aspects have been discovered only in very recent years. Hence it has become customary to divide the field of critical phenomena into an early 'classical' era and a more recent 'modern' era.

Although a wide variety of physical systems exhibit critical phenomena, we shall concentrate on the liquid–gas and magnetic critical points for the sake of simplicity.

1.1.1. *Fluid systems*

Our discussion of fluid systems will focus first on the equation of state, a functional relationship of the form $f(P, \rho, T) = 0$, which relates the thermodynamic parameters—pressure, density, and temperature. The equation of state thus defines a surface in a three-dimensional space whose coordinates are P, ρ, T; each of the points on this surface corresponds to an equilibrium state of the system. In order to aid in visualization of this $P\rho T$ surface, it is convenient to consider its projections on to the PT, $P\rho$, and ρT planes; these are shown schematically in Figs. 1.1(a), 1.2(a), and 1.3(a) respectively. We see from Fig. 1.1(a) that the projection on to the PT plane produces three separate regions, corresponding to the three familiar phases of matter—the solid, liquid, and gaseous phases. The solid and gaseous phases are in equilibrium along the *sublimation curve*, the solid and liquid phases are in equilibrium along the *fusion curve*, and the liquid and gaseous phases are in equilibrium along the *vapour pressure curve*. Each point on these three curves represents an equilibrium state in which two or more phases can coexist—and the *triple point* represents an equilibrium state in which all three phases coexist.

We notice, however, that the vapour pressure curve does not extend forever, as the fusion curve appears to do, but rather that it terminates in a point. This point is called the *critical point*, and its coordinates are

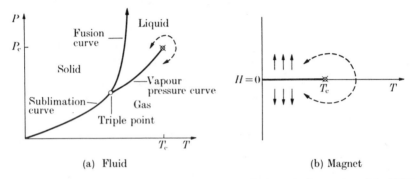

(a) Fluid   (b) Magnet

Fig. 1.1. (a) Projection of the PVT surface in the PT plane. (b) Projection of the HMT surface in the HT plane.

denoted by (P_c, ρ_c, T_c), where P_c, ρ_c, and T_c are the critical pressure, critical density, and critical temperature respectively. The fact that the vapour pressure curve terminates in a critical point means that one can convert a liquid to a gas continuously, without crossing the phase transition line, as is indicated by the dotted path shown in Fig. 1.1(a). In this sense there is no fundamental difference between the liquid and gaseous phases. It is widely believed that the fusion curve does *not* also terminate in a (second) critical point. However, thus far it has not been possible to prove the non-existence of a liquid–solid critical point.

That the vapour pressure curve terminates in a critical point was not appreciated until about a hundred years ago. Prior to that time scientists regarded certain gases as being 'permanent' in the sense that these gases could not be made to condense no matter how much pressure was applied (compressing a gas was a standard laboratory procedure for liquefaction). Presumably the work on these gases was carried out at a temperature $T > T_c$, while a prerequisite for obtaining droplets of the condensed phase is that the material be brought to a temperature lower than the critical temperature T_c. Thus, for example, helium cannot be liquefied no matter how great the applied pressure unless the temperature is reduced below the critical temperature of $5 \cdot 2K$.

In addition to the PT projection, it is also useful to consider the projections in the $P\rho$ and ρT planes; these are shown schematically in Figs. 1.2(a) and 1.3(a) respectively. These projections tell us a great deal

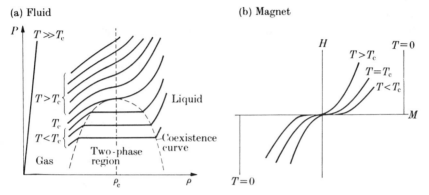

FIG. 1.2. (a) Isothermal cross-sections of the PVT surface (or, literally, the $P\rho T$ surface). The exponent δ is a measure of the degree of the critical isotherm. The fact that $\delta > 1$ reflects the fact that the compressibility K_T is infinite at T_c. The fashion in which K_T diverges at $T \to T_c$ is described by the exponents γ and γ'. (b) Isothermal cross-sections of the HMT surface. At high temperatures all isotherms approach straight lines, corresponding to the non-interacting limits, $P = \rho kT/m$ and $H = TM/\mathbf{C}$, where $\mathbf{C}$ is Curie's constant (cf. Chapter 6).

concerning the qualitative features of the critical point. From Fig. 1.2(a) and also from Fig. 1.3(a) we see that at low temperatures there is a rather large difference between the liquid and gas densities, ρ_L and ρ_G, but that as the critical temperature is approached this density difference tends to zero. The existence of a quantity which is non-zero below the critical temperature and zero above it will be seen to be a common feature associated with the critical points of a wide variety of physical systems. We say that $\rho_L - \rho_G$ is the *order parameter* for the liquid–gas critical point.

A second striking feature of Fig. 1.2(a) concerns the shape of the isotherms as the critical point is approached. At very high temperatures the ideal gas law is obeyed and the isotherms are therefore the straight lines given by the ideal gas equation of state

$$P = \rho kT/m, \qquad (1.1)$$

where k is the Boltzmann constant and m is the mass of a molecule. Hence one might suspect that the curvature of the isotherms which becomes apparent as T decreases towards the critical temperature is a manifestation of interactions among the constituent molecules of the fluid.

This is in fact the case, and we can discuss what is going on as the critical point is approached rather more easily if we utilize an analogy between a fluid and a ferromagnetic system, the lattice-gas model (see Appendix A for a more detailed discussion). According to this model, we imagine that the macroscopic volume V containing the fluid is partitioned into fixed microscopic cells whose volume v is roughly the size

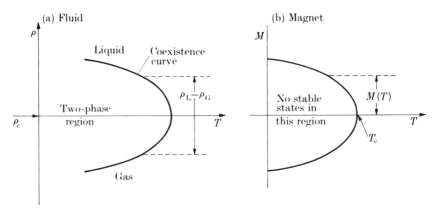

FIG. 1.3. (a) Projection of the $P\rho T$ surface in the ρT plane. Note that as $T \to T_c$ from below, the vapour density ρ_G increases and the liquid density ρ_L decreases. Sometimes one writes $\rho_c - \rho_G \sim |\epsilon|^{\beta_G}$ and $\rho_L - \rho_c \sim |\epsilon|^{\beta_L}$. However, in practice $\beta_G \simeq \beta_L$ and we may simply write $\rho_L - \rho_G \sim |\epsilon|^{\beta}$. (b) Projection of the HMT surface in the MT plane. Note that the spontaneous magnetization M plays the role of $\rho - \rho_c$. Unlike the fluid case, this projection is the same as the $H = 0$ cross-section of the HMT surface.

of the constituent molecules of the fluid. Next we construct an analogous magnetic system by considering each cell to be a lattice site on which a magnetic moment is situated. This magnetic moment is chosen to point in the upward direction if the corresponding cell of the fluid is occupied by the centre of a molecule, and the magnet is assumed to be pointing downward otherwise. Thus for temperatures well above the critical temperature, the free motion of the gas molecules of our fluid system will correspond to a rapid and random 'flipping' of the magnetic moments from one orientation (up or down) to the other. However, as the temperature is lowered toward the critical point, small 'droplets' of correlated spins appear. As the critical point is approached still closer, the droplets grow in their dimensions as shown in Fig. 1.4. In fact, it is experimentally feasible to bring a fluid close enough to its critical point for these droplets to acquire lateral dimensions on the order of the wavelength of light, whereupon the light is scattered strongly. This phenomenon, called *critical opalescence*, was discovered by Andrews just over a century ago in the course of his measurements

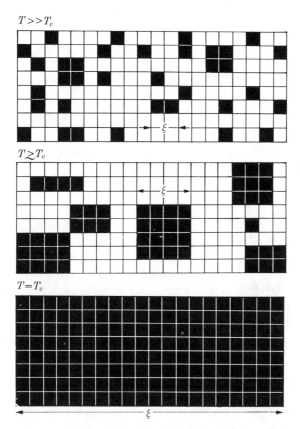

FIG. 1.4. Sketch of the lattice-gas model of a fluid system as the temperature approaches the critical temperature. Each cell is coloured black if it is occupied by the centre of a molecule and we associate an 'up' spin with this site. The correlation length ξ may be thought of as being roughly the diameter of an 'island' of aligned spins. By courtesy of D. L. Njus.

of the critical behaviour of carbon dioxide. To this day it provides one of the more striking manifestations of the critical point.

Fig. 1.4 may be misleading in that we may get the impression that the system becomes completely ordered as T approaches T_c from above (in the sense that all the moments are oriented parallel to one another). Of course this is not true; the system is completely ordered only at a temperature of absolute zero, and the net order (measured by the order parameter, which is the spontaneous magnetization in this case—cf. Fig. 1.3(b)) is zero at the critical temperature $T = T_c$. Hence Fig. 1.4 is to be interpreted as a small portion of a large system. To clarify this remark further, Fig. 1.5 shows a system consisting of a two-dimensional lattice of cells with 64 cells on each side (so that there are altogether $2^{12} = 4096$ cells). This sequence of diagrams may be interpreted as representing the approach to the critical point from the low-temperature side. Thus (a) represents the completely ordered state at $T = 0$ while in (b) the temperature of this two-dimensional lattice gas might be roughly $\frac{1}{4}T_c$. We notice that relatively few of the spins have flipped from their up-configuration to their down-configuration, and this fact corresponds to the extreme flatness of the Ising model magnetization curve (cf. Fig. 9.9). In (c) the temperature is, say, $\frac{1}{2}T_c$ and a larger number of spins have flipped orientations. In (d) the temperature is getting close to T_c, and we observe that the overturned spins appear to cluster together in small 'islands'. In (e) $T \simeq T_c$, while in (f) $T \gtrsim T_c$ and the islands of aligned moments are quite large. Notice that even though above the critical temperature the net magnetization is zero (i.e. there are roughly as many up spins as there are down spins), nevertheless there exists a considerable degree of 'order' in the system, as reflected in the large dimension of the islands of aligned spins. This order is frequently called *short-range order* in order to distinguish it from spontaneous magnetization or *long-range order*.

The phenomenon of critical opalescence is captured in the sequence of six photographs shown in Fig. 1.6. In part (a) the temperature is well above T_c and we see the single, clear phase. The fluctuations are weak and there is no appreciable light scattering. In parts (b), (c), and (d) the temperature is successively lowered and we see that the sample tube takes on a dramatic glow. In (e) the temperature is just below T_c and although a meniscus between the liquid and gaseous phases is apparent there are still liquid globules of higher density raining out of the gas phase above the meniscus, and vice versa. Finally, (f) shows complete separation of the liquid and gas phases.

The concept of density changes near the critical point is illustrated in a sequence of four photographs in Fig. 1.7. These show a sealed glass bulb which has been filled with a quantity of carbon dioxide chosen such that the average density inside the bulb is very close to the critical

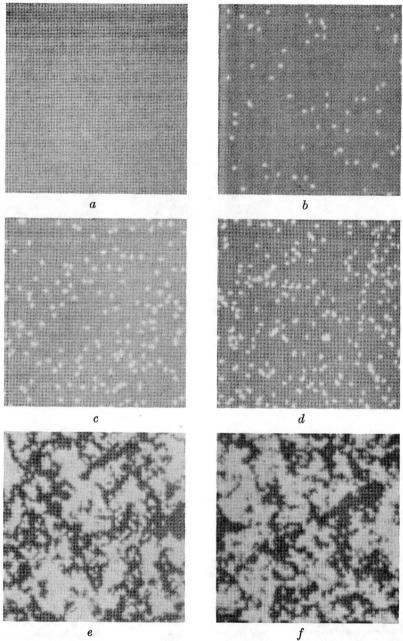

FIG. 1.5. Schematic indication of the lattice-gas model of a fluid system. Fig. 1.4 (which contains 8 × 20 cells) should be interpreted as representing a relatively small portion of this lattice (which contains 64 × 64 cells). (a) is the completely ordered state (which exists only at $T = 0$); (b), $T \simeq \frac{1}{4}T_c$; (c), $T \simeq \frac{1}{2}T_c$; (d), $T \simeq \frac{3}{4}T_c$; (e), $T \simeq T_c$; (f), $T \gtrsim T_c$. This illustration and the associated temperatures are to be regarded as purely *schematic*. In fact, the figure was constructed from a computer simulation of the time-dependent aspects of the two-dimensional Ising model and actually represents rather different phenomena (cf. Appendix E). After Ogita *et al.* (1969).

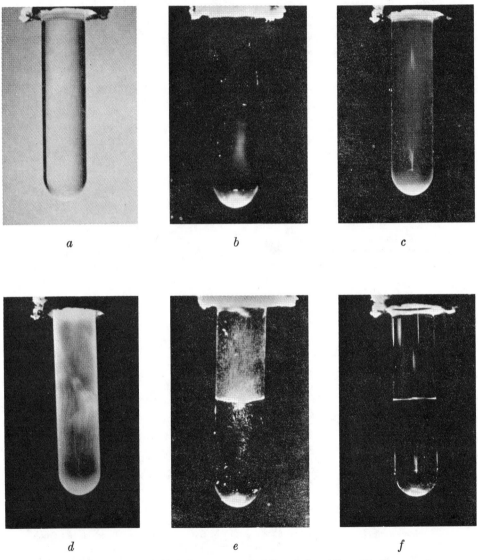

FIG. 1.6. Behaviour of a fluid as the temperature is lowered past the critical temperature: (a), $T \gg T_c$; (b), $T \gtrsim T_c$; (c), $T \simeq T_c$; (d), $T \lesssim T_c$; (e), $T < T_c$; and (f), $T \ll T_c$. The fluid shown is the binary mixture cyclohexane-aniline. After Ferrell (1968).

density ρ_c. Also inside the bulb are three balls with pre-determined densities that are slightly less than, approximately equal to, and slightly larger than ρ_c. In (a), the temperature is greater than T_c and the fluid consists of a single, relatively homogeneous 'phase' with slight density variations due to gravity. Hence the heaviest ball is at the bottom and the lightest ball is at the top of the tube. In (b) the temperature is lowered to just above T_c and we observe the critical opalescence as the entire tube takes on a milky white colour. Since the compressibility is large, the density distribution is very sensitive to pressure gradients, and the middle ball—in this photograph—is not precisely at the centre of the tube. In (c) the temperature has fallen just below T_c and a meniscus is seen to form, while in the final stage (d) the temperature is well below T_c and all three balls float at the, by now, well-defined meniscus. The fact that this meniscus forms near the centre of the tube supports the classic *law of rectilinear diameter* according to which $\frac{1}{2}(\rho_L + \rho_G) \simeq \rho_c$.

After the preceding qualitative discussion, it should come as no surprise that the isotherms in Fig. 1.2(a) indicate a significant departure from the linear behaviour predicted by the ideal gas law; we are now in a position to understand something of the nature of this deviation. Specifically, we see from Fig. 1.2(a) that the isotherms actually acquire a flat portion in the immediate vicinity of the critical point, i.e. the slope $(\partial P/\partial\rho)$ becomes zero as $T \to T_c^+$. Now the isothermal compressibility K_T of a fluid is defined to be $\rho^{-1}(\partial\rho/\partial P)_T$, and hence the flat portion of the isotherm in Fig. 1.2(a) corresponds to an isothermal compressibility which diverges to infinity as the critical point is approached. Of course an infinite value of $(\partial\rho/\partial P)_T$ means that the response of the density to a very small pressure fluctuation is infinite. Thus we might expect that this divergence in the isothermal compressibility is connected to the huge density fluctuations which we associated with critical opalescence. In fact, we shall see in Chapter 7 that the density fluctuations are mathematically related in a direct fashion to the isothermal compressibility and hence to the derivative $(\partial\rho/\partial P)_T$.

Thus far our discussion of the classical era of fluid critical phenomena has centred on experimental results. This is not to say that there were no theories, however. In fact, barely three years had passed after Andrews' classic 1869 publication on the critical point of carbon dioxide when van der Waals, in his Ph.D. dissertation, published a theoretical description of the critical region which to this day provides an intriguingly accurate description of the critical region for

temperatures that are not too close to T_c (cf. Chapter 5). The phenomenon of critical opalescence was also discussed in the early years of the
twentieth century by many theoreticians—among them von
Smoluchowski, Einstein, Ornstein, and Zernike.

1.1.2. *Magnetic systems*

It is customary to stress the analogies between magnetic and fluid
transitions for pedagogical reasons. For example, if we apply pressure
to a fluid system the density increases, while if we apply a magnetic
field H to a ferromagnetic system the magnetization M increases. Hence
in a sense H is analogous to P and M to ρ and the equilibrium $P\rho T$
surface of a fluid system corresponds, after a fashion, to the HMT
surface of a magnetic system.

Figs. 1.1(b), 1.2(b), and 1.3(b) show the projections of the HMT surface on to the HT, HM, and MT planes respectively. Much of the
qualitative discussion presented above also applies to the magnetic
system. In particular, the response function K_T is analogous to the
isothermal susceptibility $\chi_T \equiv (\partial M / \partial H)_T$, which approaches infinity
near T_c corresponding to the 'flattening' of the critical isotherm
($T \equiv T_c$) in Fig. 1.2(b). Associated with this divergent response function is an enormous increase in the fluctuations of the magnetization,
and a relation, analogous to that for the fluid system, relates χ_T to
these magnetization fluctuations (cf. Appendix A).

In the early years of the twentieth century, theoretical progress was
made in understanding magnetic transitions, and the approach taken
was not altogether unlike the classic work of van der Waals on fluid
systems. In 1907, just a few years after the pioneering experimental
work of Curie, Hopkinson, and others, Pierre Weiss proposed a theory
of ferromagnetism in which he assumed that the constituent magnetic
moments interact with one another through an artificial 'molecular
field' which is proportional to the average magnetization (see Chapter
6). More specific models of the interacting magnetic moments were
put forward some years later. These models have in common the feature
that they assume the magnetic moments to be localized on fixed lattice
sites and that they influence one another through pairwise interactions
with an energy that achieves its maximum value, J, when the moments
are parallel.

Two particular forms of the interaction are particularly interesting
to this day. In the first of these, due to Wilhelm Lenz but called the
Ising model, the magnetic moments are assumed to be classical, one-

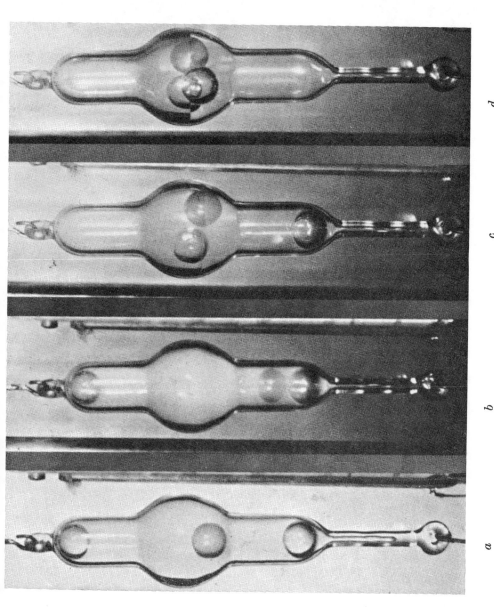

Fig. 1.7. Behaviour of carbon dioxide near the critical point. The glass tube is filled with an average density very close to the critical density ρ_c. The three balls have densities slightly less than, about equal to, and slightly greater than ρ_c. (a), $T \gg T_c$; (b), $T \gtrsim T_c$; (c) $T \lesssim T_c$; and (d), $T \ll T_c$. From Sengers and Levelt-Sengers (1968).

dimensional 'sticks' capable of only two orientations. Thus the Ising model is the magnetic model analogous to the lattice-gas model of a fluid system discussed in the previous subsection.

A second model, called the Heisenberg model, regards the magnetic moments as being related to *quantum-mechanical* three-component spin operators, and assumes that the energy is proportional to the scalar product of these operators. Although the original Weiss model of a magnetic system is quite easy to solve exactly, neither the Ising nor the Heisenberg model has yielded as yet to an exact solution for a three-dimensional lattice. Nevertheless, these models appear to represent reasonable theoretical descriptions of certain physical systems, and therefore their continued study has provided us with considerable insight regarding magnetic phase transitions.

1.2. Modern era of critical phenomena

1.2.1. *Critical point exponents*

There is no clear-cut development which delineates the beginning of the modern era of critical phenomena. Some might say that it began in the 1940s with Guggenheim's realization that the coexistence curve of a fluid system is not parabolic and with Onsager's exact solution of the two-dimensional Ising model, while others might set the date somewhat later, for example in the early 1960s when Heller, Benedek and Jacrot (together with theoreticians such as Domb, Rushbrooke, Fisher, and Marshall) came to recognize that certain entities, the critical point exponents, were significant entities worthy of special attention in their own right.

In any case, our brief introduction to the modern era will begin by considering the subject of critical-point exponents. Consider, for example, the now classic Guggenheim plot, reproduced in Fig. 1.8, which shows the temperature dependence of the liquid–gas density difference $\rho_L - \rho_G$ for eight different simple fluids (Ne, A, Kr, Xe, N_2, O_2, CO, and CH_4). We call $\rho_L - \rho_G$ the *order parameter* because it is non-zero only in the ordered phase. The fact that the data, properly normalized, fall on one and the same curve is in accord with the *law of corresponding states*. What is interesting is that the shape of the solid curve—supposedly a 'best fit' to the data—is a cubic function rather than the quadratic function that the van der Waals theory would predict.

For the analogous magnetic system the appropriate order parameter

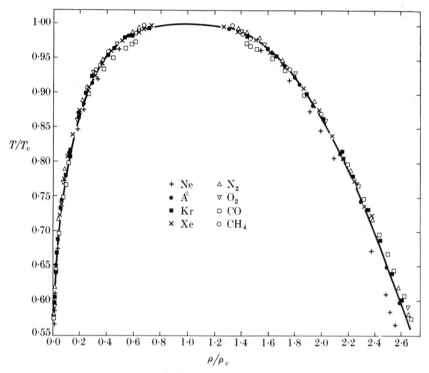

FIG. 1.8. Measurements on eight fluids of the coexistence curve (a reflection of the $P\rho T$ surface in the ρT plane analogous to Fig. 1.3). The solid curve corresponds to a fit to a cubic equation, i.e. to the choice $\beta = \frac{1}{3}$, where $\rho - \rho_c \sim (-\epsilon)^{\beta}$. From Guggenheim (1945).

is the zero-field magnetization M because M is a measure of the degree to which the magnetic moments are aligned throughout the crystal. Here again the classic Weiss theory predicts a quadratic dependence $M^2 \propto (T_c - T)$, whereas we see that $M^3 \propto (T_c - T)$ would seem to be an appropriate fit to the measurements of Heller and Benedek shown in Fig. 1.9.

It is customary to say that the order parameter varies as $(-\epsilon)^{\beta}$ where

$$\epsilon \equiv \frac{T - T_c}{T_c} \tag{1.2}$$

and where the critical-point exponent β typically has a value in the range 0·3–0·5. It is important to stress that it is not necessary to have a strict proportionality between the order parameter and $(-\epsilon)^{\beta}$ in order to be able to define a critical-point exponent. In fact, if we knew that a simple relation of the form $M = \mathscr{B}(-\epsilon)^{\beta}$ were valid, then three measurements in the critical region would suffice to determine the exponent β! In practice there are frequently correction terms, so that M might have the form $\mathscr{B}_0(-\epsilon)^\beta\{1 + B(-\epsilon)^x + \ldots\}$ with $x > 0$. Hence a more natural definition of the critical-point exponent β is

$$\beta \equiv \lim_{\epsilon \to 0} \frac{\ln M}{\ln(-\epsilon)}, \tag{1.3}$$

where the correction terms will drop out in taking the limit. In fact, critical-point exponents are frequently determined by measuring the slopes of log–log plots of experimental data, since l'Hospital's rule, together with eqn (1.3), implies that $\beta = d(\ln M)/d\{\ln(-\epsilon)\}$. Although this is a particularly quick method of determining the exponent, it requires a prior knowledge of the critical temperature, so that in practice one must frequently resort to plotting $M^{1/\beta}$ for several trial values of β until a value is found which produces a straight line.

At one time many workers believed that all materials have the same exponents. For example, we remarked above that all eight fluids shown in the Guggenheim plot, Fig. 1.8, appear to have roughly the same exponent, $\beta \simeq \frac{1}{3}$. Hence it was rather satisfying when the first accurate measurements of β for a magnetic system, those of Heller and Benedek in 1962, produced the value $\beta = 0{\cdot}335 \pm 0{\cdot}005$ (cf. Fig. 1.9), and

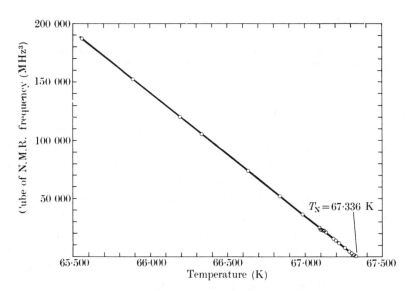

FIG. 1.9. Dependence upon temperature of the cube of the zero-field magnetization for MnF_2. Since MnF_2 is an antiferromagnet instead of a ferromagnet, the critical temperature is denoted by T_N rather than by T_c. After Heller and Benedek (1962).

subsequent measurements on other magnetic systems also appeared to yield similar values of β. However, this once-hoped-for universality has yet to be more rigorously demonstrated, and there now exists a growing list of materials for which $\beta = \frac{1}{3}$ is definitely outside the experimental error. For example, particularly accurate measurements supporting $\beta = 0.354$ for helium are shown in Fig. 1.10.

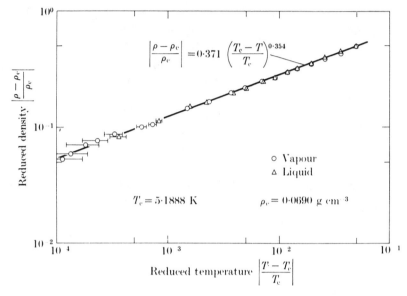

FIG. 1.10. Measurements of the coexistence curve of helium in the neighbourhood of its critical-point. The critical-point exponent β has a value of 0.354. After Roach (1968).

For some of these materials, however, the source of the discrepancy may be due to complicating factors such as the lattice compressibility. In Fig. 1.11, for example, are shown some recent data on the magnetization of CoO, which contracts suddenly on cooling through the critical temperature so that the exchange energy between neighbouring atomic moments increases. Hence when the critical temperature is approached from below, the system finds the exchange energy and hence the effective critical temperature decreasing (kT_c is generally thought to be a linear function of J, as one might imagine from dimensional analysis), and the measured critical-point exponent β is decreased below what one would expect for an incompressible lattice. Thus the value $\beta = 0.244 \pm 0.015$ is obtained from the slope of the log–log plot of the CoO data in Fig. 1.11 whereas, when corrected for this lattice contraction effect, the data indicate $\beta = 0.290 \pm 0.025$.

In Chapter 3 we shall define a good many of the commonly used critical-point exponents—suffice it to say here that there are essentially as many exponents as there are singular functions, and the Greek alphabet is fast being exhausted. Three of the most common critical-point exponents—α', β, and γ'—are defined for fluid and magnetic systems in Table 1.1. Note that minus signs are associated with the

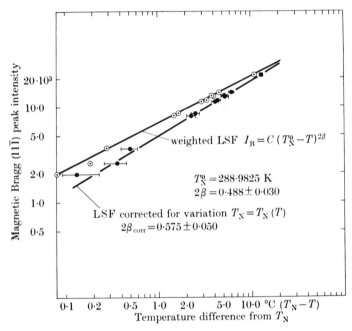

FIG. 1.11. Dependence of the logarithm of the magnetic Bragg peak intensity from neutron scattering from CoO upon the logarithm of $(T_N - T)$. This intensity is proportional to the square of the spontaneous magnetization. The upper curve is a least squares fit (LSF) to a power law, assuming that the lattice is incompressible. The lower curve is a similar least squares fit to data that have been corrected for the lattice compressibility. The critical exponent obtained from the lower curve is clearly in better agreement than the upper curve with the anticipated value of $\beta \simeq 0.3$. After Rechtin, Moss, and Averbach (1970).

exponents for response functions such as the specific heat, compressibility, and susceptibility which are expected, theoretically speaking, to diverge to infinity at the critical point; hence the exponents α' and γ' are defined such that they are positive quantities. Of course, no one has ever measured an infinite value for any of these response functions. This is not only because we never can make measurements arbitrarily close to T_c (measurements for $\epsilon < 10^{-6}$ or closer than one part in a

TABLE 1.1

Representative critical-point exponents for fluid and magnetic systems. For simplicity we have only considered here the approach to T_c from the low-temperature side. More complete tables are shown in Chapter 3

Definition	α'	β	γ'	$\alpha' + 2\beta + \gamma'$
Fluid	$C_{V=V_c} \sim (-\epsilon)^{-\alpha'}$	$\rho_L - \rho_G \sim (-\epsilon)^{\beta}$	$K_T \sim (-\epsilon)^{-\gamma'}$	—
Magnet	$C_{H=0} \sim (-\epsilon)^{-\alpha'}$	$M_{H=0} \sim (-\epsilon)^{\beta}$	$\chi_T \sim (-\epsilon)^{-\gamma'}$	—
Typical experimental values				
Fluid or magnet	0–0·2	0·3–0·5	1·1–1·4	$\lesssim 2$
Theories				
van der Waals or Weiss	0 (discontinuity)	$\frac{1}{2}$	1	2
Two dimensional Ising model	0 (log)	$\frac{1}{8}$	$\frac{7}{4}$	2

million from the critical temperature are extremely rare) but also because some sort of rounding-off of the data is frequently found, as we see, for example, in the specific-heat data shown in Fig. 1.12.

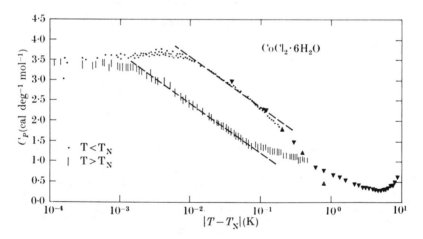

Fig. 1.12. Dependence of the specific heat upon the logarithm of $|T - T_N|$ for cobalt chloride. The data appear to be fitted fairly well by a logarithmic divergence except within a few millidegrees of T_N. After Kadanoff *et. al.* (1967).

1.2.2. *Results from model systems*

The number of model systems which have been studied as a means of gaining insight into the nature of phase transitions and critical phenomena is extremely large and therefore we shall limit our remarks here to the two models we introduced above, the Ising model and the Heisenberg model. Although both these models were proposed in the early years of this century, it is only within the last two decades that much of their richness has come to be appreciated.

The highlight in any discussion of the Ising model is perforce

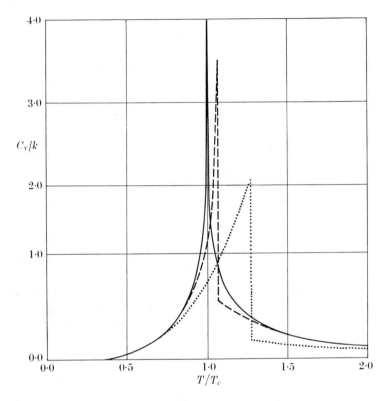

FIG. 1.13. The solid curve shows the specific heat of the two-dimensional Ising model as obtained from the exact solution of Onsager (solid curve), from the Bethe approximation (dotted curve), and the Kramers–Wannier and Kikuchi approximation (broken curve). After Domb (1960).

Onsager's solution, in 1944, for the $H = 0$ partition function of a two-dimensional lattice. From the partition function he was able to demonstrate that the specific heat possesses a logarithmic divergence at T_c when approached from either side of the transition. This result stood in dramatic contradistinction to the predictions of the mean field theory and other theories of cooperative phenomena of that day which predicted a simple discontinuity in the specific heat (cf. Fig. 1.13). In particular, it showed that Ehrenfest's classification according to which derivative of the free energy undergoes a discontinuity was, in a sense, inadequate for the description of the two-dimensional Ising model phase transition. (In this monograph we shall follow Ehrenfest in so far as we shall denote by 'first-order transitions' phase transitions in which the first derivative of an appropriate thermodynamic potential is discontinuous, while we shall denote by 'higher-order' or 'continuous' transitions the characteristic behaviour near the critical points of fluids and magnets for which the first derivatives are continuous while the second derivatives are either discontinuous or infinite. This classification, due to Fisher, is somewhat more general than Ehrenfest's.)

Because Onsager's solution is for the zero-field partition function, it is not possible to obtain directly the exponents β, γ, and γ'. The spontaneous magnetization and hence the exponent β was, however, obtained a few years later by Onsager, who announced his result in 1948 as a discussion remark following a paper of Tisza on the general theory of phase transitions. Onsager never published his derivation, however, and four years later Yang published a complete derivation that abounds in complicated elliptic integrals and represents a true mathematical *tour de force*.

The susceptibility of the two-dimensional Ising model has not yet been calculated, so Table 1.1 shows only an approximate (though an extremely reliable) value for the exponent γ'. This approximate value is not obtained from the usual sort of closed-form approximations, but rather from extrapolations based upon a method of successive approximations in which successively longer-range correlations are taken into account. As we shall see in Chapter 9, closed form approximations have so far not been capable of describing the subtle and fascinating physical feature of the critical point—namely that although the interaction between the constituent magnetic moments is generally of extremely short range (for example, extending only to neighbouring moments), this interaction nevertheless 'propagates' from one moment to the next, tending to create a preferred direction for all magnetic moments and in fact the resulting order becomes infinite in range as the critical point is approached. In particular, these approximation methods in which longer and longer correlation ranges are successively taken into account have demonstrated the crucial role of lattice dimensionality in determining critical behaviour. Hence the fact that the critical-point exponents shown in Table 1.1 for the two-dimensional Ising model do not agree with most experimental measurements on simple fluids does not necessarily indicate a breakdown of the intuitively plausible lattice-gas model.

Since no one has yet succeeded in solving the three-dimensional Ising model exactly (and since some workers have claimed that it may in fact be an insoluble problem!), we must perforce consider alternate models. Now the Heisenberg model is not even soluble for two-dimensional lattices, so this avenue of approach would appear fruitless. However, some years ago Berlin and Kac succeeded in solving, for both two- and three-dimensional lattices, a rather unphysical model called the spherical model. This model is in some respects similar to the Ising model but, in the critical region, the exponents of the spherical model agree with experiments as poorly as do those of the two-dimensional Ising model. The principal virtue to this day of the spherical model result is that it represents one of the very few non-trivial examples of a many-particle system that can be solved exactly for three dimensions. This statement is not intended to deprecate model systems which as yet have no physical system as a counterpart—there is generally much to be learned from a theoretical model when considered in its own right, and frequently the predictions of the theoretical model motivate a successful search for the appropriate physical system.

1.2.3. *Exponent inequalities and the scaling hypothesis*

At this point, the reader may be wondering why so much effort has been directed towards the investigation of the critical-point exponents and he may be led to inquire why more attention is not focused on finding a theory that fits experimental data over the entire range of temperatures. Our answer to his question has two parts. The first is that the physical phenomena that come into play—namely, the building up of extremely long-range correlations extending to perhaps a billion constituent particles—are absent when one is far from the critical point. The second part of our answer is that there exist among the exponents certain relations which arise from fundamental thermodynamic and statistical mechanical considerations and thus transcend any particular system under consideration. As an example of this, we show in Chapter 4 that a large number of inequalities among the various critical-point exponents can be obtained from elementary thermodynamic reasoning, together with certain additional assumptions. One of these inequalities states that $\alpha' + 2\beta + \gamma' \geq 2$, and Table 1.1 shows the values of the quantity $\alpha' + 2\beta + \gamma'$ for several theoretical models and for typical experimental systems. We see that

$\alpha' + 2\beta + \gamma'$ apparently equals 2 for the theoretical models and either equals or falls just short of 2 for the experimental systems. However, the equality can frequently be obtained by reconsideration of the error bounds placed on the data. Hence one might be led to conjecture that the inequality $\alpha' + 2\beta + \gamma' \geq 2$ is in fact obeyed as an equality. Attempts to prove this rigorously have thus far been unsuccessful.

An alternate approach which does predict that the inequalities be satisfied as equalities involves making an assumption that certain thermodynamic functions are homogeneous functions of their arguments. This assumption, which has come to be called the *static scaling hypothesis*, leads not only to the prediction that most of the exponent inequalities are satisfied as equalities, but also to additional predictions concerning the equation of state of a substance in the very immediate vicinity of the critical point. Many of these predictions have been partially corroborated by experimental measurement and theoretical calculation on model systems, and the scaling law approach has thereby enjoyed quite a remarkable success in providing a unified approach to many phenomena associated with the critical point. It should be pointed out, however, that certain examples still stand in contradiction to the predictions of the scaling hypothesis, and hence the ultimate validity of the hypothesis must be regarded as an open question. More important, perhaps, is the fact that the scaling hypothesis is at best unproved, and indeed to some workers represents an *ad hoc* assumption entirely devoid of physical rationale.

1.3. Phase transitions in other systems

Phenomena analogous to those described above are found to be associated with critical points in a wide variety of other physical systems. For example, many features of the superfluid, superconducting, and ferroelectric transitions are quite similar to those found for liquid–gas and magnetic critical points. In particular, the analogy between the phase transitions discussed above and the phase transition in a binary mixture (or a binary alloy) is relatively easy to describe. A binary mixture of two different fluids exhibits a critical temperature below which the two components do not mix homogeneously in all proportions, and one can define a two-component region, bounded by a coexistence curve, which is in many respects analogous to the two-phase region of a simple one-component fluid.

Analogies among phase transitions in superfluids, superconductors,

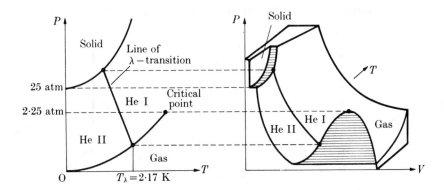

Fig. 1.14. Sketch (not to scale) of the equilibrium PVT surface of ^{4}He in the low-temperature region, and the projection of this surface on to the PT plane. Contrast this phase diagram with that of a normal fluid shown in Fig. 1.1(a). From Huang (1963).

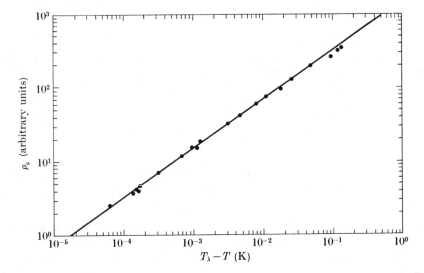

Fig. 1.15. Dependence of the logarithm of the superfluid density upon the logarithm of the temperature difference from the λ temperature for ^{4}He. The superfluid density is the square of the order parameter for the λ-transition, and one sees from the slope of this plot that the appropriate exponent has a value of about $\frac{2}{3}$ or roughly that of M^2, where M is the order parameter for the ferromagnetic transition. After Kadanoff *et. al.* (1967).

and ferroelectrics must, however, be made with care. For example, there is actually an entire line of critical points in a superfluid along which various critical phenomena take place, and as we see from the PT-projection of the PVT surface of helium shown in Fig. 1.14, this

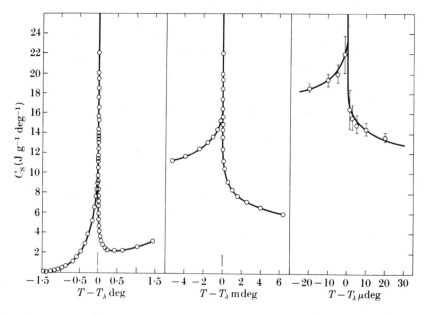

FIG. 1.16. Specific heat of ^{4}He as a function of $T - T_\lambda$ in K. Notice that the shape of the specific heat curve is rather like the Greek letter λ, whence the origin of the term 'λ-transition'. The fact that the specific heat is only about ten times its 'normal' value even at temperatures only a few microdegrees from T_λ is correlated with the fact that the critical-point exponent is extremely small (in fact, α is probably zero, corresponding to a logarithmic divergence). The width of the small vertical line just above the origin indicates the portion of the diagram that is expanded in width in the curve directly to the right. After Buckingham and Fairbank (1965).

'λ-line' terminates at the vapour pressure curve in a 'λ-point'. These phase transitions appear to be characterized by an order parameter which is non-zero below T_c and zero above (cf. the data for the order parameter in the superfluid transition in Fig. 1.15) and an anomaly in the specific heat (cf. Fig. 1.16, which shows the temperature dependence of C_S, the specific heat under saturated vapour pressure, for ^{4}He; note that the temperature scales are progressively expanded by factors of 1000).

Table 1.2 gives a brief list of phase transitions and the order parameters of each. For pedagogical reasons, we shall restrict most of our

TABLE 1.2

Partial list of phase transitions and the order parameters $\langle p \rangle$ associated with each. Adapted from Kadanoff et al. (1967)

Transition	Meaning of $\langle p \rangle$	Free choice in $\langle p \rangle$	Thermodynamic conjugate of $\langle p \rangle$
liquid–gas	$\rho - \rho_c$	$p > 0 =$ liquid $p < 0 =$ vapour (2 choices)	μ
ferromagnetic	magnetization M	if n equivalent 'easy axes' $2n$ choices	applied magnetic field, H, along easy axes
antiferromagnetic	sublattice magnetization	if n 'easy axes' $2n$ choices	not physical
Heisenberg model ferromagnet	magnetization M	direction of M (can choose any value on surface of sphere)	H
Ising model ferro-magnet	s_j	2 choices	H
superconductors	Δ (complex gap parameter)	phase of Δ	not physical
superfluid	$\langle \psi \rangle$ (condensate wave function)	phase of $\langle \psi \rangle$	not physical
ferroelectric	lattice polarization	finite number of choices	electric field
phase separation	concentration	2 choices	a difference of chemical potentials

discussion in this monograph to the simpler cases of fluid and magnetic phase transitions.

Suggested further reading
Andrews (1869).
Hopkinson (1890).
Weiss (1907).
Temperley (1956).
Brout (1965).
Mendelssohn (1966).
Landau, Akhiezer, and Lifshitz (1967).
Widom (1967).
Stanley (1971a).

2

USEFUL THERMODYNAMIC RELATIONS
FOR FLUID AND MAGNETIC SYSTEMS

THE material in this chapter is intended to serve as a brief review of those aspects of thermodynamics that will be used to describe phase transitions and critical phenomena. We shall first review the thermodynamic relations appropriate for a fluid system, following which we shall discuss the analogous relations for a magnetic system.

2.1. The thermodynamic state functions U, E, G, and A

We begin by recalling the first law of thermodynamics, which states that the differential dU, defined by the relation

$$dU \equiv dQ - dW, \qquad (2.1)$$

is exact. Here dQ is the differential quantity of heat absorbed by the system and $dW = P\,dV$ is the differential quantity of work done by the system, where P is the pressure and V is the volume. That dU is exact means that there exists a function U representing the internal energy of the system, such that the change in U, $\Delta U = \int_i^f dU$, is the same for all processes leading from a given initial state i to a given final state f. We say that U is a *state function* because, providing we choose $U = 0$ for some reference state i, U depends only on the state f. We shall usually write $U = U(S, V)$, where S is the entropy.

Three other state functions are useful—the enthalpy E, the Gibbs potential G, and the Helmholtz potential A. They are defined by the relations

$$E = E(S, P) \equiv U + PV, \qquad (2.2)$$

$$G = G(T, P) \equiv U - TS + PV, \qquad (2.3)$$

and

$$A = A(T, V) \equiv U - TS. \qquad (2.4)$$

2.2. Differential relations for the state functions: the thermodynamic square

We will frequently use the following two mathematical lemmas, which the reader can easily prove for himself.

LEMMA 1. *Let* df *be an exact differential of the function* $f(x, y)$.

Then

$$df = u(x, y)\, dx + v(x, y)\, dy \qquad (2.5)$$

with

$$u(x, y) = \left(\frac{\partial f}{\partial x}\right)_y; \qquad v(x, y) = \left(\frac{\partial f}{\partial y}\right)_x; \qquad (2.6a)$$

and

$$\left(\frac{\partial u}{\partial y}\right)_x = \frac{\partial^2 f}{\partial y\, \partial x} = \frac{\partial^2 f}{\partial x\, \partial y} = \left(\frac{\partial v}{\partial x}\right)_y. \qquad (2.6b)$$

LEMMA 2. *Let* x, y, *and* z *satisfy some functional relationship* $f(x, y, z)$ $= 0$ (e.g. P, V, and T satisfy an equation of state of the form $f(P, V, T)$ $= 0$). *Let* g *be a function of any two of* x, y, *and* z.

Then

$$\left(\frac{\partial x}{\partial y}\right)_z = \left\{\left(\frac{\partial y}{\partial x}\right)_z\right\}^{-1}, \qquad (2.7)$$

$$\left(\frac{\partial x}{\partial y}\right)_z\left(\frac{\partial y}{\partial z}\right)_x\left(\frac{\partial z}{\partial x}\right)_y = -1, \qquad (2.8)$$

$$\left(\frac{\partial x}{\partial z}\right)_g = \left(\frac{\partial x}{\partial y}\right)_g\left(\frac{\partial y}{\partial z}\right)_g. \qquad (2.9)$$

We begin with the differential expressions for the four state functions. The first relation is obtained from eqn (2.1) with the substitutions $dQ = TdS$ and $dW = PdV$; the remaining relations are obtained from (2.2)–(2.4) by writing $d(PV) = PdV + VdP$ and $d(TS) = TdS + SdT$. Thus we obtain

$$dU = T\, dS - P\, dV, \qquad (2.10a)$$

$$dE = T\, dS + V\, dP, \qquad (2.10b)$$

$$dG = -S\, dT + V\, dP, \qquad (2.10c)$$

$$dA = -S\, dT - P\, dV. \qquad (2.10d)$$

On using (2.6a), we find

$$T = \left(\frac{\partial U}{\partial S}\right)_V; \qquad -P = \left(\frac{\partial U}{\partial V}\right)_S; \qquad (2.11\text{a})$$

$$T = \left(\frac{\partial E}{\partial S}\right)_P; \qquad V = \left(\frac{\partial E}{\partial P}\right)_S; \qquad (2.11\text{b})$$

$$-S = \left(\frac{\partial G}{\partial T}\right)_P; \qquad V = \left(\frac{\partial G}{\partial P}\right)_T; \qquad (2.11\text{c})$$

$$-S = \left(\frac{\partial A}{\partial T}\right)_V; \qquad -P = \left(\frac{\partial A}{\partial V}\right)_T. \qquad (2.11\text{d})$$

A convenient mnemonic device for recalling these equations is the thermodynamic square shown in Fig. 2.1(a). On each side of the square

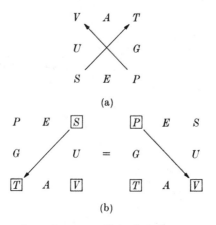

(a)

(b)

Fig. 2.1. (a) The thermodynamic square. Note that the arrows enable one to get the sign right in eqns (2.10) and (2.11). (b) Two thermodynamic squares rotated 180° as required to obtain the Maxwell relation, eqn (2.12d).

appears one of the state functions, surrounded by its two natural independent variables. The reader can easily discover how to use the square to construct the relations (2.10) and (2.11).

On applying eqn (2.6b) of Lemma 1 to each of eqns (2.10) and (2.11) we obtain the four 'Maxwell relations':

$$\left(\frac{\partial T}{\partial V}\right)_S = -\left(\frac{\partial P}{\partial S}\right)_V, \qquad (2.12\text{a})$$

$$\left(\frac{\partial T}{\partial P}\right)_S = \left(\frac{\partial V}{\partial S}\right)_P, \qquad (2.12\text{b})$$

$$\left(\frac{\partial S}{\partial P}\right)_T = -\left(\frac{\partial V}{\partial T}\right)_P, \tag{2.12c}$$

$$\left(\frac{\partial S}{\partial V}\right)_T = \left(\frac{\partial P}{\partial T}\right)_V. \tag{2.12d}$$

The Maxwell relations are also easily obtained from the thermodynamic square, providing we rotate it to the appropriate position. Thus, by rotating the square 180° (see Fig. 2.1(b)) we can read off the last relation, eqn (2.12d).

Clearly a knowledge of any one of the four state functions U, E, G, and A for all values of its natural variables ((S, V), (S, P), (T, P), and (T, V) respectively) is sufficient to determine the remaining three state functions. For example, U can be obtained from A by using

$$U = A + TS = A - T\left(\frac{\partial A}{\partial T}\right)_V = -T^2\left(\frac{\partial}{\partial T}\left[\frac{A}{T}\right]\right)_V. \tag{2.13}$$

2.3. Two basic response functions: the specific heat and the compressibility

Two types of response function will be introduced in this section: (i) the specific heats C_P and C_V, which measure the heat absorption from a temperature stimulus, and (ii) the compressibilities K_T and K_S, which measure the response of the volume to a pressure stimulus.

(i) The specific heat at constant x ($x = P$ or V) is defined by

$$C_x \equiv \left(\frac{\mathrm{d}Q}{\mathrm{d}T}\right)_x = T\left(\frac{\partial S}{\partial T}\right)_x. \tag{2.14}$$

If the volume is held constant, then

$$C_V = T\left(\frac{\partial S}{\partial T}\right)_V = \left(\frac{\partial U}{\partial T}\right)_V = -T\left(\frac{\partial^2 A}{\partial T^2}\right)_V, \tag{2.15}$$

using (2.10a) and (2.11d). Similarly, if the pressure is held constant, then $\mathrm{d}Q = \mathrm{d}U + P\,\mathrm{d}V = \mathrm{d}(U + PV) = \mathrm{d}E$, so that

$$C_P = T\left(\frac{\partial S}{\partial T}\right)_P = \left(\frac{\partial E}{\partial T}\right)_P = -T\left(\frac{\partial^2 G}{\partial T^2}\right)_P. \tag{2.16}$$

(ii) The isothermal and adiabatic compressibilities are defined by the relations

$$K_T \equiv -\frac{1}{V}\left(\frac{\partial V}{\partial P}\right)_T = \frac{1}{\rho}\left(\frac{\partial \rho}{\partial P}\right)_T = -\frac{1}{V}\left(\frac{\partial^2 G}{\partial P^2}\right)_T \tag{2.17}$$

and

$$K_S \equiv -\frac{1}{V}\left(\frac{\partial V}{\partial P}\right)_S = \frac{1}{\rho}\left(\frac{\partial \rho}{\partial P}\right)_S = -\frac{1}{V}\left(\frac{\partial^2 E}{\partial P^2}\right)_S, \qquad (2.18)$$

where $\rho \equiv mn$ is the mass density, and $n \equiv N/V$ is the average particle density.

A third response function which receives considerably less attention is the coefficient of thermal expansion, α_P, defined by

$$\alpha_P \equiv \frac{1}{V}\left(\frac{\partial V}{\partial T}\right)_P. \qquad (2.19)$$

The response functions are not all independent of one another. Two particularly useful relations among them are

$$K_T(C_P - C_V) = TV\alpha_P^2 \qquad\qquad \blacktriangleright \ (2.20a)$$

and

$$C_P(K_T - K_S) = TV\alpha_P^2. \qquad\qquad \blacktriangleright \ (2.20b)$$

To derive eqns (2.20a) and (2.20b), we begin with the relations

$$\left(\frac{\partial S}{\partial T}\right)_P = \left(\frac{\partial S}{\partial T}\right)_V + \left(\frac{\partial S}{\partial V}\right)_T\left(\frac{\partial V}{\partial T}\right)_P \qquad (2.21a)$$

and

$$\left(\frac{\partial V}{\partial P}\right)_T = \left(\frac{\partial V}{\partial P}\right)_S + \left(\frac{\partial V}{\partial S}\right)_P\left(\frac{\partial S}{\partial P}\right)_T, \qquad (2.21b)$$

which are special cases of the general result

$$\left(\frac{\partial w}{\partial y}\right)_x = \left(\frac{\partial w}{\partial y}\right)_z + \left(\frac{\partial w}{\partial z}\right)_y\left(\frac{\partial z}{\partial y}\right)_x. \qquad (2.22)$$

Hence

$$K_T(C_P - C_V) = -\frac{T}{V}\left(\frac{\partial V}{\partial P}\right)_T\left(\frac{\partial S}{\partial V}\right)_T\left(\frac{\partial V}{\partial T}\right)_P \qquad (2.23a)$$

and

$$C_P(K_T - K_S) = -\frac{T}{V}\left(\frac{\partial S}{\partial T}\right)_P\left(\frac{\partial V}{\partial S}\right)_P\left(\frac{\partial S}{\partial P}\right)_T. \qquad (2.23b)$$

Next we observe that eqn (2.8) of Lemma 2 implies that

$$\left(\frac{\partial V}{\partial P}\right)_T\left(\frac{\partial P}{\partial T}\right)_V\left(\frac{\partial T}{\partial V}\right)_P = -1. \qquad (2.24)$$

Using the Maxwell relation (2.12d) and eqn (2.7) of Lemma 2, eqn (2.24) becomes

$$\left(\frac{\partial V}{\partial P}\right)_T\left(\frac{\partial S}{\partial V}\right)_T = -\left(\frac{\partial V}{\partial T}\right)_P. \qquad (2.25)$$

On substituting eqn (2.25) into eqn (2.23a) and using the definition (2.19) of the thermal expansion coefficient α_P, we obtain the first desired result, eqn (2.20a).

Equation (2.20b) is derived in an analogous fashion. Thus by eqn (2.8)

$$\left(\frac{\partial S}{\partial T}\right)_P \left(\frac{\partial T}{\partial P}\right)_S \left(\frac{\partial P}{\partial S}\right)_T = -1, \tag{2.26}$$

which may be rewritten using eqns (2.7) and (2.12b) as

$$\left(\frac{\partial S}{\partial T}\right)_P \left(\frac{\partial V}{\partial S}\right)_P = -\left(\frac{\partial S}{\partial P}\right)_T. \tag{2.27}$$

Finally, eqn (2.23b) together with eqns (2.27) and (2.12c) result in eqn (2.20b).

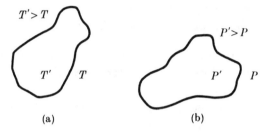

Fig. 2.2. Using the principle—often attributed to Le Chatelier—that any spontaneous change in the parameters of a system that is in stable equilibrium will give rise to processes that tend to restore the system to equilibrium, the reader can complete the following heuristic arguments that (a) $C = dQ/dT > 0$, and (b) $K \sim -dV/dP > 0$: (a) suppose there has occurred a spontaneous temperature fluctuation in which the temperature of the system is increased from T to T'; (b) suppose there occurs a spontaneous pressure fluctuation $P \to P'$. The non-negativity of these response functions also follows directly from the fluctuation–dissipation relations derived in Appendix A.

Now in order that our fluid system be thermally and mechanically stable, the specific heat and compressibility should be positive for all T (see Fig. 2.2). Hence eqns (2.20) imply that

$$C_P \geq C_V \tag{2.28a}$$

and

$$K_T \geq K_S \tag{2.28b}$$

for all temperatures. In particular, as $T \to T_c$, we shall see that $C_P \gg C_V$ and $K_T \gg K_S$. (Note that the relations (2.28) reduce to equalities only (i) when $T = 0$ or (ii) when $\alpha_P = 0$, as, for example, in the case of water at about 4 °C.) Equations (2.20) and their magnetic analogues will prove extremely useful in our discussion below. In particular, it is elementary to obtain from eqns (2.20) the following corollaries:

$$C_P/C_V = K_T/K_S, \qquad (2.29)$$

$$C_P = \frac{TV\alpha_P^2}{(K_T - K_S)}, \qquad (2.30)$$

and

$$C_V = \frac{TV\alpha_P^2 K_S}{K_T(K_T - K_S)}. \qquad (2.31)$$

2.4. Stability and convexity relations

In this section we shall see that the positivity of the specific heat and compressibility, as required for thermal and mechanical stability, implies certain convexity properties of the thermodynamic potentials A and G. These properties will be used in Chapter 4.

DEFINITION. A function $f(x)$ is a convex function of x providing the chord joining the points $f(x_1)$ and $f(x_2)$ lies above or on the curve $f(x)$ for all x in the interval $x_1 < x < x_2$.

For example, the function $f(x) = x^2$ is convex (see Fig. 2.3) as is the

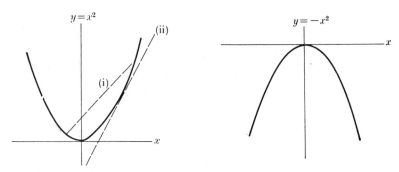

FIG. 2.3. The parabola $f(x) = x^2$ is a convex function since, for example, (i) any chord lies above or on the parabola, (ii) any tangent lies below or on the parabola, and (iii) the second derivative $f''(x)$ is positive. Therefore the parabola $f(x) = -x^2$ is a concave function.

function $f(x) = |x|$. Note that this definition does not require $f(x)$ to be differentiable. However, should the derivative $f'(x)$ exist, then it follows that a tangent to a convex function always lies below the function except at the point of tangency. If, moreover, the second derivative of a convex function exists, then $f''(x) \geq 0$ for all x.

DEFINITION. A function $f(x)$ is a concave function of x if the function $-f(x)$ is convex. Thus, for example, $f(x) = -x^2$ is concave.

In the previous section we noted that for a mechanically stable system, the specific heat and the compressibility must be positive for all temperatures. Thus we are led to the following theorem.

THEOREM. (i) *The Gibbs potential* $G(T, P)$ *is a concave function of both temperature and pressure, and* (ii) *the Helmholtz potential* $A(T, V)$ *is a concave function of the temperature and a convex function of the volume.*

Proof. We begin by noting from eqns (2.11c) and (2.11d) that

$$S = -\left(\frac{\partial G}{\partial T}\right)_P = -\left(\frac{\partial A}{\partial T}\right)_V. \tag{2.32}$$

Hence

$$\left(\frac{\partial^2 G}{\partial T^2}\right)_P = -\left(\frac{\partial S}{\partial T}\right)_P = -\frac{1}{T} C_P \le 0 \tag{2.33}$$

and

$$\left(\frac{\partial^2 A}{\partial T^2}\right)_V = -\left(\frac{\partial S}{\partial T}\right)_V = -\frac{1}{T} C_V \le 0. \tag{2.34}$$

Since the second derivatives exist (except for $T = T_c$) and are non-positive, it follows that both $G(T, P)$ and $A(T, V)$ are concave functions of temperature.

Next we observe, again from eqns (2.11c) and (2.11d), that

$$\left(\frac{\partial^2 G}{\partial P^2}\right)_T = \left(\frac{\partial V}{\partial P}\right)_T = -VK_T \le 0 \tag{2.35}$$

and

$$\left(\frac{\partial^2 A}{\partial V^2}\right)_T = -\left(\frac{\partial P}{\partial V}\right)_T = (VK_T)^{-1} \ge 0, \tag{2.36}$$

from which it follows that $G(T, P)$ is a concave function of pressure and $A(T, V)$ is a convex function of volume.

COROLLARY. *The curvature of* $G(T, P)$ *with respect to* P *(for fixed* T*) is the negative reciprocal of the curvature of* $A(T, V)$ *with respect to* V *(for fixed* T*).*

Proof. Directly from eqns (2.35) and (2.36) it follows that

$$\left(\frac{\partial^2 G}{\partial P^2}\right)_T = -\left\{\left(\frac{\partial^2 A}{\partial V^2}\right)_T\right\}^{-1}. \tag{2.37}$$

2.5. Geometrical interpretation of the Gibbs and Helmholtz potentials

First let us note from eqns (2.3) and (2.4) that $G(T, P)$ and $A(T, V)$ are simply related by $G = A + PV$. Therefore, given one of these potentials, we may easily find the other by means of the simple geometric construction shown in Fig. 2.4 (for a fixed temperature above T_c). Note that the convexity relations for A and G imply that the second derivatives with respect to any argument have the same sign for all values of the argument, thereby guaranteeing that no ambiguities will arise in this construction. More familiar quantities are the volume and

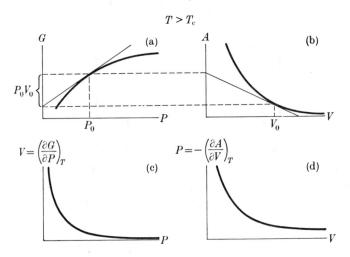

Fig. 2.4. Geometrical interpretation of the relation between the Gibbs and Helmholtz potentials at a fixed temperature $T > T_c$. This construction, adapted from Morse (1969) and Griffiths (unpublished lecture notes), is based on the identities $A = G - PV$ and $V = (\partial G/\partial P)_T$. The vertical distance between the dashed lines in (a) and (b) is the product PV, and on subtracting this from G we obtain A. Note that the construction indicated is unambiguous because $G(T, P)$ is a concave function of P (for all P) and $A(T, V)$ is a convex function of V (for all V). Also shown are the volume as a function of pressure (obtained from the pressure derivative of $G(T, P)$) and the pressure as a function of volume (obtained from the volume derivative of $A(T, V)$).

the pressure, given by eqns (2.11c) and (2.11d). These quantities are shown in Figs. 2.4(c) and 2.4(d).

By definition, a first-order phase transition is characterized by a finite discontinuity in a first derivative of the Gibbs potential $G(T, P)$. In Fig. 2.5(a) G is plotted as a function of P for a fixed temperature less than T_c. We see that at a certain pressure the tangent to the curve changes discontinuously. This discontinuous change in the first derivative of G corresponds to an infinite curvature in G and hence, by eqn (2.37), to a zero value for the curvature of A. Therefore we expect that a plot of A as a function of V will have a corresponding straight-line portion, as in Fig. 2.5(b). P–V (pressure–volume) and V–P isotherms are shown in Fig. 2.5(c) and (d) respectively.

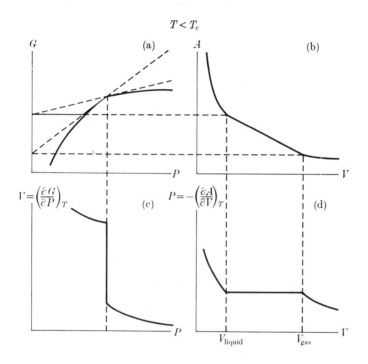

FIG. 2.5. The Gibbs and Helmholtz potentials for a fixed temperature $T < T_c$, where T_c is the transition temperature for a first-order phase transition. Also shown are the volume as a function of pressure and the pressure as a function of volume. Note that the value of the pressure at which the volume discontinuity occurs is not the critical pressure P_c but rather the value of the saturated vapour pressure $P_{sat}(T)$ given by the vapour pressure curve of Fig. 1.1a. Thus the system is seen to undergo a *first-order* transition on crossing the vapour pressure curve at constant temperature T ($T < T_c$).

Finally, in Fig. 2.6(a) we show the temperature dependence of G (or A) that would result in a phase transition with an entropy discontinuity or latent heat (first-order transition); in Fig. 2.6(b) we show the case of zero latent heat (higher-order transition). The temperature dependence of the entropy is shown in Figs. 2.6(c) and (d) for the two cases.

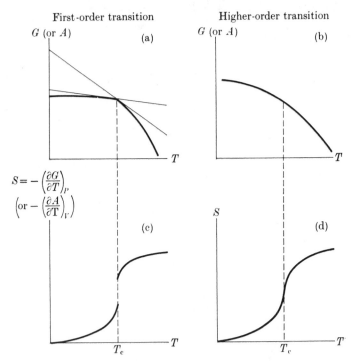

FIG. 2.6. (a) Temperature dependence of either the Gibbs potential at fixed pressure or the Helmholtz potential at fixed volume. The system shown undergoes a phase transition at $T = T_c$, accompanied by a latent heat (entropy discontinuity). (b) Same as (a) except that the phase transition has no latent heat. (c)–(d) show the entropy obtained from the temperature derivative of A or G. For this particular example, $C \sim dS/dT$ appears to diverge at T_c for both the first-order and the higher-order transition.

2.6. Analogies between fluids and magnets

Next we shall develop thermodynamic relations for magnetic systems which are analogous to those discussed above for fluid (liquid–gas) systems.

We shall assume that volume and pressure effects in our magnetic system can be neglected. Hence, for example, when we write C_H for the specific heat in constant magnetic field H, we shall assume that the volume is also held constant. Thus our thermodynamic parameters are now H, M and T instead of P, V, and T. Almost all the equations in the following section may be obtained from those in the preceding sections by making the substitutions

$$V \to -M, \tag{2.38}$$

$$P \to H. \tag{2.39}$$

The minus sign in (2.38), though bothersome, is absolutely necessary. For example, the response function $\chi \equiv \partial M / \partial H$ is positive (the magnetization increases with magnetic field), whereas the analogous derivative for the fluid system, $\partial V / \partial P$, is negative (the volume decreases with pressure).

There is a second magnet-fluid analogy, in which $M \to \rho - \rho_c$ and $H \to \mu - \mu_c$ where here μ denotes the chemical potential. This is pedagogically the sounder analogy, since μ is the thermodynamic variable conjugate to the density and the density is the order parameter. However for our purposes it is more convenient to develop the analogy stated in eqns (2.38) and (2.39).

2.7. The thermodynamic state functions for a magnetic system

The first law of thermodynamics for a magnetic system may be written either as

$$dU_a = dQ - M\,dH \qquad (2.40a)$$

or as

$$dU_b = dQ + H\,dM. \qquad (2.40b)$$

The distinction is carefully discussed by Kittel (1958), Guggenheim (1967), Callen (1960), Reif (1965), Wannier (1966), and Fay (1965). Here we shall use eqn (2.40b). The remaining three state functions are then defined, in analogy with the fluid case, as

$$E = E(S, H) \equiv U - MH, \qquad (2.41)$$

$$G = G(T, H) \equiv U - TS - MH, \qquad (2.42)$$

and

$$A = A(T, M) \equiv U - TS. \qquad (2.43)$$

2.8. Differential relations and the thermodynamic square for magnetic state functions

The substitutions $V \to -M$ and $P \to H$ are sufficient to obtain the results of this section from the corresponding expressions in § 2.2 for fluid systems. Thus the thermodynamic square is as shown in Fig. 2.7, and the differential expressions for the magnetic state functions are

$$dU = T\,dS + H\,dM, \qquad (2.44a)$$

$$dE = T\,dS - M\,dH, \qquad (2.44b)$$

$$dG = -S\,dT - M\,dH, \qquad (2.44c)$$

$$dA = -S\,dT + H\,dM. \qquad (2.44d)$$

(a)

$U = U\ (S,V)\ (dU = dQ - P\ dV)$

$E = E\ (S,P) = U + PV$

$G = G\ (T,P) = U - TS + PV$

$A = A\ (T,V) = U - TS$

(b)

$U = U\ (S,M)\ (dU = dQ + H\ dM)$

$E = E\ (S,H) = U - HM$

$G = G\ (T,H) = U - TS - HM$

$A = A\ (T,M) = U - TS$

$$dU = T\ dS - P\ dV \Rightarrow \left(\frac{\partial T}{\partial V}\right)_S = -\left(\frac{\partial P}{\partial S}\right)_V \qquad dU = T\ dS + H\ dM \Rightarrow \left(\frac{\partial T}{\partial M}\right)_S = \left(\frac{\partial H}{\partial S}\right)_M$$

$$dE = T\ dS + V\ dP \Rightarrow \left(\frac{\partial T}{\partial P}\right)_S = \left(\frac{\partial V}{\partial S}\right)_P \qquad dE = TdS - M\ dH \Rightarrow \left(\frac{\partial T}{\partial H}\right)_S = -\left(\frac{\partial M}{\partial S}\right)_H$$

$$dG = -S\ dT + V\ dP \Rightarrow \left(\frac{\partial S}{\partial P}\right)_T = -\left(\frac{\partial V}{\partial T}\right)_P \qquad dG = -S\ dT - MdH \Rightarrow \left(\frac{\partial S}{\partial H}\right)_T = \left(\frac{\partial M}{\partial T}\right)_H$$

$$dA = -S\ dT - P\ dV \Rightarrow \left(\frac{\partial S}{\partial V}\right)_T = \left(\frac{\partial P}{\partial T}\right)_V \qquad dA = -S\ dT + HdM \Rightarrow \left(\frac{\partial S}{\partial M}\right)_T = -\left(\frac{\partial H}{\partial T}\right)_M$$

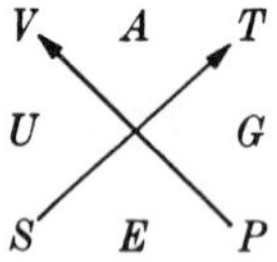

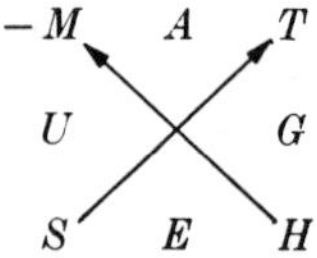

FIG. 2.7. Comparison of thermodynamic relations for (a) fluid systems and (b) magnetic systems. Note that the two are related by the substitutions $V \to -M$ and $P \to H$.

Similarly, we have

$$T = \left(\frac{\partial U}{\partial S}\right)_M; \qquad H = \left(\frac{\partial U}{\partial M}\right)_S; \tag{2.45a}$$

$$T = \left(\frac{\partial E}{\partial S}\right)_H; \qquad -M = \left(\frac{\partial E}{\partial H}\right)_S; \tag{2.45b}$$

$$-S = \left(\frac{\partial G}{\partial T}\right)_H; \qquad -M = \left(\frac{\partial G}{\partial H}\right)_T; \tag{2.45c}$$

$$-S = \left(\frac{\partial A}{\partial T}\right)_M; \qquad H = \left(\frac{\partial A}{\partial M}\right)_T. \tag{2.45d}$$

The four corresponding Maxwell relations are

$$\left(\frac{\partial T}{\partial M}\right)_S = \left(\frac{\partial H}{\partial S}\right)_M, \tag{2.46a}$$

$$\left(\frac{\partial T}{\partial H}\right)_S = -\left(\frac{\partial M}{\partial S}\right)_H, \tag{2.46b}$$

$$\left(\frac{\partial S}{\partial H}\right)_T = \left(\frac{\partial M}{\partial T}\right)_H, \tag{2.46c}$$

$$\left(\frac{\partial S}{\partial M}\right)_T = -\left(\frac{\partial H}{\partial T}\right)_M. \tag{2.46d}$$

2.9. Magnetic response functions: specific heat and susceptibility

(i) Corresponding to C_V and C_P we introduce C_M and C_H, defined by

$$C_M \equiv T\left(\frac{\partial S}{\partial T}\right)_M = \left(\frac{\partial U}{\partial T}\right)_M = -T\left(\frac{\partial^2 A}{\partial T^2}\right)_M \tag{2.47}$$

and

$$C_H \equiv T\left(\frac{\partial S}{\partial T}\right)_H = \left(\frac{\partial E}{\partial T}\right)_H = -T\left(\frac{\partial^2 G}{\partial T^2}\right)_H. \tag{2.48}$$

(ii) Instead of the isothermal and adiabatic compressibilities K_T and K_S, we now have the isothermal susceptibility

$$\chi_T \equiv \left(\frac{\partial M}{\partial H}\right)_T = -\left(\frac{\partial^2 G}{\partial H^2}\right)_T \tag{2.49}$$

and the adiabatic susceptibility

$$\chi_S \equiv \left(\frac{\partial M}{\partial H}\right)_S = -\left(\frac{\partial^2 E}{\partial H^2}\right)_S. \tag{2.50}$$

Note from eqns (2.17) and (2.18) that the 'normalizing' factor V, which occurs in the definition of the compressibilities, does not appear in the corresponding definitions of the magnetic susceptibilities.

It is now straightforward to verify the relations

$$\chi_T(C_H - C_M) = T\alpha_H{}^2 \qquad\qquad \blacktriangleright \tag{2.51a}$$

and

$$C_H(\chi_T - \chi_S) = T\alpha_H{}^2, \qquad\qquad \blacktriangleright \tag{2.51b}$$

which are analogous to eqns (2.20). Here α_H is defined by

$$\alpha_H \equiv \left(\frac{\partial M}{\partial T}\right)_H. \tag{2.52}$$

In analogy to eqn (2.29), we have

$$C_H/C_M = \chi_T/\chi_S. \tag{2.53}$$

We mention, for later reference, that eqn (2.51a) may be written in the form

$$C_H - C_M = T\alpha_M^2 \chi_T, \tag{2.54}$$

where here

$$\alpha_M \equiv \left(\frac{\partial H}{\partial T}\right)_M = -\frac{\alpha_H}{\chi_T}.$$

(2.55)

2.10. Convexity relations for magnetic systems

In the case of a magnetic system we cannot argue, as we did in § 2.3 for fluid systems, that χ and C are both non-negative. In fact we know

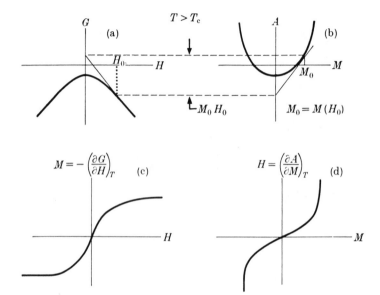

FIG. 2.8. Geometrical relation between $G(T, H)$ and $A(T, M)$ for a fixed temperature $T > T_c$. Also shown are the magnetization as a function of field (obtained from $M = -(\partial G/\partial H)_T$) and the field as a function of magnetization (obtained from $H = (\partial A/\partial M)_T$).

that $\chi < 0$ for a large number of diamagnetic materials. However, it can be shown (Griffiths 1964) that for a system whose Hamiltonian is of the form

$$\mathscr{H} = \mathscr{H}_0 - H\mathscr{M}$$

(2.56)

(where H is the magnetic field and $\mathscr{M}$ is the magnetization operator) that (i) the Gibbs potential $G(T, H)$ is a concave function of both temperature and magnetic field and (ii) the Helmholtz potential $A(T, M)$ is a concave function of the temperature and a convex function of the magnetization. It is reasonable that many real ferromagnets are closely approximated by a Hamiltonian of the form of (2.56); for example, familiar magnetic models, such as the Heisenberg and Ising models, can be written in the form (2.56).

2.11. Geometrical interpretation of the thermodynamic potentials for a magnetic system

For magnetic systems $G(T, H) = A(T, M) - MH$. Hence the potentials G and A are related to each other by geometric constructions

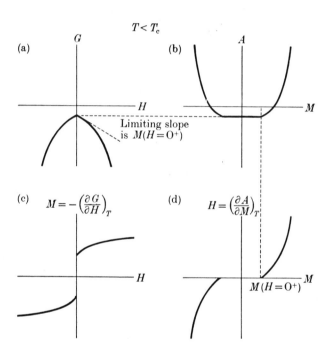

FIG. 2.9. Relation between $G(T, H)$ and $A(T, M)$ for a fixed temperature $T < T_c$. Also shown are the M–H isotherms (obtained from $G(T, H)$) and the H–M isotherms (obtained from $A(T, M)$).

analogous to Figs. 2.4 and 2.5 for the fluid case. Such constructions are shown in Fig. 2.8 for $T > T_c$ and in Fig. 2.9 for $T < T_c$. Also shown are M–H isotherms (obtained from the relation $M = -(\partial G/\partial H)_T$) and H–M isotherms (obtained from $H = (\partial A/\partial M)_T$).

Two slight differences between the magnetic and the fluid cases are worth noting here: (i) $A(T, M)$ is an even function of M, corresponding to the physical idea that $M(H)$ should be an odd function of H. We will use this fact when we consider the Landau theory in Chapter 10. (ii) The straight-line segment of the A–M isotherm is horizontal. This makes proving the Griffiths inequality easier for a magnetic system than for a fluid system (cf. Chapter 4).

Suggested further reading

Kittel (1958).
Callen (1960).
Fay (1965).
Reif (1965).
Wannier (1966).
Guggenheim (1967).
Kubo (1968).
Kittel (1969).

PART II

CRITICAL-POINT EXPONENTS AND RIGOROUS RELATIONS AMONG THEM

3

CRITICAL-POINT EXPONENTS

3.1. Definition of a critical-point exponent

IN recent years the study of critical phenomena has come to focus more and more on the values of a set of indices, called critical-point exponents, which describe the behaviour near the critical point of the various quantities of interest.

We begin with a precise and general definition of a critical-point exponent to describe the behaviour near the critical point of a general function $f(\epsilon)$, where

$$\epsilon \equiv \frac{T - T_c}{T_c} = \frac{T}{T_c} - 1 \qquad (3.1)$$

serves as a dimensionless variable to measure the difference in temperature from the critical temperature. We assume that this function $f(\epsilon)$ is positive and continuous for sufficiently small, positive values of ϵ, and that the limit

$$\lambda \equiv \lim_{\epsilon \to 0} \frac{\ln f(\epsilon)}{\ln \epsilon} \qquad \blacktriangleright (3.2)$$

exists. This limit, λ, is called the critical-point exponent to be associated with the function $f(\epsilon)$. As a shorthand notation, we frequently denote the fact that λ is the critical-point exponent for the function $f(\epsilon)$ by writing $f(\epsilon) \sim \epsilon^\lambda$.

It is important to stress that the relation $f(\epsilon) \sim \epsilon^\lambda$ does not imply the relation

$$f(\epsilon) = A\epsilon^x \qquad [x = \lambda], \qquad (3.3)$$

although of course the converse is true. In fact, it is relatively rare that the behaviour of a typical thermodynamic function is as simple as (3.3); in general we find that there are correction terms, and eqn (3.3) is replaced by a functional expression such as

$$f(\epsilon) = A\epsilon^x(1 + B\epsilon^y + \cdots) \qquad [y > 0]. \qquad (3.4)$$

We see that the definition (3.2) of a critical-point exponent does not distinguish the functional forms (3.3) and (3.4)—for both functions, $\lambda = x$.

We may rightfully question why we should focus on a quantity such as the critical-point exponent, which contains considerably less information than the complete functional form. The answer seems to lie in the experimental fact that sufficiently near the critical point the behaviour of the leading terms dominates. Therefore log–log plots of experimental data, such as that of Fig. 1.10, display straight-line behaviour sufficiently near the critical point, and hence the critical-point exponent is easily determined as the slope of this straight-line region. Hence critical-point exponents are measurable while the complete function may not be. A second reason for focusing on the critical-point exponents is that, as we shall see, there exists a large number of relations among the exponents that arise from fundamental thermodynamic and statistical mechanical considerations and thus transcend any particular system.

We conclude this section with a discussion of four simple examples (cf. Fig. 3.1). These examples are designed to illustrate the extent to which there is a simple one-to-one relationship between the magnitude of a critical-point exponent and the qualitative behaviour of the function under consideration near the critical point, $\epsilon \equiv 0$. In the first two of these examples, illustrated in Fig. 3.1(a) and (b), we show that negative values of the critical-point exponent λ defined in eqn (3.2) correspond to a function $f(\epsilon)$ which diverges to infinity at the critical point, while positive values of λ correspond to a function $f(\epsilon)$ which approaches zero. More important, perhaps, is the fact that the smaller the magnitude of λ, the 'sharper' is the behaviour of $f(\epsilon)$ in the sense that for negative λ the divergence is more sudden while for positive λ the approach to zero becomes steeper.

The intermediate case when λ is zero does not correspond to a unique type of behaviour; in fact, $\lambda = 0$ can correspond to a logarithmic divergence as in Fig. 3.1(c), to a cusp-like singularity as in Fig. 3.1(d), or to a perfectly analytic function with no anomalous behaviour worse than, say, a jump discontinuity. Since these three possibilities differ sharply from one another, we are motivated to introduce yet another sort of critical-point exponent that serves to distinguish the case of the logarithmic singularity from that of the cusp singularity. To find the exponent λ that describes the singular part of a function $f(\epsilon)$ with a

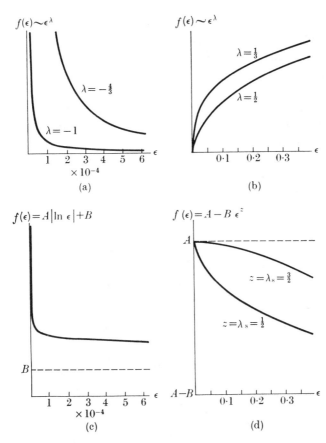

FIG. 3.1. Examples of the various types of behaviour near $\epsilon \equiv T/T_c - 1 = 0$. (a) $f(\epsilon) \sim \epsilon^\lambda$ with $\lambda < 0$. (b) $f(\epsilon) \sim \epsilon^\lambda$ with $\lambda > 0$. (c) $f(\epsilon) \sim |\ln f(\epsilon)|$, so that $\lambda = 0$. (d) $f(\epsilon) \sim$ const. $- \epsilon^{\frac{1}{2}}$ so that $\lambda = 0$, but $\lambda_s = \frac{1}{2}$.

cusp-like singularity, we first find the smallest integer j such that the derivative $\partial^j f/\partial \epsilon^j \equiv f^{(j)}(\epsilon)$ diverges as $\epsilon \to 0$. We then define

$$\lambda_s \equiv j + \lim_{\epsilon \to 0} \frac{\ln |f^{(j)}(\epsilon)|}{\ln \epsilon}. \tag{3.5}$$

For example, the function plotted in Fig. 3.1(d) is $f(\epsilon) =$ const. $- \epsilon^{\frac{1}{2}}$, and the first derivative diverges as $\epsilon^{-\frac{1}{2}}$ when $\epsilon \to 0$. Hence from eqn (3.5) it follows that $\lambda_s = 1 - \frac{1}{2} = \frac{1}{2}$ for this example.

The remaining sections of this chapter are designed to introduce the most commonly used critical-point exponents for fluid and magnetic systems. In general, the same notation for a critical exponent is used for the analogous fluid and magnetic functions, so that it is convenient to treat them together. It is important to stress that not all the exponents are defined along paths on which T is varying (cf. Table 3.1), and the discussion of this section applies to any function $f(x)$ where x might denote M or H as well as ϵ.

TABLE 3.1

Notation for measurements at constant P(or H), V(or M), T, and S

Fluid		Magnet
P = const. ($=P_c$) (critical) isobar V = const. ($=V_c$) ρ = const. ($=\rho_c$) (critical) isochore		H = const. ($=0$) (critical) isochamp M = const. ($=0$) (critical) isomag
	T = const. ($=T_c$) (critical) isotherm S = const. ($=S_c$) (critical) isentrope	

Note that measurements along the vapour pressure curve $P = P_{sat}(T)$ are *not* at constant pressure, nor at constant volume.

3.2. The critical-point exponents α, β, γ, δ, ν, and η

We begin our survey of a few selected critical-point exponents with a further analogy between fluid and magnetic phase transitions. Fig. 1.3 shows the temperature dependence of the liquid–gas density difference and the zero-field magnetization $M_0(T)$ of an idealized (single-domain) ferromagnet at constant magnetic field $H = 0$. From the shape of these curves it would seem natural to postulate temperature dependences in the neighbourhood of T_c which have the form

$$\{\rho_L(T) - \rho_G(T)\}/2\rho_c = \mathscr{B}(1 - T/T_c)^\beta[1 + \cdots] \qquad (3.6)$$

and

$$M_0(T)/M_0(0) = \mathscr{B}(1 - T/T_c)^\beta[1 + \cdots]. \qquad (3.7)$$

As a shorthand notation, we commonly write $\Delta\rho \equiv \rho_L - \rho_G \sim (-\epsilon)^\beta$ (or $M \sim (-\epsilon)^\beta$).

The normalization constants ρ_c and $M_0(0)$ are included in order that the coefficient $\mathscr{B}$ vary only slightly from system to system (generally $\mathscr{B}$ is of the order unity).

Fig. 1.2 shows PV isotherms and HM isotherms for fluid and magnetic systems respectively. The slopes of these isotherms are proportional, respectively, to the inverse isothermal compressibility K_T^{-1} and the inverse isothermal susceptibility χ_T^{-1}. Although both K_T and χ_T diverge to infinity as $T \to T_c$, we must distinguish whether the critical point is being approached from above or from below. Thus we define the (not necessarily equal) exponents γ' and γ by

$$K_T/K_T^\circ = \begin{cases} \mathscr{C}'(-\epsilon)^{-\gamma'}(1 + \cdots) & [T < T_c, \quad \rho = \rho_L(T) \quad \text{or} \quad \rho_G(T)] \\ \mathscr{C}\,\epsilon^{-\gamma}\,(1 + \cdots) & [T > T_c, \quad \rho = \rho_c] \end{cases} \tag{3.8}$$

for the fluid case, and

$$\chi_T/\chi_T^\circ = \begin{cases} \mathscr{C}'(-\epsilon)^{-\gamma'}(1 + \cdots) & [T < T_c, \quad H = 0] \\ \mathscr{C}\epsilon^{-\gamma}\,(1 + \cdots) & [T > T_c, \quad H = 0] \end{cases} \tag{3.9}$$

for the magnetic case. Here $K_T^\circ = 1/P_c^\circ \equiv m/kT_c\rho_c$ is the compressibility of an ideal gas of density ρ_c at $T = T_c$. Similarly, χ_T° is the susceptibility of a system of non-interacting magnetic moments (paramagnet) evaluated at the critical point. Measurements often come sufficiently close to T_c that K_T/K_T° is as large as 10^6. Note that we have included a minus sign in the definitions of γ' and γ in order that both exponents be positive.

The exponent δ describes the variation of $P - P_c$ with $\rho - \rho_c$ (and of H and M) along the critical isotherm, $T = T_c$:

$$(P - P_c)/P_c^\circ = \mathscr{D}|\rho/\rho_c - 1|^\delta \operatorname{sgn}(\rho - \rho_c) \quad [T = T_c], \tag{3.10}$$

$$H/H_c^\circ = \mathscr{D}|M_H(T = T_c)/M_0(T = 0)|^\delta \quad [T = T_c]. \tag{3.11}$$

Here $P_c^\circ = kT_c\rho_c/m$ is the pressure that the system would have at $\rho = \rho_c$ and $T = T_c$ if its particles did not interact; also, $H_c^\circ \equiv kT_c/m_0$, where here m_0 is the magnetic moment per spin. Thus the index δ serves to describe the degree of the critical isotherm, and from casual examination of Fig. 1.2 we might expect δ to be an odd integer such as 3 or 5. Experimentally, δ is found to have values (for different systems) which are between ~ 4 and ~ 6 and are generally non-integral—whence the complicated form of the defining equations (3.10) and (3.11). As a shorthand notation, we frequently write $\Delta P \sim (\Delta\rho)^\delta$ and $H \sim M^\delta$. The larger the value of δ, the 'flatter' the critical isotherm (and the harder it is to obtain an accurate experimental determination of δ); for example, $\delta \simeq 15$ for the two-dimensional Ising model, corresponding to an extremely flat critical isotherm.

The specific heat exponents α' and α are defined by the relations

$$C_V = \begin{cases} \mathscr{A}'(-\epsilon)^{-\alpha'}(1 + \cdots)\,[T < T_c\,\rho = \rho_L\,(T)\,\text{or}\,\rho_G(T) \ (3.12) \\ \mathscr{A}\epsilon^{-\alpha}(1 + \cdots) \quad [T > T_c, \rho = \rho_c] \hspace{1.2cm} (3.13) \end{cases}$$

for the fluid systems, and

$$C_H = \begin{cases} \mathscr{A}'(-\epsilon)^{-\alpha'}(1 + \cdots) \ [T < T_c] \\ \mathscr{A}\epsilon^{-\alpha}(1 + \cdots) \quad\ [T > T_c] \end{cases} \quad [H = 0] \quad (3.14)$$

for magnetic systems.

Note that we have departed from our strict fluid-magnet analogy $V \leftrightarrow -M$ and $P \leftrightarrow H$ in our definition (3.14). Fisher (1967) has argued

TABLE 3.2

Summary of definitions of critical-point exponents for fluid systems. Here
$$\epsilon \equiv T/T_c - 1$$

Expo-nent	Definition	Conditions ϵ	$P-P_c$	$\rho-\rho_c$	Quantity
α'	$C_V \sim (-\epsilon)^{-\alpha'}$	<0	0	0	specific heat at constant volume $V = V_c$
α	$C_V \sim \epsilon^{-\alpha}$	>0	0	0	
β	$\rho_L - \rho_G \sim (-\epsilon)^\beta$	<0	0	$\neq 0$	liquid–gas density difference (or shape of coexistence curve)
γ'	$K_T \sim (-\epsilon)^{-\gamma'}$	<0	0	$\neq 0$	isothermal compressibility
γ	$K_T \sim \epsilon^{-\gamma}$	>0	0	0	
δ	$P - P_c$ $\sim |\rho_L - \rho_G|^\delta \,\text{sgn}\,(\rho_L - \rho_G)$	0	$\neq 0$	$\neq 0$	critical isotherm
ν'	$\xi \sim (-\epsilon)^{-\nu'}$	<0	0	$\neq 0$	correlation length
ν	$\xi \sim \epsilon^{-\nu}$	>0	0	0	
η	$G(r) \sim |r|^{-(d-2+\eta)}$	0	0	0	pair correlation function (d = dimensionality)
Δ'_l	$\dfrac{\partial^l G}{\partial P^l} \equiv G^{(l)} \sim (-\epsilon)^{-\Delta} {}_l G^{(l-1)}$	<0	0	0	successive pressure derivatives of the Gibbs potential $G(T, P)$
Δ_{2l}	$\dfrac{\partial^{2l} G}{\partial P^{2l}} \equiv G^{(2l)} \sim \epsilon^{-2\Delta}{}_{2l}G^{(2l-2)}$	>0	0	0	

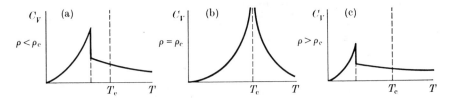

FIG. 3.2. Expected behaviour for the specific heat at densities less than, equal to, and greater than the critical density. After Heller (1967).

that in a sense $H = 0$ implies $M = 0$. This is certainly true for $T > T_c$, while for $T < T_c$ the two-phase $M = 0$ specific heat also corresponds to $H = 0$ since only then can oppositely magnetized domains coexist to yield zero total magnetization.

Suppose we measure C_V as a function of temperature for a system whose volume V_0 is not equal to V_c, for example along a path parallel to, but slightly below, the line $\rho = \rho_c$ in Fig. 1.3, corresponding to $V_0 > V_c$ or $\rho_0 < \rho_c$. Our expectations are shown in Fig. 3.2: $C_{V_0 \neq V_c}$

TABLE 3.3

Summary of definitions of critical-point exponents for magnetic systems.
Here $\epsilon \equiv T/T_c - 1$

Expo-nent	Definition	Conditions			Quantity
		ϵ	H	M	
α'	$C_H \sim (-\epsilon)^{-\alpha'}$	< 0	0	0	specific heat at constant magnetic field
α	$C_H \sim \epsilon^{-\alpha}$	> 0	0	0	
β	$M \sim (-\epsilon)^\beta$	< 0	0	$\neq 0$	zero-field magnetization
γ'	$\chi_T \sim (-\epsilon)^{-\gamma'}$	< 0	0	$\neq 0$	zero-field isothermal susceptibility
γ	$\chi_T \sim \epsilon^{-\gamma}$	> 0	0	0	
δ	$H \sim |M|^\delta \, \mathrm{sgn}\,(M)$	0	$\neq 0$	$\neq 0$	critical isotherm
ν'	$\xi \sim (-\epsilon)^{-\nu'}$	< 0	0	$\neq 0$	correlation length
ν	$\xi \sim \epsilon^{-\nu}$	> 0	0	0	
η	$\Gamma(r) \sim |r|^{-(d-2+\eta)}$	0	0	0	pair correlation function ($d =$ dimensionality)
Δ'_ℓ	$\dfrac{\partial^\ell G}{\partial H^\ell} \equiv G^{(\ell)} \sim (-\epsilon)^{-\Delta'_\ell} G^{(\ell-1)}$	< 0	0	0	successive field derivatives of the Gibbs potential $G(T, H)$
$\Delta_{2\ell}$	$\dfrac{\partial^{2\ell} G}{\partial H^{2\ell}} \equiv G^{(2\ell)} \sim \epsilon^{-2\Delta_{2\ell}} G^{(2\ell-2)}$	> 0	0	0	

will undergo a jump discontinuity at the temperature $T_0(< T_c)$ at which the isochore $\rho = \rho_0$ passes through the coexistence curve.

Experimentally, either the values of the exponents α' and α are zero or they are small positive numbers ($\alpha \simeq \frac{1}{8}$ is the three-dimensional Ising model prediction). The case $\alpha = 0$ corresponds in some measurements to a logarithmic singularity, and in others to a cusp-like singularity (in which case $\alpha_s \neq 0$).

The exponents ν', ν, and η refer to the behaviour of the pair correlation function $G(r)$ in the critical region. Since a rather extensive discussion of $G(r)$ will be presented in Chapter 7, at this time we give only a brief definition of the exponents themselves. The correlation length ξ is a measure of the range of the correlation function. It is assumed that

$$\xi = \begin{cases} \xi'_0(1 - T/T_c)^{-\nu'} & [T < T_c, H = 0] \\ \xi_0(T/T_c - 1)^{-\nu} & [T > T_c, H = 0] \end{cases}. \qquad (3.15)$$

In many cases the pair correlation function at $T = T_c$ falls off to zero with distance r with the simple power law form

$$G(r) \sim \frac{W}{r^{d-2+\eta}} \qquad [T = T_c; \quad P = P_c \quad (\text{or } H = 0)], \qquad (3.16)$$

thereby defining the critical-point exponent η. Here d is the dimensionality of the system.

The exponent definitions are summarized for fluid and magnetic systems in Tables 3.2 and 3.3 respectively.

3.3. Numerical values of critical-point exponents

Values of the exponents α', α, β, γ', γ, δ, ν', ν, and η are given in Table 3.4 for selected fluid and magnetic systems, and for a few theoretical models. Despite the fact that the critical temperatures change substantially from material to material (see, e.g. Table 3.5), we notice a distinct similarity among the values found experimentally for each exponent. At one time many workers believed that all materials have the same exponents, though this once-hoped-for universality has not been borne out by more recent experiments. For example, many of the early measurements on magnetic systems indicated that β had a value extremely close to $\frac{1}{3}$, but very recently Ho and Litster (1969) have shown unequivocally that $\beta = 0.368 \pm 0.005$ for the insulating ferromagnet $CrBr_3$.

What can theoretical models say about the critical-point exponents? First of all, a given model predicts a given set of critical-point

TABLE 3.4

Values of critical-point exponents for selected systems

System	$T < T_c$				$T = T_c$		$T > T_c$		
	α'	β	γ'	ν'	δ	η	α	γ	ν
Fluids									
CO_2	$\sim 0{\cdot}1$	$0{\cdot}34$	$\sim 1{\cdot}0$	—	$4{\cdot}2$	—	$\sim 0{\cdot}1$	$1{\cdot}35$	—
Xe	$< 0{\cdot}2$	$0{\cdot}35$	$\sim 1{\cdot}2$	$0{\cdot}57$	$4{\cdot}4$	—	—	$1{\cdot}3$	—
Magnets									
Ni	$\alpha'_s = -0{\cdot}3$	$0{\cdot}42$	—	—	$4{\cdot}22$	—	0	$1{\cdot}35$	—
EuS	$\alpha'_s = -0{\cdot}15$	$0{\cdot}33$	—	—	—	—	$0{\cdot}05$	—	—
$CrBr_3$	—	$0{\cdot}368$	—	—	$4{\cdot}3$	—	—	$1{\cdot}215$	—
Soluble Models									
classical	0 (disc)	$\frac{1}{2}$	1	$\frac{1}{2}$	3	0	0 (disc)	1	$\frac{1}{2}$
Ornstein–Zernike	—	—	—	—	5	0	$\alpha_s = -1$	2	1
$d = 3$ spherical model	—	$\frac{1}{2}$	—	—	5	0	$\alpha_s = -1$	2	1
$d = 2$ Ising model	0 (log)	$\frac{1}{8}$	$\sim \frac{7}{4}$	1	~ 15	$\frac{1}{4}$	0 (log)	$\sim \frac{7}{4}$	1
Approximations									
$d = 3$ Ising model (fluid?)	$\sim \frac{1}{8}$ or $\sim \frac{1}{16}$	$\sim \frac{5}{16}$	$\sim \frac{5}{4}$ or $\sim \frac{21}{16}$	—	~ 5	$\sim 0{\cdot}041$	$\sim \frac{1}{8}$	$\sim \frac{5}{4}$	$\sim 0{\cdot}638$
$d = 3$ Heisenberg model (magnet?)	—	$(\sim 0{\cdot}345?)$	—	—	~ 5	$(\sim 0{\cdot}03?)$	$\alpha_s \sim -0{\cdot}1$	$\sim 1{\cdot}4$	$\sim 0{\cdot}70$

TABLE 3.5

Critical-point parameters (critical temperature, critical pressure, and critical density) for selected fluids

Fluid	T_c (K)	P_c (atm)	ρ_c (g cm^{-3})
Water	647·5	218·5	0·325
Alcohol	516·6	63·1	0·28
Ether	467·0	35·5	0·26
Carbon dioxide	304·20	72·8	0·46
Xenon	289·75	57·64	1·105
Oxygen	154·6	49·7	0·41
Argon	150·8	48·34	0·53
Hydrogen	32·98	12·76	0·0314
Helium-4	5·19	2·25	0·069
Helium-3	3·324	1.15	0·04178

exponents, independent of the location of the critical point. Moreover, the van der Waals theory of a fluid, the molecular field theory of a magnet, and the Landau theory (at least in their original forms) all give the same values for each of the exponents. These theories we call *classical* to suggest the fact that until fairly recently they were essentially the only theories of critical-point exponents. However, it is clear from Table 3.4 (and Fig 3.3) that these classical theories fail to predict the observed values of the exponents. Also failing in this regard are the Ornstein–Zernike theory and the various exactly soluble models such as the two-dimensional Ising model, the spherical model, and the ideal Bose gas.

On the other hand, the predictions of the three-dimensional Ising model and the three-dimensional classical Heisenberg model do appear to mirror to a certain extent the data on, respectively, fluid and magnetic systems. Certainly a first goal of any theory of critical phenomena is to find such theoretical models and, moreover, to understand which features of the models are relevant in determining the values of the critical-point exponents and which are not. Even if we should eventually succeed in this first goal, we will still be left with the question, 'Isn't there some underlying theory of all these exponents which tells us how they hang together?' A second goal, then, is to study relations among the various exponents. One step in this direction is taken in the following chapter, where we shall demonstrate certain rigorous inequalities which relate the various exponents. In Chapter 11 we shall study in some detail the recent homogeneous function or scaling law theory of exponents which, among other results, predicts that the inequalities be satisfied as equalities.

3.4. The exponents Δ and θ

We introduce in this section additional critical-point exponents that will be needed elsewhere.

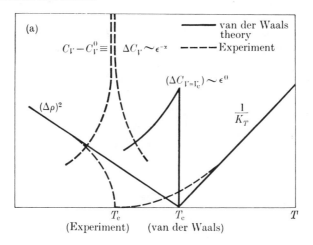

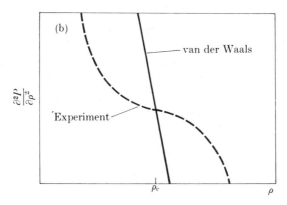

FIG. 3.3. Schematic comparison of the predictions of the classic van der Waals theory with typical experimental measurements—on fluid systems—of the exponents α (specific heat), β (coexistence curve), γ (isothermal compressibility), and δ (critical isotherm). Experiments of the sort shown in (b) have been carried out on xenon by Habgood and Schneider (1954).

3.4.1. *The gap exponents Δ'_ℓ and $\Delta_{2\ell}$*

Recall that for a magnetic system, the spontaneous magnetization and the zero-field susceptibility are proportional, respectively, to the first and second derivatives of the Gibbs potential $G(T, H)$ with respect to the magnetic field H (and evaluated at $H = 0$). In 1963 Essam and Fisher suggested that one might consider higher field derivatives of $G(T, H)$, thereby defining a sequence of exponents Δ'_ℓ called gap exponents. Thus

$$(\partial G/\partial H)_T \equiv G^{(1)} \sim (1 - T/T_c)^{-\Delta'_1} G^{(0)} \qquad (3.17)$$

and

$$(\partial^2 G/\partial H^2)_T \equiv G^{(2)} \sim (1 - T/T_c)^{-\Delta'_2} G^{(1)}. \qquad (3.18)$$

In general, then,

$$(\partial^\ell G/H^\ell)_T \equiv G^{(\ell)} \sim (1 - T/T_c)^{-\Delta'_\ell} G^{(\ell-1)}. \qquad \blacktriangleright \ (3.19)$$

In eqns (3.17)–(3.19), $H = 0$ and $T \to T_c^+$.

Since $G^{(1)} \sim M \sim (1 - T/T_c)^\beta$, and the specific heat $C_{H=0} \sim \partial^2 G/\partial T^2$ diverges as $(1 - T/T_c)^{-\alpha'}$, it follows that $G^{(0)} \sim (1 - T/T_c)^{2-\alpha'}$ and hence

$$\Delta'_1 = 2 - \alpha' - \beta. \qquad \blacktriangleright \ (3.20)$$

Similarly, the fact that $G^{(2)} \sim \chi_T \sim (1 - T/T_c)^{-\gamma'}$ means that

$$\Delta'_2 = \beta + \gamma'. \qquad \blacktriangleright \ (3.21)$$

For the molecular field theory, $\Delta'_\ell = \frac{3}{2}$ for all ℓ (e.g. $\Delta'_1 = 2 - \alpha' - \beta = \frac{3}{2}$, $\Delta'_2 = \beta + \gamma' = \frac{3}{2}, \cdots$). For the two-dimensional Ising model, $\alpha' = 0$ and $\beta = \frac{1}{8}$ are rigorous results. Hence $\Delta'_1 = \Delta'_2 \equiv \frac{15}{8}$ exactly providing $\gamma' = \frac{7}{4}$. Numerical studies of Essam and Fisher (1963) find $\Delta'_3 = 1\cdot87 \pm 0\cdot05 \simeq \frac{15}{8}$, so that it is certainly plausible that $\Delta'_\ell = \frac{15}{8}$ for all ℓ for the two-dimensional Ising model. Note that the conjectured equality $\Delta'_1 = \Delta'_2$ (or $2 - \alpha' - \beta = \beta + \gamma'$) serves to relate the exponents α', β, and γ':

$$\alpha' + 2\beta + \gamma' = 2. \qquad (3.22)$$

The conjectured equality (3.22) is valid for the molecular field theory and for the two-dimensional Ising model, and it is not inconsistent with numerical studies on the three-dimensional Ising model.

Above T_c, all odd field derivatives of $G(T, H)$ are zero when evaluated at $H = 0$. Hence we generalize (3.19) to

$$(\partial^{2\ell} G/\partial H^{2\ell})_T \equiv G^{(2\ell)} \sim (T/T_c - 1)^{-2\Delta_{2\ell}} G^{(2\ell-2)}. \qquad (3.23)$$

Numerical calculations for the Ising model give

$$\Delta_4 \simeq \Delta_6 \simeq \Delta_8 \simeq \frac{15}{8} \qquad (3.24)$$

for two-dimensional lattices and

$$\Delta_4 \simeq \Delta_6 \simeq \Delta_8 \simeq 1{\cdot}56 \pm 0{\cdot}03 \simeq \tfrac{25}{16} \qquad (3.25)$$

for three-dimensional lattices (Essam and Hunter, 1968). For the three-dimensional Heisenberg model,

$$\Delta_4 \simeq \Delta_6 \simeq \Delta_8 \simeq 1{\cdot}81 \qquad (3.26)$$

for spin quantum number $S = \tfrac{1}{2}$ and

$$\Delta_4 \simeq \Delta_6 \simeq \cdots \simeq 1{\cdot}73 \qquad (3.27)$$

for $S = \infty$, the classical Heisenberg model (Baker, Gilbert, Eve, and Rushbrooke 1967, Stephenson and Wood 1968). Thus we see that these model systems lend support to the general conjecture that the gap exponents are equal for all orders. It should be remarked that the gap exponents are largely of theoretical interest at the present time, since the higher derivatives of the Gibbs potential have not yet been measured experimentally.

3.4.2. *The vapour pressure curvature exponent θ*

The specific heat C_V may be written, directly from eqn (2.10c), as

$$C_V = TV(\partial^2 P/\partial T^2)_V - TN(\partial^2 \mu/\partial T^2)_V, \qquad (3.28)$$

where $\mu \equiv G/N$ is the chemical potential. Yang and Yang (1964) pointed out that if C_V is singular at $T = T_c$, then from eqn (3.28) $(\partial^2 P/\partial T^2)_V$ or $(\partial^2 \mu/\partial T^2)_V$—or both—will be singular. We define the exponent θ by

$$\left(\frac{\partial^2 P}{\partial T^2}\right) \sim (1 - T/T_c)^{-\theta}. \qquad (3.29)$$

Thus θ is a measure of the degree of divergence (if any) of the curvature of the vapour pressure curve as $T \to T_c$ from below (see Fig. 1.1). Yang and Yang argue that for the lattice-gas model of a fluid $(\partial^2 \mu/\partial T^2)_{V=V_c}$ approaches zero at the critical point, so that $\theta = \alpha'$. However, for real gases the curvatures of μ and P might both be singular, so that θ might differ from α'. Recently Moldover (1969) has found experimentally that for ^4He $(\partial^2 \mu/\partial T^2)_V$ is less singular than $(\partial^2 P/\partial T^2)_V$ and may even be constant near T_c. In any case, the specific heat singularity in ^4He appears to be dominated by $(\partial^2 P/\partial T^2)_V$.

3.5. Useful relations among critical-point exponents

To conclude this chapter, we present general relations among critical-point exponents that will be of use, for example, in proving inequalities among the various critical-point exponents in Chapter 4.

LEMMA 3. *Let* $f(x) \sim x^\lambda$, $g(x) \sim x^\varphi$, *and let* $f(x) \leq g(x)$ *for sufficiently small positive* x. *Then* $\lambda \geq \varphi$.

The proof of Lemma 3 is quite straightforward, but for the sake of completeness we shall reproduce it here. The assumption $f(x) \leq g(x)$ implies that $\ln f(x) \leq \ln g(x)$. Now for $x < 1$, $\ln x < 0$, and

$$\frac{\ln f(x)}{\ln x} \geq \frac{\ln g(x)}{\ln x}, \tag{3.30}$$

where the inequality sign is reversed. The relation (3.30) is valid as $x \to 0$, whence from the fundamental definition of a critical point exponent, eqn (3.2), it follows that $\lambda \geq \varphi$.

Lemma 3 is basic to the proofs of exponent inequalities so it is perhaps useful to keep some simple examples in mind. In particular, we note that Lemma 3 is valid regardless of the signs of λ and φ. For example, suppose $f(x) = 1/x$ and $g(x) = 1/x^{4/3}$. Since $f(x) < g(x)$ for $0 < x < 1$, the hypotheses of Lemma 3 are fulfilled, and $\lambda = -1$, $\varphi = -\frac{4}{3}$ so that indeed $\lambda > \varphi$ (cf. Fig. 3.1(a)).

Additional relations that are useful in proving exponent inequalities have been given by Griffiths (1965b).

Suggested further reading
Guggenheim (1945).
Fisher (1967).
Heller (1967).

4

EXPONENT INEQUALITIES

THE only rigorous relations thus far proposed among the critical-point exponents are a set of inequalities. In this chapter we shall present in detail the proofs for magnetic systems of a few representative examples of such inequalities; most of the remaining inequalities that have been proved thus far are discussed in §4.3 and are listed in Table 4.1.

4.1. The Rushbrooke and Coopersmith inequalities

4.1.1. The Rushbrooke inequality: $H = 0$, $T \to T_c^-$

We begin by recalling from §2.9 that the specific heat C_M must be positive so that eqn (2.51a) leads to the inequality

$$C_H \geq T \left\{ \left(\frac{\partial M}{\partial T} \right)_H \right\}^2 \Big/ \chi_T. \tag{4.1}$$

Using the definitions of the critical-point exponents from Chapter 3,

$$C_H \sim (-\epsilon)^{-\alpha'}, \quad \chi_T \sim (-\epsilon)^{-\gamma'}, \quad \text{and} \quad (\partial M/\partial T)_H \sim (-\epsilon)^{\beta-1}$$

and using Lemma 3 (§3.5), we see that (4.1) implies that $-\alpha' \leq 2(\beta - 1) + \gamma'$ or equivalently,

$$\alpha' + 2\beta + \gamma' \geq 2. \qquad \blacktriangleright \tag{4.2}$$

The relation (4.2) is generally called the Rushbrooke inequality (Rushbrooke 1963), although actually it was first conjectured, on the basis of heuristic arguments, in the form of an equality (Essam and Fisher 1963). Using the definitions of the gap exponents Δ'_ℓ presented in §3.4, eqn (4.2) may be written equivalently as

$$\Delta'_2 \geq \Delta'_1. \tag{4.3}$$

After the style of Onsager, Rushbrooke first proposed the inequality (4.2) as a discussion remark at the end of a seminar by someone else, namely a seminar which presented the latest (1963) estimates of the exponents α', β, and γ' obtained from the detailed Padé approximant analyses: $\alpha' = 0$, $\beta = \frac{5}{16}$, $\gamma' = \frac{5}{4}$. (Numerical approximation methods, useful for models that have not been solved exactly, are discussed in Chapter 9.) Rushbrooke calmly questioned the speaker as to whether he had any explanation as to why the quantity $\alpha' + 2\beta + \gamma'$ $(=\frac{15}{8})$ fell short of 2. The speaker replied that he knew of no reason why they should add up to 2 in the first place, whereupon Rushbrooke produced a two-line derivation of the now famous inequality (4.2). The first efforts were directed toward finding some loop-hole in Rushbrooke's thermodynamic argument. Not surprisingly, thermodynamics held up under scrutiny and the errors were then presumed to lie in the numerical approximations. Subsequent arguments based upon series extrapolations now favour the value $\alpha' = \frac{1}{8}$, so that (4.2) would appear to be obeyed as an equality (Gaunt 1967, Gaunt and Domb 1968, Garelick and Essam 1968, Guttmann and Thompson 1969).

For the two-dimensional Ising model it has been proved rigorously (Onsager 1944) that $\alpha' = 0$ (corresponding to a logarithmic singularity in the zero-field specific heat) and $\beta = \frac{1}{8}$ (Yang 1952). Hence the Rushbrooke inequality implies that $\gamma' \geq \frac{7}{4}$. Using arguments that fall barely short of being fully rigorous, it has been shown that in fact $\gamma' = \gamma = \frac{7}{4}$ (Fisher 1964). Hence for the two-dimensional Ising model the Rushbrooke inequality would appear to be satisfied as an equality, as it is for the classical theories for which $\alpha' = 0$ (corresponding to a jump discontinuity) $\beta = \frac{1}{2}$, and $\gamma' = 1$.

For the relatively few materials on which α', β, and γ' have all been measured, the experimental values often 'fail to add up to two' unless the error bars are taken into account. Thus experimental results, as well as theoretical work on the Ising model for $d = 2$ and 3, suggest that the Rushbrooke inequality may well be obeyed as an equality. This is, in fact, one of the predictions of the (non-rigorous) *scaling law theory of exponents*, which we shall develop in Chapter 11.

We now show that (4.2) fails to be an equality if and only if the quantity

$$R \equiv \lim_{\epsilon \to 0} C_M/C_H \qquad (4.4)$$

has the value unity (Rushbrooke 1965). We begin by writing eqn (2.51a) in the form

$$1 - C_M/C_H = T\alpha_H^2/C_H\chi_T. \qquad (4.5)$$

(a) Consider first the case $R = 1$, and define the positive exponent x by the (assumed) dependence near $T = T_c$,

$$C_M/C_H \sim R - (-\epsilon)^x(1 + \cdots) = 1 - (-\epsilon)^x(1 + \cdots). \qquad (4.6)$$

Then, on substituting into (4.5), we obtain

$$1 - \{1 - (-\epsilon)^x(1 + \cdots)\} = (-\epsilon)^{\alpha' + 2\beta + \gamma' - 2}(1 + \cdots), \qquad (4.7)$$

or

$$2 + x = \alpha' + 2\beta + \gamma'. \qquad (4.8)$$

(b) Next consider the case $R \equiv 1 - y < 1$ (i.e. $y > 0$). Then near T_c eqn (4.5) leads to

$$1 - \{1 - y(1 + \cdots)\} = y(1 + \cdots) = (-\epsilon)^{\alpha' + 2\beta + \gamma' - 2}(1 + \cdots) \qquad (4.9)$$

and

$$2 = \alpha' + 2\beta + \gamma'. \qquad (4.10)$$

4.1.2. *The Coopersmith inequality:* $T = T_c$, $H \to 0^+$

The Rushbrooke inequality was obtained by setting $H = 0$ in (4.1) and letting $T \to T_c$. Suppose instead that we set $T = T_c$ and then let $H \to 0$. In this case, $M \sim H^{1/\delta}$ so that $\chi_T \sim H^{1/\delta - 1}$. Next introduce the exponents φ and ψ by means of the definitions

$$C_H \sim H^{-\varphi} \sim M^{-\varphi\delta} \qquad [T = T_c] \qquad (4.11)$$

and

$$S(H) \sim -H^{\psi} \sim M^{\psi\delta} \qquad [T = T_c]. \qquad (4.12)$$

Finally, we use the Maxwell relation (2.46c) to replace $(\partial M/\partial T)_H$ on the right-hand side of (4.1) by $(\partial S/\partial H)_T \sim -H^{\psi - 1}$. Hence (4.1) becomes, for $T = T_c$,

$$H^{-\varphi}(1 + \cdots) \geq H^{2(\psi - 1) - (1/\delta - 1)}(1 + \cdots) \qquad (4.13)$$

or

$$\varphi + 2\psi - 1/\delta \geq 1. \qquad \blacktriangleright \ (4.14)$$

Equation (4.14) was first derived by Coopersmith (1968a). Unfortunately the exponents φ and ψ are not readily measured experimentally.

The exponent ζ is defined by means of the relation

$$S \sim -M^{1+\zeta} \qquad (T = T_c). \qquad (4.15)$$

From eqn (4.12) we have that

$$\zeta = \psi\delta - 1, \qquad (4.16)$$

so that the Coopersmith inequality may be written in terms of ζ as

$$\varphi + (2\zeta + 1)/\delta \geq 1 \quad \text{or} \quad \varphi\delta + 2\zeta + 1 \geq \delta. \qquad (4.17)$$

Many authors prefer to use ζ instead of ψ.

The mean field approximation (cf. Chapter 6) predicts that $\varphi = 0$,

$\delta = 3$, and $\psi = \frac{2}{3}$ (so that $\zeta = \psi\delta - 1 = 1$); hence the Coopersmith inequality reduces to an equality for this model. Unfortunately, the new exponents φ and ψ are not known for most model systems, thereby precluding extensive tests of the Coopersmith inequality.

4.2. The Griffiths inequality

The proof of the Griffiths inequality (Griffiths 1965a, 1965b),

$$\alpha' + \beta(1 + \delta) \geq 2, \qquad (4.18)$$

illustrates the utility of the convexity relations described in Chapter 2.

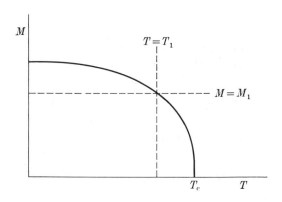

Fig. 4.1. Zero-field magnetization as a function of temperature. M_1 is the value of the magnetization when $T = T_1$. This diagram is useful in proving the Griffiths inequality (4.18).

We consider here a magnetic system as the proof is simpler than for a fluid. Choose an arbitrary temperature $T = T_1$ which is less than T_c and let M_1 denote the value of the zero-field magnetization at temperature T_1 (cf. Fig. 4.1), that is, $M_1(T_1)$ is just the spontaneous magnetization which is assumed to vanish as $(T_c - T_1)^\beta$ when $T_1 \to T_c$. Now since for all $M < M_1$, $(\partial A/\partial M)_T = H = 0$ (see eqn (2.45d)), it follows that

$$A(T_1, M) = A(T_1, 0), \qquad M \leq M_1(T_1). \qquad (4.19)$$

Similarly, the fourth Maxwell relation for magnetic systems, eqn (2.46d), implies that

$$S(T_1, M) = S(T_1, 0), \qquad M \leq M_1(T_1). \qquad (4.20)$$

Now define the functions

$$A^*(T, M) \equiv \{A(T, M) - A_c\} + (T - T_c)S_c \qquad (4.21)$$

and

$$S^*(T, M) \equiv S(T, M) - S_c, \qquad (4.22)$$

where $A_c = A(T_c, M)$ and $S_c \equiv S(T_c, M)$. With these definitions,

$$S^*(T, M) = -\left(\frac{\partial A^*}{\partial T}\right)_M. \qquad (4.23)$$

The function A^* is shown as a function of T and M in Fig. 4.2. Note from (4.21) that A^* has the same convexity properties as does A, since the addition of a term linear in T cannot affect the second derivatives.

We next utilize the convexity properties of A^* to obtain our first inequality. Choose some temperature $T_1 < T_c$ and, as before, let

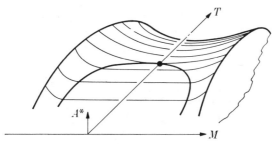

FIG. 4.2. General appearance of the function $A^*(T, M)$; note that the point $(T = T_c, M = 0)$ is a saddle point. After Griffiths (1965b).

$M_1(T_1)$ denote the saturation magnetization corresponding to T_1. Then consider the function $A^*(T, M)$ for constant $M = M_1$ (cf. Fig. 4.3). Draw a tangent to this curve at the point $T = T_1$; the equation of the tangent is

$$f(T) = A^*(T_1, M_1) + (T - T_1)\left(\frac{\partial A^*}{\partial T}\right)_{T=T_1}. \qquad (4.24)$$

Then since A^* is a concave function of T for constant M, all points on the A^* curve must lie below (or on) the tangent curve (4.24); in particular, $A^*(T, M_1)$ for $T = T_c$ must lie below (4.24), that is, $A^*(T_c, M_1) \leq f(T_c)$ or, on using (4.23) and (4.24),

$$A^*(T_c, M_1) \leq A^*(T_1, M_1) - (T_c - T_1)S^*(T_1, M_1). \qquad (4.25)$$

Because of the relations (4.19) and (4.20), eqn (4.25) is equivalent to

$$A^*(T_c, M_1) \leq A^*(T_1, 0) - (T_c - T_1)S^*(T_1, 0). \qquad (4.26)$$

Next observe that the concavity property of A^* implies $A^*(T_1, 0) \leq A^*(T_c, 0)$ since the tangent to $A^*(T, M_1)$ is horizontal at $T = T_c$. Finally, we note from the definition (4.21) that $A^*(T_c, 0) = 0$; hence (4.26) implies that

$$A^*(T_c, M_1) \leq -(T_c - T_1)S^*(T_1, 0). \qquad (4.27)$$

Assuming that $H(M_1, T_c) \sim M_1^\delta$ (recall eqn (3.11)), the relation $H = (\partial A^*/\partial M)_T$ implies

$$A^*(T_c, M_1) \sim M_1^{\delta+1} \sim (T_c - T_1)^{\beta(\delta+1)}. \qquad (4.28)$$

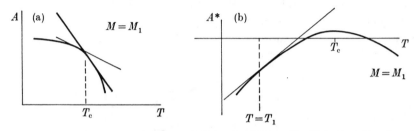

FIG. 4.3. (a) $A(T, M_1)$ as a function of T for fixed $M = M_1$. Note that the slope is always negative, as is required for the entropy to be positive. (b) $A^*(T, M_1)$ as a function of T for $M = M_1$. The slope of this plot is given by

$$-(\partial A^*/\partial T)_M = S^*(T, M_1) \equiv S(T, M_1) - S_c.$$

Consider next the right-hand side of (4.27). Since T is slowly varying near T_c, we may write

$$\frac{C}{T} = \left(\frac{\partial S}{\partial T}\right)_{M=0} \sim (T_c - T)^{-\alpha'}, \qquad (4.29)$$

or

$$S^*(T_1, 0) \equiv S(T_1, 0) - S(T_c, 0) \sim -(T_c - T_1)^{1-\alpha'}. \qquad (4.30)$$

Hence the right-hand side of (4.27) varies as $(T_c - T_1)^{2-\alpha'}$; comparing with (4.28) and using Lemma 3 (§ 3.5) yields

$$\beta(\delta + 1) \geq 2 - \alpha', \qquad \blacktriangleright (4.31)$$

which may be recognized as the Griffiths inequality (4.18).

Additional complications arise for a fluid system because a plot of A as a function of V has a straight-line portion whose slope, given by $-P$, changes with temperature as $T \rightarrow T_c$, and the derivation of the Griffiths inequality for the fluid case is rather more difficult (Griffiths 1965b).

Note from Table 3.4 that (4.31) holds as an equality for the classical theories. Observe, further, that for the two-dimensional Ising model, it provides the rigorous result $\delta \geq 15$, which is consistent with the value $\delta = 15 \pm 0.08$ provided by numerical approximation methods (Gaunt, Fisher, Sykes, and Essam 1964).

The Griffiths inequality may be tested by comparison with experiments on a system such as helium-4 for which all three exponents α', β, and δ have been measured. Specifically, Roach and Douglass (1967) and

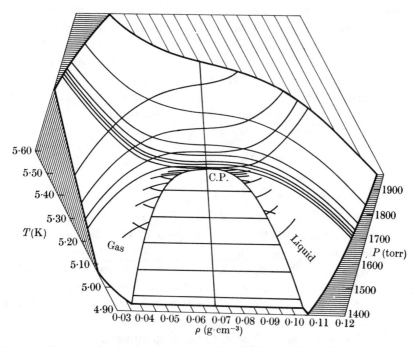

Fig. 4.4. Pressure–density–temperature surface of ^{4}He near the critical point, showing paths along which Roach and Douglass obtained their data. After Roach (1968). (Note that 1 atm = 760 torr.)

Roach (1968) have attempted to map out the $P\rho T$ surface of ^{4}He by making measurements along the paths shown in Fig. 4.4. They obtained

$$\beta = 0.354 \pm 0.010 \tag{4.32}$$

and

$$3.8 \leq \delta \leq 4.1. \tag{4.33}$$

In order to test the Griffiths inequality, Roach and Douglass calculated $\alpha' = 0.017 \pm 0.008$ by analysing specific heat data taken by Moldover

(1969). Using these values of the exponents, we obtain $1 \cdot 660 \leq \alpha' + \beta(\delta + 1) \leq 1 \cdot 881$, that is

$$\{\alpha' + \beta(\delta + 1)\}_{\max} = 1 \cdot 881 < 2. \tag{4.34}$$

Thus the experimental results as reported would appear to be inconsistent with the requirements of thermodynamics. However, in any situation such as this there is always one easy way out, namely to conclude that the experimental measurements did not extend in temperature sufficiently close to T_c that the limiting behaviour could be ascertained. If this is the case, thermodynamics has told us something very useful indeed: the experiment should be repeated. In the particular case of the Roach–Douglass helium data, several other possible interpretations of the discrepancy (4.34) have been suggested (Moldover 1969, Coopersmith 1968b). In particular, Moldover (1969) has recently published a detailed account of his own specific heat measurements and he concludes that $\alpha' \sim 0 \cdot 15$, whereupon

$$\{\alpha' + \beta(\delta + 1)\}_{\max} = 2.01 > 2. \tag{4.35}$$

4.3. More inequalities

In the preceding sections we discussed in detail three of the more easily proved inequalities that relate various of the critical-point exponents. These inequalities are of particular utility for reasons indicated above and also because each has as a counterpart an equality predicted by the non-rigorous scaling-law theory of exponents. There exist several more useful inequalities which we shall collect in this section. Since most of these remaining inequalities cannot be demonstrated easily, we shall refer the interested reader to the original work for details of the proofs.

4.3.1. *The remaining Griffiths inequalities*

What is generally termed *the* Griffiths inequality, (4.18), is actually only one of many inequalities proved by Griffiths. For example, two other inequalities Griffiths (1965b, 1968) has proved are

$$\gamma' \geq \beta(\delta - 1) \tag{4.36}$$

and

$$\gamma(\delta + 1) \geq (2 - \alpha)(\delta - 1). \tag{4.37}$$

The remaining Griffiths inequalities are listed in Table 4.1 (Nos. 3–9). Most of these require for their 'proofs' certain plausible assumptions

TABLE 4.1

Rigorous inequalities among critical-point exponents for magnetic systems. Assumptions are listed in Table 4.2; exponents not defined in Table 4.2 are defined in Table 3.2. Adapted from Griffiths (1965b).

	Inequality	Assumptions
1.	$\alpha' + 2\beta + \gamma' \geq 2$	—
2.	$\alpha' + \beta(\delta + 1) \geq 2$	—
3.	$(2 - \alpha')\zeta + 1 \geq (1 - \alpha')\delta$	B
4.	$\gamma'(\delta + 1) \geq (2 - \alpha')(\delta - 1)$	B, C_1
5.	$\gamma' \geq \beta(\delta - 1)$	A, C_1
6.†	$\delta \geq \delta_s$	—
7.†	$\delta_s \geq \min[\delta, \zeta + \sigma]$	B
8.†	$(2 - \alpha)\sigma \geq \delta_s + 1$	B
9.†	$\gamma(\delta_s + 1) \geq (2 - \alpha)(\delta_s - 1)$	B, C_2
10.	$\varphi + 2\psi - 1/\delta \geq 1$	—
11.	$d(\delta - 1)/(\delta + 1) \geq 2 - \eta$	D–F
12.	$d\gamma'/(2 - \alpha') \geq d\gamma'/(2\beta + \gamma') \geq 2 - \eta$	D–F
13.	$(2 - \eta)\nu \geq \gamma$	D–F
14.	$d\alpha' \geq 2 - \eta_E$	D–F
15.	$2d\varphi\delta/(\delta + 1 + \varphi\delta) \geq d\varphi/(\psi + \varphi) \geq 2 - \eta_E$	D–F
16.	$d\nu' \geq 2 - \alpha'$	other
17.	$d\nu \geq 2 - \alpha$	other

† If $\alpha \geq \alpha'$, inequalities 3, 6, 7, and 8 lead to the result $\delta_s = \delta$ (Griffiths 1965b).

that have not been established yet for general systems; these auxiliary assumptions are given in Table 4.2. For example, (4.37) is obtained by making essentially the following three assumptions:

$$(\partial M/\partial T)_H \leq 0 \quad \text{for} \quad H \geq 0, \tag{4.38}$$

$$(\partial^2 M/\partial H^2)_T \leq 0 \quad \text{for} \quad H \geq 0, \tag{4.39}$$

and

$$\alpha \geq \alpha'. \tag{4.40}$$

Assumption (4.38) has recently been proved for Ising ferromagnets, and is plausible for other systems (cf. Fig. 4.5(a)). Assumption (4.39) states that the M–H isotherms are concave and, although it has not been proved either generally or for any particular systems, it is certainly plausible (cf. Fig. 4.5(b)). Assumption (4.40) states that the specific heat singularity is not stronger when T_c is approached from below than from above. It is valid for the Ising model if $d = 2$ (and probably also valid if $d = 3$). Thus the inequality (4.37) may well hold for some systems. One must note the difference in proving a general thermodynamic

TABLE 4.2

Assumptions referred to in Table 4.1

A. $(\partial H/\partial T)_M = -(\partial S/\partial M)_T \geq 0$ for $M \geq 0$. Cf. Fig. 4.5(a).

B. $(\partial S/\partial M)_T \leq 0$ for $M \geq 0$; $T = T_c$.

C_1. $(\partial H/\partial M)_T$ is a non-decreasing function of M (≥ 0) for $T \leq T_c$. Cf. Fig. 4.5(b)

C_2. $(\partial H/\partial M)_T$ is a non-decreasing function of M (≥ 0) for $T \geq T_c$. Cf. Fig. 4.5(b)

D. *Positivity of the correlation functions*—for all temperatures T and any non-negative field (i.e. $H \geq 0$), $\Gamma_1 > 0$, $\Gamma_2(r) - \Gamma_1^2 \geq 0$, and $\Gamma_4(r_1,r_2,r_3) - \Gamma_2(r_1)\Gamma_2(r_3 - r_2) \geq 0$, where here Γ_j denotes the j-spin correlation function (and hence Γ_1 is proportional to the spontaneous magnetization).

E. *Monotonicity with magnetic field*—for any fixed value of T, $\Gamma_1(T, H)$, $\Gamma_2(T, H)$, and $\Gamma_4(T, H)$ are monotonic non-decreasing functions of H for $H \geq 0$ (in particular, for $T > T_c$, $\Gamma_1 = 0$ provided $H = 0$).

F. *Monotonicity with temperature*—for any fixed (non-negative) value of H, $\Gamma_1(T, H)$, $\Gamma_2(T, H)$, and $\Gamma_4(T, H)$ are monotonic non-increasing functions of T.

Exponent definitions:

    (a) $S_C - S \sim M^{\zeta+1}$ $T = T_c$

    (b) $T - T_c \sim M^\sigma$ $S = S_c$

    (c) $H \sim M^{\delta_s}$ $S = S_c$

    (d) $C_H \sim H^{-\varphi}$ $T = T_c$

    (e) $S(H) \sim -H^\psi$ $T = T_c$ $[\zeta \equiv \psi\delta - 1]$

    (f) $\Gamma_{EE} \sim r^{-(d-2+\eta_E)}$ $T = T_c$

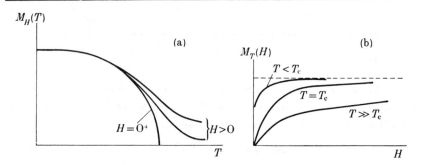

FIG. 4.5. (a) General form of the magnetization curves required by assumption (4.38), that M be a non-increasing function of T for fixed non-negative H. (b) Assumption (4.39) requires that the M–H isotherms be concave.

relation, such as (4.1) and (4.18), and one valid for particular systems for which auxiliary assumptions such as eqns (4.38)–(4.40) must be satisfied.

Notice that if we further assume that $2 - \alpha - \gamma > 0$ (as is generally true), then (4.37) implies

$$\frac{2 - \alpha + \gamma}{2 - \alpha - \gamma} \geq \delta, \qquad (4.41)$$

which is a useful upper bound on the critical isotherm exponent δ.

For example, for the three-dimensional classical Heisenberg model, one set of numerical estimates is ($\alpha = -0.10$, $\gamma = 1.40$) and another set is ($\alpha = -\frac{1}{16}$, $\gamma = \frac{11}{8}$). Equation (4.41) predicts $\delta \leq 5$ for both sets of data (Stanley 1971b). We can combine (4.41) and (4.31) to obtain the relation

$$\frac{2 - \alpha + \gamma}{2 - \alpha - \gamma} \geq \delta \geq \frac{2 - \alpha' - \beta}{\beta} \qquad (4.42)$$

or equivalently,

$$\beta(2 - \alpha + \gamma) \geq (2 - \alpha' - \beta)(2 - \alpha - \gamma). \qquad (4.43)$$

Note that (4.42) and (4.43) contain both high-temperature and low-temperature exponents. Note also from Table 3.4 that both (4.42) and (4.43) are satisfied as equalities for the classical theories.

For the $d = 2$ Ising model $\alpha' = \alpha = 0$; also assumption (4.38) has been proved rigorously. Hence the only assumptions that are still needed to obtain the value $\delta = 15$ from (4.42) are (4.39) and $\gamma = \frac{7}{4}$, which is almost rigorous.

Note that the values $\alpha = \frac{1}{8}$, $\gamma = \frac{5}{4}$, and $\delta = 5.2$ are inconsistent with (4.41), while the values of $\alpha' = \frac{1}{16}$, $\alpha = \frac{1}{8}$, $\beta = \frac{5}{16}$, and $\gamma = \frac{5}{4}$ are inconsistent with (4.43). These values were at one time the best numerical estimates for the $d = 3$ Ising model (Baker and Gaunt 1967). Note also that we can restore the inequality (4.41) by setting $\delta = 5$ and the inequality (4.43) by setting $\alpha' = \frac{1}{8}$. Of course there are many other sets of values that are capable of satisfying (4.41) and (4.43) but current estimates support decreasing δ to 5 and increasing α' to $\frac{1}{8}$ (Gaunt 1967, Gaunt and Domb 1968, Garelick and Essam 1968). With these values the inequalities (4.31), (4.41), and (4.43) become equalities; also in this case both the Rushbrooke inequality (4.2) and the inequality (4.36) predict that $\gamma' \geq \frac{5}{4}$. It is therefore still possible that $\gamma' = \frac{21}{16}$, as is favoured by some recent work (Guttmann and Thompson 1969).

4.3.2. The Buckingham–Gunton inequalities

We begin by recalling that the critical exponent η, as defined in eqn (3.16), characterizes the decay with distance r of the pair correlation function $\Gamma_2(r)$. Buckingham and Gunton (1969) and Fisher (1969) have proved the following two inequalities relating η and the dimensionality d of the system (see also Stell 1968a):

$$d\frac{\delta - 1}{\delta + 1} \geq 2 - \eta \qquad (4.44)$$

and

$$\frac{d\gamma'}{2\beta + \gamma'} \geq 2 - \eta. \tag{4.45}$$

A corollary to (4.45) is easily obtained from the Rushbrooke inequality, $2 - \alpha' \leq 2\beta + \gamma'$,

$$\frac{d\gamma'}{2 - \alpha'} \geq 2 - \eta. \tag{4.46}$$

The proofs of (4.44) and (4.45) depend on three further assumptions concerning the behaviour of the correlation functions of the system.

D. *Positivity of the correlation functions*—for all temperatures T and any non-negative field (i.e. $H \geq 0$),

$$\Gamma_1 \geq 0, \tag{4.47}$$

$$\Gamma_2(r) - \Gamma_1^2 \geq 0, \tag{4.48}$$

and

$$\Gamma_4(r_1, r_2, r_3) - \Gamma_2(r_1)\Gamma_2(r_3 - r_2) \geq 0, \tag{4.49}$$

where Γ_j denotes the j-spin correlation function (and hence Γ_1 is proportional to the spontaneous magnetization.)

E. *Monotonicity with magnetic field*—for any fixed value of T, $\Gamma_1(T, H)$, $\Gamma_2(T, H)$, and $\Gamma_4(T, H)$ are monotonic non-decreasing functions of H for $H \geq 0$ (in particular, for $T > T_c$, $\Gamma_1 = 0$ provided $H = 0$).

F. *Monotonicity with temperature*—for any fixed (non-negative) value of H, $\Gamma_1(T, H)$, $\Gamma_2(T, H)$ and $\Gamma_4(T, H)$ are monotonic non-increasing functions of T.

Assumptions D–F have been proved rigorously for the Ising model with arbitrary ferromagnetic interactions of arbitrary range and with arbitrary spin quantum number S (Kelly and Sherman 1968, Griffiths 1967a, 1968). For other magnetic models and for real magnets, assumptions D–F are at least intuitively plausible. For fluid systems, on the other hand, assumption D (positivity) is certainly not valid; rather, $\Gamma(r)$ oscillates in sign. Moreover, it is not yet known whether or not the spin correlation function develops a monotonically-decaying tail as $T \to T_c$ (Fisher 1969). Hence the Buckingham–Gunton inequalities have not been proved for fluid systems—but neither have they been disproved!

The inequality (4.44) is especially useful since it provides for η a lower bound which depends only on δ. For example, if $\delta = 5$ for a three-dimensional system (as is true for the spherical model and as is extremely likely for both the Ising and Heisenberg models), then (4.44) requires that $\eta \geq 0$. Moreover, many experimental measurements of δ yield a value *smaller* than 5, whence (4.44) implies that η must be *larger* than zero. Note, however, that if $\delta = 4 \cdot 3$ for a system (as for $CrBr_3$—cf. Table 3.4), then we cannot conclude that η must be larger than $0 \cdot 132$ unless the assumptions D–F could be verified for the system.

We observe that the inequalities (4.44)–(4.46) differ from the inequalities we have considered earlier in this chapter in that they are not satisfied by the van der Waals and mean field theories. They are, however, satisfied (as equalities!) by the two-dimensional Ising model, providing we choose $\delta = 15$ and $\gamma' = \frac{7}{4}$ (Fisher 1964).

4.3.3. *The Fisher inequalities*

Fisher (1969) has recently extended and refined the arguments for the Buckingham–Gunton inequalities. He has also proved, subject to the same assumptions (D–F above), the following inequalities:

$$(2 - \eta)\nu \geq \gamma \tag{4.50}$$

$$d\alpha' \geq 2 - \eta_E, \tag{4.51}$$

and

$$\frac{d\varphi\delta}{1 + \zeta + \varphi\delta} \equiv \frac{d\varphi}{\varphi + \psi} \geq 2 - \eta_E, \tag{4.52}$$

where the exponent η_E characterizes the decay of the *energy–energy* correlation function at $T = T_c$; ν, η, φ, ψ, and ζ are defined, respectively, in eqns (3.15), (3.16), (4.11), (4.12), and 4.16.

We can easily derive the following corollary to (4.52),

$$\frac{2d\varphi}{1 + \varphi + 1/\delta} \geq 2 - \eta_E, \tag{4.53}$$

from the Coopersmith equality (4.14), $(\varphi + \psi) \geq \frac{1}{2}(1 + 1/\delta + \varphi)$.

As we can see from the exponents presented in Table 3.4 the inequality (4.50) is satisfied as an equality for (a) the classical theories, (b) the $d = 2$ Ising model, and (c) the $d = 3$ spherical model. Unlike the other inequalities discussed in this chapter, (4.51) and (4.52) are not satisfied as equalities when the scaling hypothesis (cf. Chapter 11) is made.

4.3.4. *The Josephson inequalities*

Josephson (1967) has proved the inequalities

$$d\nu' \geq 2 - \alpha' \tag{4.54}$$

and

$$d\nu \geq 2 - \alpha, \tag{4.55}$$

subject to certain additional assumptions which we shall not discuss here. Observe that if we could prove the analogue of (4.50) for $T < T_c$, $(2 - \eta)\nu' \geq \gamma'$, then combination with (4.46) would produce the first Josephson inequality (4.54).

The Josephson inequalities, (4.54) and (4.55), fail for the classical theories for $d \geq 3$ (which predict the same values of the critical exponents regardless of the lattice dimensionality d). Observe from Table 3.4 that the Ising model satisfies the Josephson inequalities as equalities for $d = 2$, and as inequalities for $d = 3$. Similarly, the spherical model satisfies (4.55) as an equality.

Suggested further reading

Fisher (1964).
Griffiths (1965*b*).
Fisher (1967).
Fisher (1969).
Griffiths, Hurst, and Sherman (1970).

PART III

CLASSICAL THEORIES OF COOPERATIVE PHENOMENA

5

THE VAN DER WAALS THEORY OF LIQUID–GAS PHASE TRANSITIONS

VAN DER WAALS presented in his doctoral thesis 'On the continuity of the liquid and gaseous states' (1873) an approximate equation of state for a fluid system. His work provided one of the earliest qualitative explanations of critical phenomena, although we shall see that many quantitative predictions are not borne out by detailed experimental work.

5.1. Heuristic derivation of the van der Waals equation of state

The simplest theory of a fluid system corresponds to assuming that the particles of the system interact so weakly with one another that these interactions may be neglected altogether. Such a treatment results in the *ideal gas law*,

$$PV = NkT = n\mathscr{R}T, \tag{5.1}$$

where N is the number of molecules and $n \equiv N/N_A$ is the number of moles in the system, N_A is Avogadro's number, k is the Boltzmann constant, and $\mathscr{R} \equiv kN_A$ is the ideal gas constant.

For a gas at very low density and/or high temperatures eqn (5.1) is fairly well obeyed. Therefore it might seem reasonable to attempt to describe an interacting gas at normal densities by means of simple corrections to the non-interacting result (5.1). To this end, van der Waals attempted to take into account (i) the non-zero size of the molecules and (ii) the attractive force between the molecules.

(i) The ideal gas law assumes that the volume of the molecules may be neglected; for example, the equation $P = n\mathscr{R}T/V$ means that as we increase the pressure at fixed temperature, the volume decreases without limit. Actually there is a limit if we think of the molecules as being like rigid spheres; i.e. there is a minimum volume

$V_{min} = N \times$ volume of a rigid sphere molecule.

Of course real molecules are not rigid spheres, yet it may be a reasonable approximation to replace V in eqn (5.1) by $V - V_{min}$, where V_{min} is regarded as a phenomenological parameter. Now in order that $V - V_{min}$ be an extensive quantity, V_{min} must be linear in N. We call the proportionality constant ℓ/N_A and therefore make the substitution $V \to V - (\ell/N_A)N$ into eqn (5.1), with the result

$$P = n\mathscr{R}T/(V - n\ell). \tag{5.2}$$

Equation (5.2) is sometimes called the Clausius equation of state.

(ii) The attractive force between the molecules of a gas clearly leads to a decrease in the pressure that the gas would otherwise exert on the walls of the container. This decrease ΔP has contributions from the decrease in momentum per molecule, and from the decrease in the number hitting the walls. If we assume that both contributions are proportional to the density of molecules, then $\Delta P \propto (\rho/m)^2$.

We call the proportionality constant a/N_A^2 and therefore subtract the term $(a/N_A^2) (N/V)^2$ from the value of the pressure given by eqn (5.2) obtaining

$$P = n\mathscr{R}T/(V - n\ell) - an^2/V^2. \tag{5.3}$$

Equation (5.3) is the van der Waals equation of state and it is usually written in the form

$$(P + an^2/V^2) (V - n\ell) = n\mathscr{R}T, \qquad \blacktriangleright (5.4)$$

or else in the form

$$(P + a/\overline{V}^2) (\overline{V} - \ell) = \mathscr{R}T, \tag{5.5}$$

where $\overline{V} \equiv V/n$ is the volume per mole or specific volume. Clearly the van der Waals equation of state reduces to the non-interacting result (5.1) in the limit of low density, i.e. when both a and ℓ are negligible compared to $\overline{V}$. It is important to stress that a and ℓ are phenomonological parameters characteristic of the fluid. Values of a and ℓ that provide a fairly satisfactory fit of (5.4) to the thermodynamic data (at least far from T_c) are listed in Table 5.1 for a few selected substances.

TABLE 5.1

The van der Waals constants a and ℓ for selected fluid systems, together with the ratio $Z_c \equiv P_c V_c / \mathscr{R} T_c$. The van der Waals constants appear in the equation of state (5.4); because of the dimensions of a and ℓ used here, we should measure P in atm, V in litres, and take $\mathscr{R} \simeq 0.08206$ litre-atm/ mole-degree. Note that a is a measure of the attractive force between particles, and ℓ measures the non-zero volume of the molecules. Since both a and ℓ vary with temperature, the values shown here are only approximate; they are obtained from the CRC Handbook, 45th edn. The systems are listed in order of increasing Z_c. The van der Waals theory predicts $Z_c = \frac{3}{8} = 0.375$, while in the non-interacting limit $Z = 1$.

Fluid	a (l^2 atm mol^{-2})	ℓ (l mol^{-1})	$Z_c = P_c V_c / \mathscr{R} T_c$
Water, H_2O	5·464	0·0305	0·230
Sulphur dioxide, SO_2	6·714	0·0564	0·269
Ethylene, C_2H_4	4·471	0·0571	0·270
Acetylene, C_2H_2	4·390	0·0514	0·274
Carbon dioxide, CO_2	3·592	0·0427	0·275
Ethane, C_2H_6	5·489	0·0638	0·285
Xenon, Xe	4·194	0·0511	0·288
Methane, CH_4	2·253	0·0428	0·290
Nitrogen, N_2	1·390	0·0391	0·291
Argon, Ar	1·345	0·0322	0·291
Oxygen, O_2	1·370	0·0318	0·292
Carbon monoxide, CO	1·485	0·0399	0·294
Hydrogen, H_2	0·2444	0·0266	0·304
Helium, ^{4}He	0·03412	0·0237	0·308
van der Waals theory	—	—	0·375

5.2. van der Waals isotherms and the Maxwell construction

Note that the van der Waals equation of state (5.4) may be written, for 1 mole ($n = 1$), as

$$V^3 - (\ell + \mathscr{R}T/P) V^2 + \frac{a}{P} V - \frac{a\ell}{P} = 0, \qquad (5.6)$$

so that for a given value of P and T there will correspond in general three values of V instead of only one. For small T all three roots of (5.6) will be real (see Figs. 5.1, 5.2(a)). As we increase the temperature these roots move together and for $T > T_c$ one pair of roots becomes complex; at very large T, eqn (5.6) becomes $V^3 - (\mathscr{R}T/P)V^2 = 0$, which is just the ideal gas law, eqn (5.1).

For all subcritical ($T < T_c$) isotherms there will be a region in which the slope $\partial V/\partial P$ is positive corresponding to the unphysical occurrence of a negative isothermal compressibility $K_T \equiv -V^{-1} (\partial V/\partial P)_T$.

This dramatic breakdown of the van der Waals theory for $T < T_c$ was recognized from the outset, and less than two years had passed

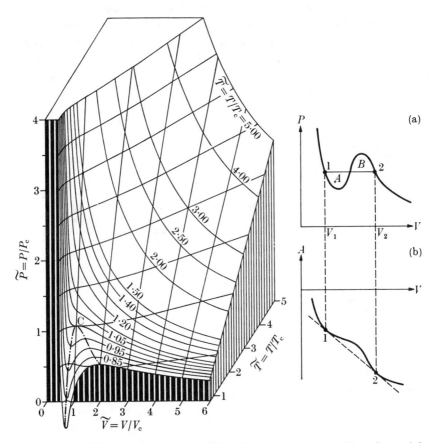

FIG. 5.1. The PVT surface for the van der Waals theory. Any point on this surface satisfies the van der Waals equation of state (5.4) and hence is predicted to correspond to an equilibrium state of the system. The reader should determine that portion of the surface that corresponds to unstable states. After Kubo (1968).

FIG. 5.2. (a) A single subcritical (or $T < T_c$) isotherm, as given by (5.4). (b) The Maxwell construction restores the convexity of the Helmholtz potential $A(T, V)$ as a function of V. After Huang (1963).

from the date of van der Waals' thesis when Maxwell proposed an *ad hoc* remedy which is commonly called the Maxwell equal-area construction. Following Huang (1963), we construct in Fig. 5.2(b) the

Helmholtz potential $A(T_0, V)$ for a fixed temperature $T_0 < T_c$, and we find that the region in which $\partial P/\partial V > 0$ corresponds to a region where $\partial^2 A/\partial V^2 < 0$ (cf. eqn (2.36)). Now $A(T_0, V)$ must be a convex function of V for all V in order that the system be stable. The Maxwell construction consists in replacing the Helmholtz potential in the entire region $V_1 < V < V_2$ (which is somewhat larger than the unstable region referred to above) by the dashed line shown in Fig. 5.2(b). This modification of the Helmholtz potential corresponds to a horizontal straight line connecting points 1 and 2 of the P–V isotherm of Fig. 5.2(a). It is easy to show that the horizontal line is at just such a value of the pressure that the areas labelled A and B of Fig. 5.2 (a) are equal, and for this reason the Maxwell construction is sometimes called the equal-area construction. An equivalent formulation of the Maxwell construction focusses instead on the Gibbs potential—whose dependence upon pressure the reader can easily construct from Fig. 5.2 using the methods developed in Fig. 2.4. One finds that the Gibbs potential becomes a three-valued function in the neighbourhood of the phase transition, and that the Maxwell construction is equivalent to deleting all but the lowest branch of this function.

5.3. The van der Waals critical point: P_c, V_c, and T_c

The P–V isotherm for $T = T_c$ is characterized by a horizontal tangent and an inflection point when $P = P_c$ and $V = V_c$. Hence we may obtain the three parameters P_c, V_c, and T_c (which define the critical point) by solving simultaneously the three equations:

(i) $(\partial P/\partial V)_{T=T_c} = 0$,

(ii) $(\partial^2 P/\partial V^2)_{T=T_c} = 0$, and

(iii) the van der Waals equation (5.4),

where in all three equations we set $P = P_c$, $V = V_c$, and $T = T_c$.

An alternative procedure is to observe that we may equate coefficients of identical powers of V in the two equations:

$$(V - V_c)^3 = V^3 - 3V^2 V_c + 3V V_c^2 - V_c^3 = 0 \tag{5.7}$$

and

$$V^3 - (\ell + \mathscr{R}T_c/P_c) V^2 + (a/P_c)V - a\ell/P_c = 0 \tag{5.8}$$

Equation (5.8) merely states that for $P = P_c$ and $T = T_c$, all three roots of eqn (5.7) are the same and in fact are equal to the critical volume V_c. We thereby obtain the three equations:

$$-3V_c = -(\ell + \mathscr{R}T_c/P_c),$$ (5.9)

$$+3V_c^2 = a/P_c,$$ (5.10)

and

$$-V_c^3 = -a\ell/P_c.$$ (5.11)

Equations (5.10) and (5.11) may be combined to yield

$$P_c = a/27\ell^2$$ (5.12)

and

$$V_c = 3\ell.$$ (5.13)

Finally, substitution of (5.12) and (5.13) into (5.9) results in

$$\mathscr{R}T_c = 8a/27\ell.$$ (5.14)

Table 3.5 lists values of P_c, ρ_c, and T_c for a few representative fluids.

5.4. The law of corresponding states

We begin by observing that eqns (5.12)–(5.14) provide three relations among the two parameters a and ℓ which characterize a given fluid. Hence there must be one relation involving P_c, V_c, and T_c which is independent of the parameters a and ℓ; clearly

$$Z_c \equiv P_c V_c/\mathscr{R}T_c = (a/27\ell^2)(3\ell)(27\ell/8a) = \tfrac{3}{8} = 0\cdot375 \quad (5.15)$$

Typical values of Z_c are listed in the last column of Table 5.1. Note that for an ideal gas (which has no critical point),

$$Z \equiv PV/\mathscr{R}T = 1$$ (5.16)

for all values of P, V, and T so that the van der Waals theory has achieved some improvement over the non-interacting result, although from Table 5.1 we see that $Z_c < \tfrac{3}{8}$ for real fluids.

Suppose we multiply both sides of the van der Waals equation by $(27\ell/a)$ and then use eqns (5.12)–(5.14) to eliminate the parameters a and ℓ; we thereby obtain

$$(\tilde{P} + 3/\tilde{V}^2)(3\tilde{V} - 1) = 8\tilde{T}, \qquad \blacktriangleright \ (5.17)$$

where

$$\tilde{P} \equiv P/P_c, \quad \tilde{V} \equiv V/V_c, \quad \text{and} \quad \tilde{T} \equiv T/T_c.$$ (5.18)

This new form of the van der Waals equation, (5.17), means that if we measure pressure, volume, and temperature, respectively, in units of P_c, V_c, and T_c, then the equation of state is the same for all substances. Thus any two fluids with the same values of $\tilde{P}$, $\tilde{V}$, and $\tilde{T}$ may be said to be in corresponding states; eqn (5.17) is generally called the law of corresponding states.

One might expect that the law of corresponding states is obeyed by real gases only to the extent that the van der Waals equation (5.4) is obeyed. However, experimentally we find remarkable confirmation of the (more general) notion that materials should behave similarly pro-

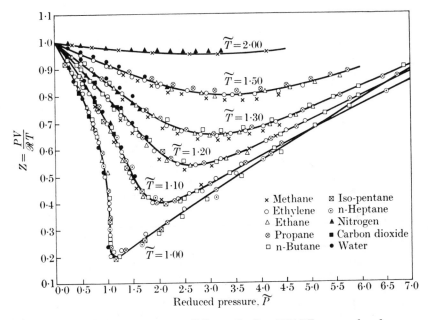

FIG. 5.3. Dependence of the compressibility ratio $Z \equiv PV/\mathscr{R}T$ upon reduced pressure $\tilde{P}$ for different reduced temperatures $\tilde{T}$. The fact that the data for a wide variety of fluids fall on identical curves supports the law of corresponding states. After Su (1946).

vided that P, V, and T are measured in units of P_c, V_c, and T_c. For example, in Fig. 5.3 we display measurements of the compressibility ratio $Z \equiv PV/\mathscr{R}T$ for a variety of fluids. Even though these materials do not obey the van der Waals equation near T_c (e.g. the value of Z_c is not $\frac{3}{8}$), the data for all materials nevertheless lie on the same curve for a given reduced temperature $\tilde{T}$. See also Fig 1.8. We shall discuss behaviour analogous to the law of corresponding states in Chapter 11.

5.5. Critical-point exponents for the van der Waals theory

One of the simpler methods of obtaining critical-point exponents for the classical van der Waals theory is to begin with eqn (5.17), the van der Waals equation in terms of reduced variables $\tilde{P}$, $\tilde{V}$, and $\tilde{T}$. We then expand each variable about the critical point; thus, on defining

$$p \equiv \tilde{P} - 1 \equiv (P - P_c)/P_c, \tag{5.19}$$

$$v \equiv \tilde{V} - 1 \equiv (V - V_c)/V_c, \tag{5.20}$$

$$\epsilon \equiv \tilde{T} - 1 \equiv (T - T_c)/T_c, \tag{5.21}$$

we find that eqn (5.17) becomes

$$\{(1 + p) + 3(1 + v)^{-2}\}\{3(1 + v) - 1\} = 8(1 + \epsilon). \tag{5.22}$$

Multiplying both sides by $(1 + v)^2$, we get

$$\{4 + 2v + v^2 + p(1 + 2v + v^2)\}\{2 + 3v\} = 8(1 + \epsilon)(1 + 2v + v^2), \tag{5.23}$$

and on expanding both sides and combining terms, we get

$$2p(1 + 7v/2 + 4v^2 + 3v^3/2) = -3v^3 + 8\epsilon(1 + 2v + v^2). \tag{5.24}$$

To obtain the critical isotherm exponent δ and the coefficient $\mathscr{D}$ defined in eqn (3.10), we set $\epsilon = 0$ in (5.24),

$$p = -\tfrac{3}{2}v^3(1 + 7v/2 + \cdots)^{-1} = -\tfrac{3}{2}v^3(1 - 7v/2 + \cdots). \tag{5.25}$$

The leading term on the right-hand side is cubic; hence $\delta = 3$. To obtain the coefficient $\mathscr{D}$, we observe that

$$\frac{P - P_c}{P_c^\circ} = \frac{P_c}{P_c^\circ}\frac{P - P_c}{P_c} = \frac{3}{8}\left(-\frac{3}{2}v^3\right) = \frac{-9}{16}v^3 \tag{5.26}$$

where we have used the fact that $P_c^\circ = \mathscr{R}T_c/V_c = \tfrac{8}{3}P_c$ from eqn (5.15). Thus from (3.10) and (5.26) we see that

$$\mathscr{D} = \frac{9}{16}. \tag{5.27}$$

Consider next the behaviour of K_T as $T \to T_c^+$ along the critical isochore ($v = 0$). From (2.17), (5.19), and (5.20),

$$(-VK_T)^{-1} = \left(\frac{\partial P}{\partial V}\right)_T = \frac{P_c}{V_c}\left(\frac{\partial p}{\partial v}\right)_T = \frac{P_c}{V_c}(-6\epsilon) \tag{5.28}$$

where the last equality in (5.28) follows from (5.24) On using $(K_T^\circ)^{-1} = P_c^\circ = \frac{8}{3} P_c$, we find

$$\frac{K_T}{K_T^\circ} = \frac{8}{3}\left(\frac{1}{6\epsilon}\right) = \frac{4}{9}\frac{1}{\epsilon}. \tag{5.29}$$

Hence $\gamma = 1$ and the coefficient $\mathscr{C}$ defined in eqn (3.8) has the value $\frac{4}{9}$. We can show in a similar fashion that when we approach the critical temperature from above, we obtain the values $\gamma' = 1$ and $\mathscr{C}' = \frac{2}{9}$. Thus the inverse compressibility is linear in ϵ, but it rises twice as fast below T_c as above T_c (cf. Fig. 5.4). This result can be remembered easily by means of the following mnemonic device: when the system is below

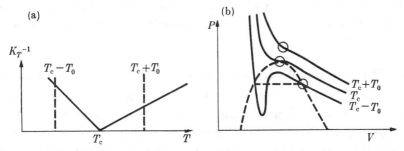

FIG. 5.4. (a) Inverse isothermal compressibility predicted by the van der Waals theory. Note that $\gamma' = \gamma = 1$, yet $\mathscr{C}'(=\frac{2}{9}) < \mathscr{C}(=\frac{4}{9})$. (b) A PV projection for three isotherms, $T = T_c$, and $T = T_c \pm T_0$. The circles indicate the regions where the slope is proportional to the inverse compressibility. That $\mathscr{C}' < \mathscr{C}$ is often said to correspond to the fact that the lower circle is further from the critical point than the upper circle.

the critical point, it is removed both in temperature *and* in density, whereas when it is above T_c, we are on the critical isochore and hence are removed only in temperature.

Similarly, we can calculate the shape of the coexistence curve and we find that the exponent β has the value $\beta = \frac{1}{2}$, while the coefficient $\mathscr{B}$ defined in eqn (3.6) has the value $\mathscr{B} = 2$.

Finally, the specific heat predicted by the van der Waals theory is

$$\Delta C_V \equiv C_V - C_V^\circ = \begin{cases} \frac{9}{2}Nk\left(1 - \frac{28}{25}\epsilon + \cdots\right) & [T < T_c] \\ 0 & [T > T_c] \end{cases} \tag{5.30}$$

where $C_V^\circ = \frac{3}{2}Nk$ is the constant volume specific heat predicted for the non-interacting (high-temperature) limit or ideal gas. Thus $\alpha' = \alpha = 0$. Note that the slope of C_V vs. T is finite as $T \to T_c$ from below, so that $\alpha_s' = 0$ as well.

It is clear from Table 3.4 and Fig. 3.3 that the predictions of this classical theory disagree with most experimental results on fluid and magnetic systems near the critical point. On the other hand, we see that the $d = 3$ Ising and Heisenberg models agree rather better with experiment. It is by now generally believed that the classical exponent values are to be associated with the assumption, inherent in all of the classical theories, that the interparticle forces are of extremely long range. In fact, the essential results of the van der Waals theory can be derived by making the approximation that each particle moves in a mean field due to all the other particles. Correspondingly, in the next chapter we show that the molecular field (or mean field) theory of magnetism can be derived by making the assumption that each spin interacts equally with all the other spins in the system. Presumably these rather drastic assumptions are unrealistic for real fluid and magnetic systems.

5.6. The van der Waals equation of state as a mean field theory

In this section we show that the van der Waals equation of state also results if we approximate the interactions among the particles comprising the fluid by an effective potential of the form shown in Fig 5.5(a). The technique of approximating the effects of molecular interactions by means of the assumption that a particle moves in such

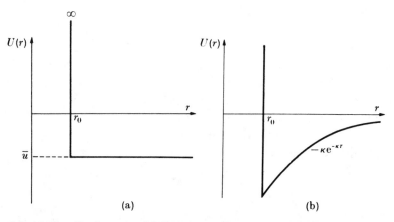

FIG. 5.5. (a) The effective potential $U(r)$ assumed in the Reif-derivation of the van der Waals equation of state consists of a hard core of radius r_0 and a long-range tail for which the potential is approximately constant. (b) The effective potential $U(r) = \infty$ for $r < r_0$ and $U(r) = -\kappa e^{-\kappa r}$ for $r > r_0$. A system of particles interacting with this potential behaves, in the limit $\kappa \to 0$, as a van der Waals gas.

a mean field due to all the other particles is a trick that we shall use again in the following chapter to develop the so-called mean field theory of magnetism. Our present mean field theory of fluids follows Reif (1965). Assume, then, that the particles are independent, yet that they move about in an effective potential of the form

$$U(r) = \begin{cases} \infty & r < r_0 \\ \bar{u} & \text{otherwise} \quad [\bar{u} < 0]. \end{cases} \tag{5.31}$$

Then the pressure in the fluid is, from (2.11d),

$$P = -\left(\frac{\partial A}{\partial V}\right)_T = kT\left\{\frac{\partial}{\partial V}(\ln Z)\right\}_T, \tag{5.32}$$

where the partition function Z is given by the product of the partition functions of the N independent particles; defining $\beta \equiv 1/kT$, we have

$$Z^{1/N} \propto \int d\mathbf{p} \int d\mathbf{r} \exp\left[-\beta\{p^2/2m + U(r)\}\right]. \tag{5.33}$$

Now the momentum integral factors out; its value is independent of V and hence cannot contribute to the pressure in (5.32). Because of the form of $U(r)$ assumed in eqn (5.31), the spatial integral may be written as

$$Z^{1/N} \propto V_{ex} e^{-\infty} + (V - V_{ex}) e^{-\beta\bar{u}}, \tag{5.34}$$

where V_{ex} represents the volume excluded by the hard core in eqn (5.31). Since the first term on the right-hand side of (5.34) is zero, the expression for the pressure in eqn (5.32) becomes simply

$$P = NkT \frac{\partial}{\partial V}\{\ln(V - V_{ex}) - \beta\bar{u}\}. \tag{5.35}$$

It is easy to see that $V_{ex} \propto N$. We call the proportionality constant (ℓ/N_A); thus

$$V_{ex} = (\ell/N_A)N = \ell n. \tag{5.36}$$

Similarly, one can argue that $\bar{u} \propto -N/V$. Calling the constant of proportionality (a/N_A^2), we have

$$\bar{u} = -(a/N_A^2)\left(\frac{N}{V}\right) = -an^2/NV. \tag{5.37}$$

On substituting (5.36) and (5.37) into (5.35) and carrying out the indicated volume differentiation, we obtain

$$P = NkT\left(\frac{1}{V - \ell n} - \beta an^2/NV^2\right) = \frac{n\mathcal{R}T}{V - \ell n} - \frac{an^2}{V^2}, \tag{5.38}$$

which is precisely the van der Waals equation of state, (5.3).

A more precise formulation of the van der Waals theory as the limit of an infinite-range potential has been formulated in recent years by Kac, Uhlenbeck, and Hemmer (1963). These authors consider the potential

$$U(r) = \begin{cases} \infty & r < r_0 \\ -\kappa e^{-\kappa r} & r > r_0 \end{cases} \tag{5.39}$$

where the parameter κ determines the range of the potential (cf. Fig. 5.5(b)). The integral $\int_{r_0}^{\infty} U(r) \, dr$ is simply $-\exp(-\kappa r_0)$, and when we take the limit $\kappa \to 0$, the potential becomes both infinite in range and infinitely weak. Kac *et al.* have shown that in this limit, the model becomes essentially the same as the van der Waals theory, with one noteworthy improvement—there are no unstable regions (i.e., the 'Maxwell construction' of § 5.2 is no longer necessary). The interested reader is referred to this important work for the details of the argument.

Suggested further reading

Huang (1963).
Kac, Uhlenbeck, and Hemmer (1963).
Baker (1963b).
Reif (1965).
Widom (1967).
Kittel (1969).

6

THE MEAN FIELD THEORY OF MAGNETIC PHASE TRANSITIONS

In 1907—two years after the appearance of Langevin's theory of paramagnetism—Pierre Weiss proposed a phenomenological theory of ferromagnetism in which he assumed that the spins interact with one another through a molecular field proportional to the average magnetization. It was only some years later, after the Heisenberg exchange interaction had been proposed, that the molecular field was given an interpretation in terms of pair-wise exchange interactions J_{ij} between spins S_i and S_j situated on sites i and j in the lattice. Fairly recently this mean field theory has received yet another interpretation. It has been shown that if we consider a model system in which every magnetic moment interacts with every other moment—with an equal strength—then the properties of this model are identical to those of the mean field theory. In this chapter we shall consider each of these three interpretations of the mean field theory in turn:

  (i) the Weiss molecular field,
 (ii) the molecular field theory as an approximation to the Heisenberg model, and
(iii) the solution of an infinite interaction range model.

We shall see that the critical-point exponents predicted by the molecular field theory are identical to those of the van der Waals theory. This should come as little surprise, as we saw in § 5.6 that the van der Waals theory could be interpreted as an infinite interaction range model also. Most systems in nature have relatively strong, short-range interactions, and we shall see in Chapter 9 that the nature of the cooperative phenomena observed depends crucially on the fashion in which these interactions 'propagate order' from one particle to another. Hence we will come to appreciate why it is that the mean field theory provides an inadequate description of phenomena in the critical region.

We begin our discussion with the derivation of the partition function and the thermodynamic properties of a system for the limit in which the constituent magnetic moments interact so weakly with one another that these interactions may be neglected altogether. We shall find that the techniques illustrated in this calculation for the 'perfect para-magnet' prove useful in deriving the predictions of the mean field theory.

6.1. The non-interacting magnetic system

The Hamiltonian for a system of N non-interacting magnetic moments in an external magnetic field $\mathbf{H}$ is

$$\mathscr{H} = -g\mu_B \sum_{i=1}^{N} \mathbf{S}_i \cdot \mathbf{H}, \tag{6.1}$$

where the product $\mathbf{S}_i \cdot \mathbf{H}$ may assume the values

$$\mathbf{S}_i \cdot \mathbf{H} = m_i H, \qquad (m_i = -S, -S + 1, \ldots, 0, \ldots, S - 1, S) \tag{6.2}$$

and where we have used the relation $-\boldsymbol{\mu} \equiv g\mu_B\mathbf{S}$ between the magnetic moment $\boldsymbol{\mu}$ and the spin $\mathbf{S}$. In eqn (6.1) g is the Landé factor and $\mu_B \equiv e\hbar/2mc$ is the Bohr magneton; since the combination $g\mu_B$ occurs in many of the formulas to be presented, we shall introduce the notation

$$\bar{\mu} \equiv g\mu_B. \tag{6.3}$$

The partition function or *Zustandsumme* is obtained by weighting each state by the appropriate Boltzmann factor from (6.1) and then summing over all $(2S + 1)^N$ states of the system. Thus we have

$$Z = \sum_{m_1 = -S}^{S} \ldots \sum_{m_N = -S}^{S} \exp\left(x \sum_{i=1}^{N} m_i\right), \tag{6.4}$$

where

$$x \equiv \bar{\mu}H/kT. \tag{6.5}$$

The summations in eqn (6.4) are particularly elementary for the case of a two-level system $(S = \tfrac{1}{2})$, and we have

$$Z = \prod_{i=1}^{N} \left\{ \sum_{m_i = -\frac{1}{2}}^{\frac{1}{2}} \exp\left(xm_i\right) \right\}$$

$$= \prod_{i=1}^{N} 2 \cosh\left(\tfrac{1}{2}x\right)$$

$$= 2^N \cosh^N\left(\tfrac{1}{2}x\right). \tag{6.6}$$

For general S, the solution becomes

$$Z = [\![\exp(-xS)\,[1 - \exp\{(2S + 1)x\}]/\{1 - \exp(x)\}\,]\!]^N$$

$$= \left[\frac{\sinh\{(S + \tfrac{1}{2})x\}}{\sinh(x/2)}\right]^N. \tag{6.7}$$

Thus it is possible to obtain a closed form expression for the partition function in the non-interacting limit.

We say that we can 'solve' the paramagnetic system in the sense that since we can obtain the partition function, we can also obtain a great many thermodynamic functions that are directly related to the partition function. For example, the Gibbs potential $G(T, H)$ is, directly from eqn (6.7) (Wannier 1966),

$$G(T, H) = -kT \ln Z$$

$$= -NkT \ln\left[\frac{\sinh\{(S + \tfrac{1}{2})x\}}{\sinh \tfrac{1}{2}x}\right]. \tag{6.8}$$

The magnetization is related to the Gibbs potential by eqn (2.45c), whence

$$M(T, H) = -\left(\frac{\partial G}{\partial H}\right)_T = NkT\,\frac{\partial}{\partial H}\ln Z$$

$$= M_0\, B_S(Sx), \tag{6.9}$$

where

$$M_0 \equiv M(T = 0, H = 0) = NS\bar{\mu} = NSg\mu_B \tag{6.10}$$

is the maximum value of the magnetization, and

$$B_S(y) \equiv \frac{2S + 1}{2S}\coth\left(\frac{2S + 1}{2S}y\right) - \frac{1}{2S}\coth\left(\frac{1}{2S}y\right) \tag{6.11}$$

is called the Brillouin function. This function, which relates the magnetization of the system to the applied field, is plotted in Fig. 6.1. Note that for $S = \tfrac{1}{2}$ (that is, the single electronic spin value),

$$B_{\frac{1}{2}}(\tfrac{1}{2}x) = 2\coth(x) - \coth(x/2)$$

$$= \frac{\coth^2(x/2) + 1}{\coth(x/2)} - \coth(x/2)$$

$$= \tanh(x/2). \tag{6.12}$$

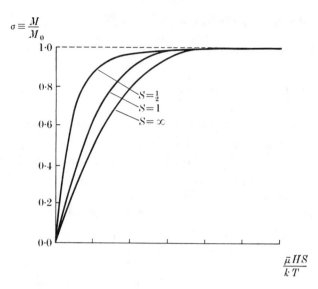

FIG. 6.1. Dependence of the reduced magnetization $\sigma \equiv M/M_0 \equiv M(T, H)/M(0, 0)$ upon $\bar{\mu}HS/kT$ for a magnetic system in the non-interacting limit, as given by eqn (6.9). The curves are those of the Brillouin function $B_S(Sx)$, which reduces to the hyperbolic tangent and to the Langevin function in the limits $S = \frac{1}{2}$ and $S = \infty$ respectively.

6.2. The assumption of an effective molecular field

We note from Fig. 6.1 that $M = 0$ when $H = 0$; we say that the spontaneous magnetization $M(T, H = 0)$ of the paramagnet is zero. This prediction is not supported by observation—in fact, for all ferromagnetic materials there is some temperature below which there exists a spontaneous magnetization.

The basic assumption of the mean field theory is that the interactions among the spins give rise to a magnetic field H_m in addition to the external field H. If one assumes that H_m is proportional to the magnetization, $H_m = \lambda M(T, H)$, then the effective field seen by each spin is

$$H_{\text{eff}} = H + \lambda M(T, H). \qquad (6.13)$$

The parameter λ in (6.13) is called the molecular field parameter. The derivation of the partition function proceeds as before, and all the above expressions remain valid providing we replace the magnetic field H (which enters into the parameter x of eqn (6.5)) by the effective field of eqn (6.13).

In particular, eqn (6.9) becomes

$$M = M_0 B_S \{\beta\bar{\mu}S (H + \lambda M)\}, \qquad (6.14)$$

where M denotes $M(T, H)$, $M_0 \equiv M(T = 0, H = 0)$ as in (6.10), and $\beta \equiv 1/kT$. For $H = 0$, eqn (6.14) reduces to

$$M = M_0 \, B_S(\beta\bar{\mu}S\lambda M).   \qquad \blacktriangleright \quad (6.15)$$

Equation (6.15) is an implicit equation in the sense that M appears in the argument of the Brillouin function as well as on the left-hand side of the equation. We consider therefore a graphical solution, and refer the reader to Fig. 6.2 which shows both sides of (6.15) plotted as functions of M. Equation (6.15) possesses the trivial solution $M = 0$ for all values of T, but there exists a second solution with $M \neq 0$, providing

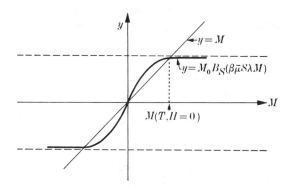

FIG. 6.2. Plotted as functions of M are the left-hand and the right-hand sides of eqn (6.15). The solution $M = 0$ exists for all T, but the solution $M \neq 0$ exists only for T sufficiently small that the initial slope of the Brillouin function is larger than 1.

the initial slope of the right-hand side of (6.15) is larger than the initial slope of the left-hand side. To study the initial slope, we consider the small-argument expansion of the Brillouin function,

$$B_S(y) = \frac{S + 1}{3S} \, y - \frac{S + 1}{3S} \, \frac{2S^2 + 2S + 1}{30S^2} \, y^3 + \cdots \qquad (6.16)$$

Therefore the initial slope of a graph of the right-hand of (6.15) as a function of M is

$$M_0 \left(\frac{S + 1}{3S} \right) \frac{\bar{\mu}S\lambda}{kT} = \mathbf{C}\,\frac{\lambda}{T}, \qquad (6.17)$$

where we have used eqn (6.10) to eliminate M_0 and where

$$\mathbf{C} \equiv \frac{N\bar{\mu}^2 \, S(S + 1)}{3k} \qquad (6.18)$$

is called the Curie constant. We see from (6.17) that non-trivial solutions exist for $T < \lambda C$, whence we have the result that the critical temperature for the molecular field theory is given by

$$T_c = \lambda C. \tag{6.19}$$

Thus T_c is proportional to the molecular field parameter λ and approaches zero when λ approaches zero. This should not be surprising, since for $\lambda = 0$ we recover the paramagnetic or non-interacting result, $T_c = 0$.

6.3. Critical-point exponents for the mean field theory

We begin by considering the case $S = \frac{1}{2}$, so that the equation of state in the mean field theory is, from (6.14) and (6.12),

$$M = M_0 \tanh\{\tfrac{1}{2}\beta\bar{\mu}\,(H + \lambda M)\}. \tag{6.20}$$

If we introduce the reduced variables

$$\sigma \equiv \frac{M}{M_0} \equiv \frac{M(T, H)}{M(0, 0)} \tag{6.21}$$

and $\tilde{T} \equiv T/T_c$, then (6.20) becomes

$$\sigma = \tanh\left(\frac{1}{2}\frac{\bar{\mu}H}{kT} + \sigma/\tilde{T}\right). \qquad \blacktriangleright \tag{6.22}$$

In order to facilitate the calculation of the critical-point exponents, it is convenient to write the equation of state (6.22) in the form

$$h \equiv \tanh\left(\frac{\bar{\mu}H}{2kT}\right) = \frac{\sigma - \tanh\,(\sigma/\tilde{T})}{1 - \sigma\tanh\,(\sigma/\tilde{T})}, \tag{6.23}$$

where we have made use of the trigonometric identity

$$\tanh\,(x + y) = \frac{\tanh x + \tanh y}{1 + (\tanh x)\,(\tanh y)}. \tag{6.24}$$

Near the critical point ($H = 0$, $M = 0$, $T = T_c$) the arguments of all the hyperbolic tangent functions in eqn (6.23) are small, and we can therefore use the expansion

$$\tanh x = x - \tfrac{1}{3}x^3 + \tfrac{2}{15}x^5 + \cdots \tag{6.25}$$

to obtain

$$h = \sigma(1 - 1/\tilde{T}) + \sigma^3\{1/(3\tilde{T}^3) + (1 - 1/\tilde{T})/\tilde{T}\} + \mathcal{O}(\sigma^5). \tag{6.26}$$

Most of the critical-point exponents for $S = \frac{1}{2}$ are obtainable from (6.26).

6.3.1. Magnetization exponent β

From eqn (6.23) we see immediately that in zero field, $h = 0$, and for $T \lesssim T_c$ eqn (6.26) becomes

$$\sigma^2 = \frac{T_c/T - 1}{T_c^3/(3T^3) + (T_c/T)(1 - T_c/T)} + \cdots$$

$$\cong 3\left(\frac{T}{T_c}\right)^2 \frac{T_c - T}{T_c}. \tag{6.27}$$

Thus the square of the zero-field magnetization vanishes linearly with $T_c - T$, so we have $\beta = \frac{1}{2}$, just as in the van der Waals theory of a fluid system. Notice, however, that the coefficient $\mathscr{B}$ defined in eqn (3.7) has the value $\sqrt{3}$ here, in contrast to the van der Waals prediction $\mathscr{B} = 2$.

6.3.2. Critical isotherm exponent δ

To obtain the curvature of the M–H isotherm at the critical temperature, we set $\tilde{T} = 1$ in (6.26) and consider both H and M to be small. Thus we have

$$\tfrac{1}{2}\beta_c\bar{\mu}H + \mathcal{O}(H^3) = \sigma^3/3 + \mathcal{O}(\sigma^5), \tag{6.28}$$

where $\beta_c \equiv 1/kT_c$. Hence $\delta = 3$, just as in the van der Waals theory. However the coefficient again differs, and we find that $\mathscr{D} = \frac{2}{3}$ (cf. eqn (5.27)).

6.3.3. Specific heat exponents α and α'

For $S = \frac{1}{2}$, a straightforward though somewhat lengthy calculation leads to the expression

$$A(T, M) = NkT\,[-\ln 2 + \tfrac{1}{2}\ln(1 - \sigma^2) + \tfrac{1}{2}\sigma\ln\{(1 + \sigma)/(1 - \sigma)\}$$
$$- \sigma^2/2\tilde{T}]. \tag{6.29}$$

From (6.29) and (2.47), we see that $C_M = -T(\partial^2 A/\partial T^2)_M = 0$ for all T. We can find C_H either from (2.48), $C_H = -T(\partial^2 G/\partial T^2)_H$, from (2.51a), $C_H = T\chi_T^{-1}\{(\partial M/\partial T)_H\}^2$, or from (2.54), $C_H = T\chi_T\{(\partial H/\partial T)_M\}^2$. All three calculations lead to the same expression: for $H = 0$, $C_H = 0$ for $T > T_c$, while for $T \lesssim T_c$

$$C_H = \tfrac{3}{2}Nk\{1 - \mathcal{C}(T_c - T)/T_c + \cdots\}. \tag{6.30}$$

Thus the limiting slope of C_H vs. T remains finite as $T \to T_c$, and we have the results $\alpha'_s = \alpha_s = 0$, as well as $\alpha' = \alpha = 0$. Further, we see that

$$C_H \to 3Nk/2 \quad \text{as} \quad T \to T_c^-, \tag{6.31}$$

so that there is a jump discontinuity in the specific heat of magnitude $\Delta C_H = 3Nk/2$ for $S = \frac{1}{2}$ (cf. Fig. 6.3).

6.3.4. *Susceptibility exponents γ and γ'*

We begin by observing that the zero-field isothermal susceptibility $\chi_T \equiv (\partial M/\partial H)_T$ satisfies the relation

$$\chi_T = \left(\frac{\partial M}{\partial \sigma}\right)_T \left(\frac{\partial \sigma}{\partial h}\right)_T \left(\frac{\partial h}{\partial H}\right)_T = (\tfrac{1}{2}N\bar\mu)(\tfrac{1}{2}\beta\bar\mu)\left(\frac{\partial \sigma}{\partial h}\right)_T = \frac{C}{T}\left(\frac{\partial \sigma}{\partial h}\right)_T, \tag{6.32}$$

where we have used (6.18) for the Curie constant for $S = \frac{1}{2}$. Hence we

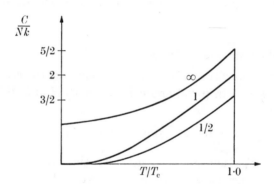

Fɪɢ. 6.3. Dependence of C_H for $H = 0$ upon reduced temperature $\tilde{T} \equiv T/T_c$. Notice that although there is a simple jump discontinuity for all values of S (i.e. $\alpha = \alpha' = 0$), the magnitude of the discontinuity depends weakly upon S. Notice also that the behaviour for the classical limit ($S = \infty$) is only slightly different from the finite-S results in the critical region, but is qualitatively different at low temperatures. After Mattis (1965).

differentiate both sides of (6.26) with respect to h, obtaining, for $T \simeq T_c$

$$1 = (\partial\sigma/\partial h)\{(1 - 1/\tilde{T}) + 3\sigma^2(1/3\tilde{T}^3) + \mathcal{O}(\sigma^4)\} \tag{6.33}$$

or, on using (6.32),

$$\chi_T = \frac{C}{T}\left\{\frac{\epsilon}{\tilde{T}} + \frac{\sigma^2}{\tilde{T}^3} + \mathcal{O}(\sigma^4)\right\}^{-1}. \tag{6.34}$$

Now for $T > T_c$, $\sigma = 0$ when $H = 0$ and (6.34) reduces to

$$\chi_T = \frac{C}{T}\left(\frac{T_c}{T}\frac{T - T_c}{T_c}\right)^{-1} = \frac{C}{T - T_c}. \tag{6.35}$$

Hence $\gamma = 1$ and the coefficient $\mathscr{C}$ defined in eqn (3.9) has the value unity also (cf. Fig. 6.4). However for $T < T_c$, we must substitute $\sigma^2 \simeq -3\epsilon$ from (6.27), whereupon

$$\chi_T \simeq \frac{1}{2} \frac{\mathbf{C}}{T} (-\epsilon)^{-1}. \tag{6.36}$$

Thus we find $\gamma' = 1$ but the coefficient $\mathscr{C}'$ is only half as large below T_c as above T_c ($\mathscr{C}' = \frac{1}{2}$).

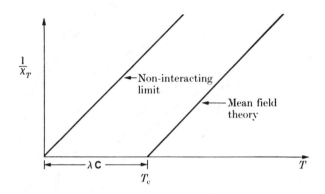

FIG. 6.4. Dependence upon temperature of the inverse isothermal susceptibility χ_T^{-1} for the non-interacting limit (perfect paramagnet) and for the mean field theory. Only the behaviour for $T > T_c$ is shown (cf. Figs. 5.4 and 10.2).

6.3.5. The gap exponents Δ_ℓ and Δ_ℓ'

The gap exponents were defined in § 3.4 in terms of successive field derivatives of the Gibbs potential. Thus, for example, we have

$$\Delta_1' = 2 - \alpha' - \beta = \tfrac{3}{2}, \tag{6.37}$$

$$\Delta_2' = \beta + \gamma' = \tfrac{3}{2}; \qquad \Delta_2 = \tfrac{1}{2}(\gamma + 2 - \alpha) = \tfrac{3}{2}. \tag{6.38}$$

To obtain Δ_3', we need to differentiate the Gibbs potential three times, or, equivalently, we need to find the field derivative of the susceptibility evaluated at $H = 0$. From (6.34) we have

$$\left(\frac{\partial \chi_T}{\partial h}\right)_T = \frac{\mathbf{C}}{T} (\epsilon + \sigma^2)^{-2} \left\{ 2\sigma \left(\frac{\partial \sigma}{\partial h}\right)_T + \mathcal{O}(\sigma^3) \right\}. \tag{6.39}$$

For $T > T_c$, $H = 0$ implies $\sigma = 0$. Hence the right-hand side of (6.39) reduces to zero, as we would already have suspected because the odd-order field derivatives of the free energy vanish for $T > T_c$.

For $T < T_c$, $(\partial\sigma/\partial h) = (T/\mathbf{C})\chi_T \simeq \frac{1}{2}(-\epsilon)^{-1}$ from (6.36) and $\sigma \simeq$ $3^{\frac{1}{2}}(-\epsilon)^{\frac{1}{2}}$ from (6.27). Hence

$$-G^{(3)} = \left(\frac{\partial\chi_T}{\partial H}\right)_T \propto \frac{\mathbf{C}}{T}(\epsilon - 3\epsilon)^{-2}(-3\epsilon)^{1/2}(-\epsilon)^{-1} \sim (-\epsilon)^{-5/2} \quad (6.40)$$

and, since $-G^2 = \chi_T \sim (-\epsilon)^{-1}$, using (3.19) we have the result

$$\Delta_3' = \tfrac{3}{2}. \quad (6.41)$$

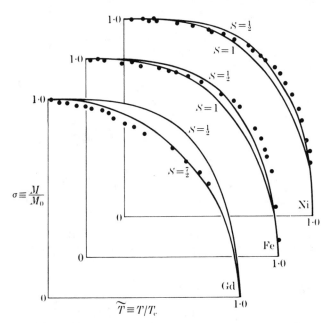

FIG. 6.5. Dependence of reduced magnetization $\sigma \equiv M/M_0 \equiv M(T, H)/M(0, 0)$ upon reduced temperature $\tilde{T} \equiv T/T_c$. The fact that there is a slightly different curve for each value of the spin quantum number S means that this law of corresponding states is valid only for a given value of S. The solid circles represent typical experimental data for Gd ($S \simeq \frac{7}{2}$), Fe ($S \simeq 1$), and Ni ($S \simeq \frac{1}{2}$). After Martin (1967).

6.3.6. *Law of corresponding states*

For general angular momentum quantum number S we can use (6.18) and (6.19) to eliminate the molecular field parameter λ from the equation of state (6.14) with the result

$$\sigma = B_S\left(\tilde{H} + \frac{3S}{S+1}\frac{\sigma}{\tilde{T}}\right), \quad (6.42)$$

where $\tilde{H} \equiv \beta\bar{\mu}SH$ is a dimensionless magnetic field. The critical-point exponents are the same for all values of angular momentum quantum number S, although the coefficients do depend upon S. For example, we find

$$\sigma^2 = \frac{10}{3} \frac{(S + 1)^2}{S^2 + (S + 1)^2} (-\epsilon) \tag{6.43}$$

and

$$\Delta C = \tfrac{5}{2} Nk \frac{(2S + 1)^2 - 1}{(2S + 1)^2 + 1}. \tag{6.44}$$

Notice that eqn (6.42) is analogous to the law of corresponding states, eqn (5.17), in the van der Waals theory. However, (6.42) predicts that even if we measure H, M, and T in the proper units, two materials will behave differently if they have different spin quantum number S (see Fig. 6.5).

6.4. The mean field theory as an approximation for the Heisenberg model

In the preceding section we obtained a set of critical indices for the molecular field theory without assuming anything concerning the nature of the coupling constant λ. Further, the theory does not give a microscopic argument for the introduction of the effective molecular field. This can be done, for example, through the Heisenberg model for a magnetic system.

The basic idea of the Heisenberg model is that there exists an exchange interaction between the magnetic moments $\mathbf{S}_i$ and $\mathbf{S}_j$ localized on sites i and j, which can be represented by the form

$$E_{\text{ex}} = -2 J_{ij} \mathbf{S}_i \cdot \mathbf{S}_j. \tag{6.45}$$

The minus sign is included in eqn (6.45) in order to conform to the convention that if the exchange parameter J_{ij} in (6.45) is positive the configuration in which the spins S_i and S_j are parallel is energetically favoured. (The case in which J_{ij} is negative favours anti-parallel alignment of $\mathbf{S}_i$ and $\mathbf{S}_j$, giving rise to an antiferromagnetic ordering.) In order to render manageable the complexity of calculations based upon (6.45), it is often assumed that $J_{ij} = 0$ except when sites i and j are neighbouring sites of the lattice. However, within the framework of the molecular field approximation it is possible to consider interaction potentials J_{ij} of arbitrary range.

The Hamiltonian for the entire system is then

$$\mathscr{H} = -\sum_{i=1}^{N} \sum_{j=1}^{N} J_{ij} \mathbf{S}_i \cdot \mathbf{S}_j - \bar{\mu} \sum_{i=1}^{N} \mathbf{S}_i \cdot \mathbf{H}. \qquad \blacktriangleright \tag{6.46}$$

The mean field approximation to the Heisenberg Hamiltonian of eqn (6.46) is one of a large class of cluster approximations (Smart 1966) for which we treat only the interactions among the spins within a cluster exactly, whereas for the remaining spins in the system we set $S_{iz} = \langle S_z \rangle$ and $S_{ix} = S_{iy} = 0$. For the molecular field approximation, we choose the cluster to consist of a single spin, so that the cluster Hamiltonian is simply

$$\mathscr{H}_i = \left(-2 \sum_j J_{ij} \langle S_z \rangle - \bar{\mu} H \right) S_{iz}. \tag{6.47}$$

Clearly the dynamics described by (6.47) are those of a single spin situated in a magnetic field of magnitude

$$H_{\mathrm{eff}} = H + 2(\hat{J}_0/\bar{\mu}) \langle S_z \rangle = H + 2(\hat{J}_0/N\bar{\mu}^2)M, \tag{6.48}$$

where $\hat{J}_0 \equiv \sum_j J_{ij}$, with $J_{ij} \equiv 0$ for $i = j$.

Comparing this equation with (6.13), we see that the molecular field coupling constant λ is given by

$$\lambda = 2\hat{J}_0/N\bar{\mu}^2. \tag{6.49}$$

Hence it follows from (6.19) that the critical temperature is

$$T_c = \frac{2}{3k} \hat{J}_0 S(S + 1). \qquad \blacktriangleright \tag{6.50}$$

We see that for the case of nearest-neighbour interactions ($J_{ij} = J$ only when site j is one of the q sites that are nearest neighbours of site i), $\hat{J}_0 = qJ$ and the critical temperature is given by the expression

$$kT_c/J = \tfrac{2}{3}qS(S + 1). \qquad \blacktriangleright \tag{6.51}$$

When we compare eqns (6.50) and (6.51) with experimental results on materials that are thought to be described fairly accurately by the Heisenberg model and also when we compare with numerical calculations (to be described in Chapter 9), we find that eqns (6.50) and (6.51) significantly overestimate the value of T_c—by as much as 50 per cent (see Fig. 6.6).

What is perhaps more disconcerting about (6.51) is that it predicts that two lattices will have the same value of the critical temperature if they have the same coordination number (i.e. the same value of q). For example, the simple cubic lattice is predicted to have the same value of T_c as the plane triangular lattice, since for both lattices $q = 6$ (cf. Fig. 6.7). Thus the mean field theory is evidently too crude to take into account the dimensionality of the lattice, which is believed to be one of the crucial features determining critical behaviour.

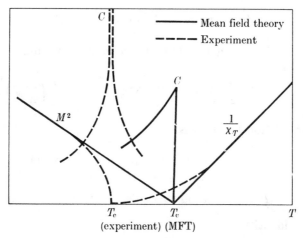

FIG. 6.6. Schematic comparison of typical experimental measurements on a Heisenberg ferromagnet (such as EuS) with the predictions of the molecular field theory (cf. Fig. 3.3). Note that the curve for $1/\chi_T$ is shown only for $T > T_c$.

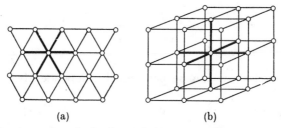

(a)   (b)

FIG. 6.7. (a) The plane triangular lattice, and (b) the simple cubic lattice. The fact that both lattices have the same coordination number ($q = 6$) means that within the mean field approximation they are predicted to have the same critical temperature (cf. (6.51)). This prediction is at odds with the more accurate approximation procedures to be discussed in Chapter 9, and it is currently believed that the critical temperature depends rather strongly upon lattice dimensionality. After Ziman (1964).

6.5. Equivalence of the mean field theory and an infinite interaction range

In this section we show that the mean field theory results follow from a model which assumes that each spin interacts equally with all the other spins in the lattice. The argument (Kac 1968) is most easily illustrated for an 'Ising' interaction,

$$\mathscr{H} = (-2J/N) \sum_{1 \leq i < j \leq N} s_i s_j. \qquad \blacktriangleright \quad (6.52)$$

Here the spins s_i are one-dimensional unit vectors, i.e. $s_i = \pm 1$ for all i. We wish to calculate the Gibbs potential per spin $\bar{G}(T, H)$, in the thermodynamic $(N \to \infty)$ limit,

$$\bar{G}(T, H) = -kT \lim_{N \to \infty} (N^{-1} \ln Z_N). \tag{6.53}$$

Our argument, following Kac (1968), begins by rewriting eqn (6.52) in the form

$$\mathcal{H} = -\frac{J}{N} \sum_{i=1}^{N} \sum_{j=1}^{N} s_i s_j + \frac{J}{N} \sum_{i=1}^{N} s_i^2 = J - \frac{J}{N} \left(\sum_{i=1}^{N} s_i \right)^2. \tag{6.54}$$

Hence the partition function is

$$Z_N = e^{-\mathcal{J}} \sum_{s_1 = -1}^{1} \cdots \sum_{s_N = -1}^{1} \exp \left\{ \mathcal{J} \left(\sum_{i=1}^{N} s_i N^{-\frac{1}{2}} \right)^2 \right\}, \tag{6.55}$$

where $\mathcal{J} \equiv J/kT \equiv \beta J$ is a dimensionless coupling constant.

Using the identity

$$\exp (a^2) = (2\pi)^{-\frac{1}{2}} \int_{-\infty}^{\infty} \exp \left(-\tfrac{1}{2} x^2 + 2^{\frac{1}{2}} ax \right) dx, \tag{6.56}$$

we can write the partition function in the form

$$Z_N = (2\pi)^{-\frac{1}{2}} e^{-\mathcal{J}} \int_{-\infty}^{\infty} e^{-x^2/2} \sum_{s_1 = -1}^{1} \cdots \sum_{s_N = -1}^{1} \exp \left\{ x(2\mathcal{J}/N)^{\frac{1}{2}} \sum_{i=1}^{N} s_i \right\} dx. \tag{6.57}$$

The summations now can be performed independently, and we have

$$Z_N = (2\pi)^{-\frac{1}{2}} e^{-\mathcal{J}} \int_{-\infty}^{\infty} e^{-x^2/2} \{ 2 \cosh (2\mathcal{J}/N)^{\frac{1}{2}} x \}^N dx. \tag{6.58}$$

If we substitute $x \equiv y N^{\frac{1}{2}}$, we obtain

$$Z_N = (N/2\pi)^{\frac{1}{2}} e^{-\mathcal{J}} 2^N \int_{-\infty}^{\infty} \{ e^{-y^2/2} \cosh (2\mathcal{J})^{\frac{1}{2}} y \}^N dy, \tag{6.59}$$

which suggests application of the saddle-point method. Thus

$$Z_N \propto N^{\frac{1}{2}} e^{-\mathcal{J}} 2^N \max_{-\infty < y < \infty} \{ e^{-y^2/2} \cosh (2\mathcal{J})^{\frac{1}{2}} y \}^N \tag{6.60}$$

and, from (6.53) the Gibbs potential is

$$\bar{G}(T, H) = -kT\{ \ln 2 + \ln F(\mathcal{J}) \}, \tag{6.61}$$

where

$$F(\mathcal{J}) = \max_{-\infty < y < \infty} \{ e^{-y^2/2} \cosh (2\mathcal{J})^{\frac{1}{2}} y \}. \tag{6.62}$$

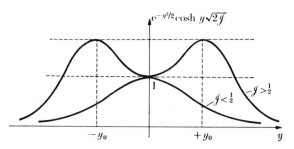

Fig. 6.8. A graph of the function which appears in the curly brackets of eqn (6.62), for two different temperature parameters $\mathscr{J} \equiv J/kT$. For $\mathscr{J} < \frac{1}{2}$ (or $T > 2J/k$), the maximum occurs at $y = 0$ and $\bar{G}(T,H) = -kT\ln 2$ from eqn (6.61). For $\mathscr{J} > \frac{1}{2}$ (or $T < 2J/k$), the maximum occurs at $y = y_0$ and there is an additional contribution to the Gibbs potential $\bar{G}(T, H)$. This figure is adapted from Kac (1968).

To find the maximum in eqn (6.62), we calculate the maximum of the logarithm of the bracketed factor,

$$f(y) = -\tfrac{1}{2}y^2 + \ln \cosh (2\mathscr{J})^{\frac{1}{2}}y. \tag{6.63}$$

Setting $f'(y) = 0$ results in the equation

$$y = (2\mathscr{J})^{\frac{1}{2}} \tanh (2\mathscr{J})^{\frac{1}{2}}y. \tag{6.64}$$

Equation (6.64) is recognized as being essentially the molecular field equation for $H = 0$; in particular, for $\mathscr{J} < \frac{1}{2}$ (or $T > 2J/k$), the only solution of (6.64) is $y_0 = 0$, whereas for $\mathscr{J} > \frac{1}{2}$ (or $T < 2J/k$) there exists a solution $y_0 \neq 0$ (see Fig. 6.8).

Suggested further reading
Weiss (1907).
Van Vleck (1945).
Ziman (1964).
Kittel and Shore (1965).
Brout (1965).
Smart (1966).
Kac (1968).

THE PAIR CORRELATION FUNCTION AND THE ORNSTEIN–ZERNIKE THEORY

IN this chapter we shall consider a third classical theory—the Ornstein–Zernike theory. In the course of our development, we shall introduce the pair correlation function for a fluid system. This function plays a particularly important role in almost all discussions of critical phenomena.

7.1. The density–density correlation function for a fluid system

Consider a fluid confined in a macroscopic volume V. If the fluid is in equilibrium and can exchange particles with an external reservoir, the probability density for the system to have N particles in a particular configuration of position and momenta $(\mathbf{r}_1, \ldots, \mathbf{r}_N, \mathbf{p}_1 \ldots, \mathbf{p}_N)$ is given by the grand canonical ensemble to be

$$P_N(\mathbf{r}_1, \ldots \mathbf{r}_N, \mathbf{p}_1, \ldots \mathbf{p}_N) = \frac{\exp\{-\beta U_N(\mathbf{r}_1, \ldots \mathbf{r}_N, \mathbf{p}_1, \ldots \mathbf{p}_N) + \beta\mu N\}}{N!\, h^{3N}\, \mathscr{L}},$$

(7.1)

where $\beta \equiv 1/kT$, h is Planck's constant, μ is the chemical potential, $U_N(\mathbf{r}_1, \mathbf{r}_2, \ldots \mathbf{r}_N, \mathbf{p}_1, \ldots \mathbf{p}_N)$ is the energy of the system, and

$$\mathscr{L} = \sum_{N=0}^{\infty} \frac{e^{\beta\mu N}}{N!\, h^{3N}} \int d\mathbf{r}_1 \ldots d\mathbf{r}_N\, d\mathbf{p}_1 \ldots d\mathbf{p}_N \times$$
$$\times \exp\{-\beta U_N(\mathbf{r}_1, \ldots \mathbf{r}_N, \mathbf{p}_1, \ldots \mathbf{p}_N)\} \quad (7.2)$$

is the grand partition function.

What will concern us here are the average values of variables when they are weighted with the probability factor (7.1). Let us first consider the density

$$n(\mathbf{r}) \equiv \sum_{i=1}^{N} \delta(\mathbf{r} - \mathbf{r}_i) \tag{7.3}$$

at a point $\mathbf{r}$ in the fluid; here $\mathbf{r}_i$ is the spatial coordinate of the ith particle. On using (7.1) we obtain for the average value of the density the expression

$$\langle n(\mathbf{r}) \rangle = \mathscr{Z}^{-1} \sum_{N=0}^{\infty} (N! h^{3N})^{-1} \int d^N \mathbf{r} \, d^N \mathbf{p} \, n(\mathbf{r}) \exp(-\beta U_N + \beta \mu N), \quad (7.4)$$

where we have adopted the shorthand notations $U_N = U_N(\mathbf{r}_1, \ldots \mathbf{r}_N,$ $\mathbf{p}_1, \ldots \mathbf{p}_N)$, $d^N \mathbf{r} = d\mathbf{r}_1 \, d\mathbf{r}_2 \ldots d\mathbf{r}_N$ and $d^N \mathbf{p} = d\mathbf{p}_1 \, d\mathbf{p}_2 \ldots d\mathbf{p}_N$. In a uniform system, $\langle n(\mathbf{r}) \rangle$ is independent of position and is therefore simply

$$\langle n(\mathbf{r}) \rangle = \left\langle \frac{N}{V} \right\rangle \equiv n. \quad (7.5)$$

A quantity that reflects in a more detailed fashion the microscopic properties of the system is

$$\langle n(\mathbf{r}) \, n(\mathbf{r}') \rangle = \mathscr{Z}^{-1} \sum_{N=0}^{\infty} (N! h^{3N})^{-1} \int d^N \mathbf{r} \, d^N \mathbf{p} \, n(\mathbf{r}) \, n(\mathbf{r}') \times$$
$$\times \exp(-\beta U_N + \beta \mu N). \quad (7.6)$$

The quantity $\langle n(\mathbf{r}) n(\mathbf{r}') \rangle$ is proportional to the probability of finding a particle at the position $\mathbf{r}$ if we know that there is a particle at the position $\mathbf{r}'$. In this sense $\langle n(\mathbf{r}) n(\mathbf{r}') \rangle$ is a measure of a conditional probability.

A function of more immediate interest is

$$G(\mathbf{r}, \mathbf{r}') \equiv \langle \{n(\mathbf{r}) - \langle n(\mathbf{r}) \rangle\} \{n(\mathbf{r}') - \langle n(\mathbf{r}') \rangle\} \rangle, \quad (7.7)$$

which measures the correlations of the fluctuations of the density from its average value. We shall call $G(\mathbf{r}, \mathbf{r}')$ the density–density correlation function or, when no confusion should arise, simply the correlation function. Since our system is assumed to be 'spatially uniform' (i.e. translationally invariant), we shall write $G(\mathbf{r}, \mathbf{r}') \to G(\mathbf{r} - \mathbf{r}')$. Furthermore, $\langle n(\mathbf{r}) \rangle = \langle n(\mathbf{r}') \rangle$, and eqn (7.7) may be written in the equivalent form

$$G(\mathbf{r} - \mathbf{r}') = \langle n(\mathbf{r}) n(\mathbf{r}') \rangle - n^2. \qquad \blacktriangleright \quad (7.8)$$

As $|\mathbf{r} - \mathbf{r}'| \to \infty$, the probability of finding a particle at $\mathbf{r}'$ becomes independent of what is happening at $\mathbf{r}$; that is to say, the densities become uncorrelated. Hence

$$\langle n(\mathbf{r}) n(\mathbf{r}') \rangle \to \langle n(\mathbf{r}) \rangle \langle n(\mathbf{r}') \rangle = n^2 \quad [\text{as } |\mathbf{r} - \mathbf{r}'| \to \infty], \quad (7.9)$$

and from eqn (7.8),

$$G(\mathbf{r} - \mathbf{r}') \to 0 \quad [\text{as } |\mathbf{r} - \mathbf{r}'| \to \infty]. \quad (7.10)$$

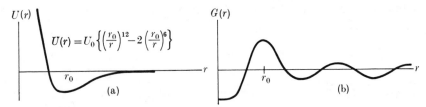

FIG. 7.1. Lennard–Jones model of a fluid. (a) Two-body potential $U(r)$. (b) Density–density correlation function $G(r)$.

The dependence upon $|\mathbf{r} - \mathbf{r}'|$ of $G(\mathbf{r} - \mathbf{r}')$ is shown in Figs. 7.1 and 7.2 for two different interparticle potentials $U(\mathbf{r})$, a Lennard–Jones potential and a hard-sphere potential.

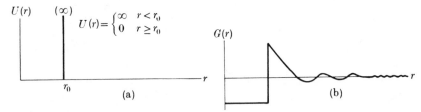

FIG. 7.2. Hard-sphere model of a fluid. (a) Two-body potential $U(r)$. (b) Density–density correlation function $G(r)$.

7.2. Relation between density fluctuations, the isothermal compressibility, and the density–density correlation function

In this section we shall show that the isothermal compressibility K_T can be directly related to the fluctuations in the total number of particles in the system—which in turn is directly related to $G(\mathbf{r} - \mathbf{r}')$. We shall see that the divergence of K_T as $T \to T_c$ is equivalent to the correlation function $G(\mathbf{r} - \mathbf{r}')$ becoming very long range (i.e. $G(\mathbf{r} - \mathbf{r}')$ approaches zero very slowly as $|\mathbf{r} - \mathbf{r}'|$ increases without limit). We first note that the fluctuation in the total number of particles is

$$\langle (N - \langle N \rangle)^2 \rangle = \langle N^2 \rangle - \langle N \rangle^2$$

$$= \mathscr{Z}^{-1} \sum_{N=0}^{\infty} (N! h^{3N})^{-1} \int d^N\mathbf{r} \, d^N\mathbf{p} \, N^2$$
$$\exp\left(-\beta U_N + \beta\mu N\right)$$

$$- \{\mathscr{Z}^{-1} \sum_{N=0}^{\infty} (N! h^{3N})^{-1} \int d^N\mathbf{r} \, d^N\mathbf{p} \, N$$
$$\exp\left(-\beta U_N + \beta\mu N\right)\}^2. \quad (7.11)$$

Equation (7.11) can be written in the form

$$\left\langle (N - \langle N \rangle)^2 \right\rangle = (kT)^2 \left(\frac{\partial^2 \ln \mathscr{Z}}{\partial \mu^2} \right)_{T, V}, \tag{7.12}$$

which the reader may verify by direct differentiation. Since

$$\frac{PV}{kT} = \ln \mathscr{Z} \tag{7.13}$$

(Huang 1963), we can write (7.12) in the form

$$\left\langle (N - \langle N \rangle)^2 \right\rangle = (kT)^2 \left\{ \frac{\partial^2 (PV/kT)}{\partial \mu^2} \right\}_{T, V}$$
$$= kTV \left(\frac{\partial^2 P}{\partial \mu^2} \right)_{T, V}. \tag{7.14}$$

Now

$$\left(\frac{\partial P}{\partial \mu} \right)_{T, V} = \frac{\langle N \rangle}{V} = n, \tag{7.15}$$

whence (7.14) becomes

$$\left\langle (N - \langle N \rangle)^2 \right\rangle = kTV \left\{ \frac{\partial(\langle N \rangle / V)}{\partial \mu} \right\}_{T, V}$$
$$= -\frac{\langle N \rangle \, kTV}{V^2} \left(\frac{\partial V}{\partial \mu} \right)_{T, N}. \tag{7.16}$$

The second line of eqn (7.16) follows from application of the identity of eqn (2.22). Next we introduce the isothermal compressibility,

$$K_T = -\frac{1}{V} \left(\frac{\partial V}{\partial P} \right)_{T, N}$$
$$= -\frac{1}{\langle N \rangle} \left(\frac{\partial V}{\partial \mu} \right)_{T, N}, \tag{7.17}$$

whence eqn (7.16) becomes

$$\left\langle (N - \langle N \rangle)^2 \right\rangle = \frac{\langle N \rangle^2 kT}{V} K_T$$
$$= \langle N \rangle n kT \, K_T. \tag{7.18}$$

Since the compressibility of an ideal gas is given by

$$K_T^\circ = \frac{1}{nkT}, \tag{7.19}$$

we can write (7.18) in the desired form,

$$\frac{K_T}{K_T^\circ} = \frac{\langle (N - \langle N \rangle)^2 \rangle}{\langle N \rangle}. \tag{7.20}$$

To relate the compressibility to the density–density correlation function $G(\mathbf{r} - \mathbf{r}')$, we observe that

$$\langle (N - \langle N \rangle)^2 \rangle = \left\langle \int d\mathbf{r} \left\{ n(\mathbf{r}) - \langle n(\mathbf{r}) \rangle \right\} \int d\mathbf{r}' \left\{ n(\mathbf{r}') - \langle n(\mathbf{r}') \rangle \right\} \right\rangle. \tag{7.21}$$

If we introduce the definition (7.7) of $G(\mathbf{r} - \mathbf{r}')$ into (7.21), we obtain

$$\langle (N - \langle N \rangle)^2 \rangle = \int d\mathbf{r} \int d\mathbf{r}' \, G(\mathbf{r} - \mathbf{r}')$$
$$= V \int d\mathbf{r}'' \, G(\mathbf{r}''), \tag{7.22}$$

where the last equality in (7.22) follows from the uniformity (translational invariance) of the system. Combining eqns (7.20) and (7.22), we have

$$\frac{K_T}{K_T^\circ} = n^{-1} \int d\mathbf{r} \, G(\mathbf{r}). \qquad \blacktriangleright \tag{7.23}$$

From eqn (7.23) we see that a divergent compressibility near T_c corresponds mathematically to an increase in the range of the pair correlation function $G(\mathbf{r} - \mathbf{r}')$. Sufficiently close to T_c the correlation length becomes as large as the wavelength of light and the density inhomogeneities scatter light strongly; this is the phenomenon of critical opalescence, which is displayed in Figs. 1.6 and 1.7.

In summary, then, we see from eqns (7.20), (7.22), and (7.23) that the three phenomena that are observed near the critical point,

(i) increase in the density fluctuations,
(ii) increase in the compressibility, and
(iii) increase in the range of the density–density correlation function

are all interrelated phenomena. In particular, eqn (7.23) is the analogue for fluid systems of the fluctuation–dissipation relation (A.20) which relates the isothermal susceptibility χ_T to the summation over all spins of the two-spin correlation function of a magnetic system.

7.3. The structure factor: relation between the pair correlation function and the scattering of electromagnetic radiation

In this section we relate the above discussion of the correlation function $G(\mathbf{r})$ to the intensity distribution of elastically-scattered electromagnetic radiation. Imagine that we irradiate our critical fluid with a monoenergetic beam of X-rays, neutrons, or light. Assume that the scattering takes place quasielastically, that is, assume that the energy of the radiation is much larger than the typical excitation energies of the system. Hence we can write for the momentum transfer vector $\mathbf{q} \equiv \mathbf{k}_s - \mathbf{k}_0$ the expression (see Fig. 7.3),

$$q = |\mathbf{q}| = 2k \sin \frac{\theta}{2}; \qquad k \equiv |\mathbf{k}_0| \simeq |\mathbf{k}_s|. \tag{7.24}$$

We wish to calculate $I(\mathbf{q})$, the intensity of the scattered radiation. Now

$$I(\mathbf{q}) = \left\langle \left| \sum_{j=1}^{N} a_j(\mathbf{q}) \right|^2 \right\rangle, \tag{7.25}$$

where $a_j(\mathbf{q})$ is the scattering amplitude for scattering from particle j.

FIG. 7.3. Scattering geometry used to determine eqn (7.24).

Now we know that the scattering from two different particles in the system is simply related by a phase factor; thus

$$a_j(\mathbf{q}) = a_1(\mathbf{q}) e^{-i\mathbf{q} \cdot (\mathbf{r}_j - \mathbf{r}_1)}. \tag{7.26}$$

Hence

$$I(\mathbf{q}) = \left\langle \left| a_1(\mathbf{q}) \sum_{j=1}^{N} e^{-i\mathbf{q} \cdot (\mathbf{r}_j - \mathbf{r}_1)} \right|^2 \right\rangle \tag{7.27}$$

$$= |a_1(\mathbf{q})|^2 \left\langle \left| \sum_{j=1}^{N} e^{-i\mathbf{q} \cdot \mathbf{r}_j} \right|^2 \right\rangle. \tag{7.28}$$

Now if there were no correlation between the particles, the scattering intensity $I^\circ(\mathbf{q})$ would be given by

$$I^\circ(\mathbf{q}) = N|a_1(\mathbf{q})|^2. \tag{7.29}$$

Hence from eqns (7.28) and (7.29)

$$\frac{I(\mathbf{q})}{I^\circ(\mathbf{q})} = N^{-1} \left\langle \sum_{i,j} e^{-i\mathbf{q} \cdot (\mathbf{r}_i - \mathbf{r}_j)} \right\rangle, \tag{7.30}$$

which may be written as

$$\frac{I(\mathbf{q})}{I^{\circ}(\mathbf{q})} = \frac{1}{N} \int d\mathbf{r} \int d\mathbf{r}' \left\langle \sum_{ij} \delta(\mathbf{r} - \mathbf{r}_i) \, \delta(\mathbf{r}' - \mathbf{r}_j) \, e^{-i\mathbf{q}\cdot(\mathbf{r}-\mathbf{r}')} \right\rangle$$

$$= \frac{1}{N} \int d\mathbf{r} \int d\mathbf{r}' \, e^{-i\mathbf{q}\cdot(\mathbf{r}-\mathbf{r}')} \left\langle \left\{ \sum_{i=1}^{N} \delta(\mathbf{r} - \mathbf{r}_i) \right\} \left\{ \sum_{j=1}^{N} \delta(\mathbf{r}' - \mathbf{r}_j) \right\} \right\rangle.$$

$$(7.31)$$

On introducing the number density $n(\mathbf{r})$ through (7.3), (7.31) becomes

$$\frac{I(\mathbf{q})}{I^{\circ}(\mathbf{q})} = \frac{1}{N} \int d\mathbf{r} \int d\mathbf{r}' \, e^{-i\mathbf{q}\cdot(\mathbf{r}-\mathbf{r}')} \langle n(\mathbf{r}) \, n(\mathbf{r}') \rangle$$

$$(7.32)$$

$$= \frac{1}{N} \int d\mathbf{r} \int d\mathbf{r}' \, e^{-i\mathbf{q}\cdot(\mathbf{r}-\mathbf{r}')} \{ G(\mathbf{r} - \mathbf{r}') + n^2 \},$$

where the second equality in (7.32) follows from (7.8). Equation (7.32) may be written as

$$\frac{I(\mathbf{q})}{I^{\circ}(\mathbf{q})} = \frac{V}{N} \int d\mathbf{r}'' \, e^{-i\mathbf{q}\cdot\mathbf{r}''} \, G(\mathbf{r}'') + \frac{V^2}{N} n^2 \, \delta(\mathbf{q}). \qquad (7.33)$$

The last term in eqn (7.33) contributes only when $\mathbf{q} = 0$, and hence, by (7.24), only for forward scattering ($\theta = 0$). Therefore it is customary to omit this term in many expressions, and to simply write

$$\frac{I(\mathbf{q})}{I^{\circ}(\mathbf{q})} = \frac{1}{n} \int d\mathbf{r} \, e^{-i\mathbf{q}\cdot\mathbf{r}} \, G(\mathbf{r}) \equiv \frac{1}{n} S(\mathbf{q}), \qquad \blacktriangleright \quad (7.34)$$

where the second equality in (7.34) serves to define the structure factor $S(\mathbf{q})$ as the spatial Fourier transform of the density–density correlation function. Equation (7.34) tells us that the intensity of radiation scattered through a wave vector $\mathbf{q}$ is changed from the value $I^{\circ}(\mathbf{q})$ (predicted if the particles of the fluid were not interacting) by an amount proportional to the Fourier transform of the density–density correlation function. In particular, as the critical point is approached, the integral in (7.34) becomes extremely large for small values of $\mathbf{q}$. This dependence upon $\mathbf{q}$ is displayed in Fig. 7.4 for scattering from a binary alloy at several temperatures in the critical region.

7.4. Ornstein–Zernike theory of the scattering amplitude

The first qualitative explanation of the huge increase of forward scattering called critical opalescence was due to Ornstein and Zernike (1914). Two particularly readable reviews of their original work and of

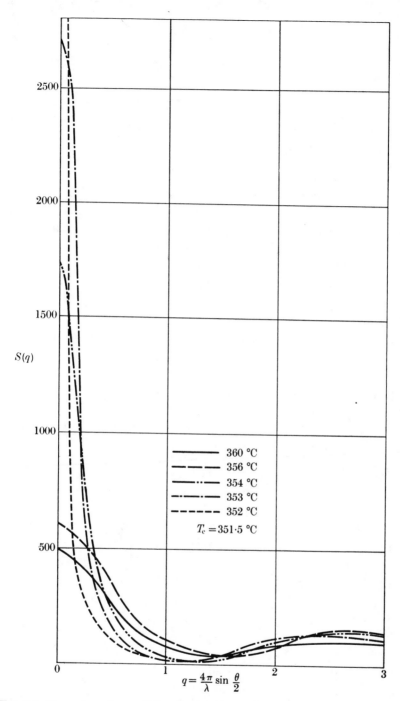

FIG. 7.4. The q dependence of the structure factor $S(q)$ for several temperatures just above the critical temperature $T_c = 351\cdot5°C$. After the work of Münster and Sagel (1958) on Al–Zn alloy. The 'lattice-gas' model of a simple fluid (cf. §1.1) also provides a useful model for a binary alloy.

numerous subsequent developments are Münster (1966) and Fisher (1964).

In order to develop the Ornstein–Zernike theory, we must make use of the density–density correlation function $G(\mathbf{r} - \mathbf{r}')$ introduced in eqn (7.7). We begin by substituting eqn (7.3) for the number density $n(\mathbf{r})$ into eqn (7.8), with the result

$$G(\mathbf{r} - \mathbf{r}') = \left\langle \sum_{i=1}^{N} \sum_{j=1}^{N} \delta(\mathbf{r} - \mathbf{r}_i)\,\delta(\mathbf{r}' - \mathbf{r}_j) \right\rangle - n^2. \qquad (7.35)$$

Notice that the terms with $i = j$ are not excluded in eqn (7.35). This is reflected in the fact that $G(\mathbf{r} - \mathbf{r}')$ includes not only correlation between different particles, but also the correlation of a particle with itself. Therefore it is convenient to partition $G(\mathbf{r} - \mathbf{r}')$ into two terms,

$$G(\mathbf{r} - \mathbf{r}') \equiv n\,\delta(\mathbf{r} - \mathbf{r}') + n^2\,\Gamma(\mathbf{r} - \mathbf{r}'), \qquad (7.36)$$

where the first term is the contribution to $G(\mathbf{r} - \mathbf{r}')$ due to correlation of a particle with itself and the second term is the contribution to $G(\mathbf{r} - \mathbf{r}')$ due to correlation between different particles. Notice that the presence of the factor n^2 in the second term implies that the function $\Gamma(\mathbf{r} - \mathbf{r}')$ is dimensionless. We next introduce the 'direct correlation function', $C(\mathbf{r})$, by means of its Fourier transform, $\hat{C}(\mathbf{q}) = \int C(\mathbf{r}) \exp(-i\mathbf{q} \cdot \mathbf{r})\,d\mathbf{r}$, where $\hat{C}(\mathbf{q})$ is defined by

$$\hat{C}(\mathbf{q}) \equiv \frac{\hat{\Gamma}(\mathbf{q})}{1 + n\hat{\Gamma}(\mathbf{q})}; \qquad (7.37)$$

here $\hat{\Gamma}(\mathbf{q}) = \int \Gamma(\mathbf{r}) \exp(-i\mathbf{q} \cdot \mathbf{r})\,d\mathbf{r}$ is the Fourier transform of the function $\Gamma(\mathbf{r})$. Note that $\hat{C}(\mathbf{q})$ thus defined does not depend very strongly on temperature: at high temperatures where $\hat{\Gamma}(\mathbf{q})$ is almost zero, $\hat{C}(\mathbf{q}) \sim \hat{\Gamma}(\mathbf{q})$, while as $T \to T_c$ where $\hat{\Gamma}(\mathbf{q} = 0) \to \infty$, $\hat{C}(\mathbf{q} = 0) \equiv \int C(\mathbf{r})\,d\mathbf{r} \sim n^{-1}$. Thus this new function $C(\mathbf{r})$ must remain comparatively short-range even at T_c! It is perhaps convenient to think of $C(\mathbf{r})$ as being some function with a range similar to that of the interaction potential $U(\mathbf{r})$, and then to imagine that $\Gamma(\mathbf{r})$ becomes long-range near T_c as a result of the propagation of this direct correlation—see for example, Fig. 7.5. This intuitive picture is supported by the Fourier transform of eqn (7.37),

$$\Gamma(\mathbf{r} - \mathbf{r}') = C(\mathbf{r} - \mathbf{r}') + n\int C(\mathbf{r} - \mathbf{r}'')\,\Gamma(\mathbf{r}'' - \mathbf{r}')\,d\mathbf{r}''. \qquad (7.38)$$

Equation (7.38) has come to be called the Ornstein–Zernike integral equation, and of course it may equally well be chosen as the defining equation for $C(\mathbf{r})$, since (7.37) and (7.38) are equivalent (as may be seen from the convolution theorem).

7.4.1. The Ornstein–Zernike assumption

Our goal is to calculate the structure factor $S(\mathbf{q}) = n + n^2\,\hat{\Gamma}(\mathbf{q})$, since this gives us the scattered intensity $I(\mathbf{q})$. We can obtain the qualitative form of $S(\mathbf{q})$ by following Ornstein and Zernike and assuming that

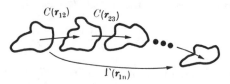

FIG. 7.5. Schematic diagram to illustrate the difference between the total correlation function $\Gamma(\mathbf{r})$ (which can be long-range) and the direct correlation function $C(\mathbf{r})$ (which is short-range).

$\hat{C}(\mathbf{q})$ may be expanded in a Tayor series about $q = 0$ for all temperatures right up to T_c. Thus, for an isotropic system,

$$\hat{C}(q) = \hat{C}(0) + \sum_{\ell=0}^{\infty} \hat{c}_\ell(n,\,T)q^\ell. \tag{7.39}$$

The coefficients $\hat{c}_\ell(n,\,T)$ in eqn (7.39) are given by Maclaurin's expansion,

$$\hat{c}_\ell(n,\,T) = \frac{1}{\ell!}\left\{\frac{\partial^\ell}{\partial q^\ell}\hat{C}(q)\right\}_{q=0} \propto \frac{i^\ell}{\ell!}\int_{-1}^{1}\mu^\ell\,\mathrm{d}\mu\int_0^\infty r^{\ell+2}\,C(r)\,\mathrm{d}r, \tag{7.40}$$

where the first integral in eqn (7.40) arises from the angular integration and is zero for odd values of the integer ℓ. Although we know from the definition of $\hat{C}(q)$ that $C(r)$ is integrable at $T = T_\mathrm{c}$, we have no reason to expect that all of the moments of $C(r)$ are also integrable at T_c. Yet let us follow Ornstein and Zernike and assume for now that the coefficients $\hat{c}_\ell(n,\,T_\mathrm{c})$ are finite for all ℓ. Then it follows at once from the defining equation for $\hat{C}(q)$ that for small q, the scattered intensity is very nearly a Lorentzian. To see this, notice from (7.36) and (7.37) that

$$\frac{1}{n}S(q) = 1 + n\hat{\Gamma}(q) = \frac{1}{1 - n\hat{C}(q)}. \qquad \blacktriangleright \ (7.41)$$

Hence

$$\frac{n}{S(q)} = 1 - n\hat{C}(q) = 1 - n\{\hat{C}(0) + \hat{c}_2(n, T)\, q^2 + \mathcal{O}(q^4)\}$$

$$= \hat{c}_2(n, T) \left\{ \frac{1 - n\hat{C}(0)}{\hat{c}_2(n, T)} - n\, q^2 + \mathcal{O}(q^4) \right\}$$

$$= R^2\{\kappa_1^2 + q^2 + \mathcal{O}(q^4)\}, \tag{7.42}$$

where

$$R^2 \equiv -n\hat{c}_2(n, T) \propto \int r^2 C(r)\, d\mathbf{r} \tag{7.43}$$

is directly related to the second moment of $C(r)$, and

$$\kappa_1^2 \equiv \frac{1 - n\hat{C}(0)}{R^2} \tag{7.44}$$

is related to the zeroth moment. From eqn (7.42) we would predict that a plot of inverse scattering intensity, $I^\circ(q)/I(q) = n/S(q)$, against q^2 for various temperatures would result in a family of curves which for sufficiently small q^2 are linear, with slopes given by $R^2 = -n\hat{c}_2(n, T)$ and intercepts by

$$R^2\, \kappa_1^2 = 1 - n\hat{C}(0) = \{1 + n\hat{\Gamma}(0)\}^{-1} = nS(0)^{-1} = K_T^\circ/K_T. \tag{7.45}$$

Note that $\kappa_1^2 \sim \epsilon^\gamma$, assuming R^2 is not singular at $T = T_c$. The slope parameter R is sometimes called the Debye persistence length but here we regard it as phenomenological parameter since we have seen no way in which $C(r)$ and hence R^2 may be calculated. A plot of inverse scattering intensity against q^2 (called an Ornstein–Zernike–Debye or OZD plot) is given in Fig. 7.6 for argon. That the limiting slopes of the curves in Fig. 7.6 change only slightly as $T \to T_c$ is consistent with our assumption that $C(r)$ remains reasonably short-range in the critical region.

7.4.2. The Ornstein–Zernike approximation

Next we make what we shall call the OZ approximation—we truncate the expression (7.42) by neglecting all terms of order q^4 and higher. In this approximation (which presumably is good for small q) $I(q)$ is indeed a Lorentzian,

$$\frac{I(q)}{I^\circ(q)} = \frac{S(q)}{n} = \frac{R^{-2}}{\kappa_1^2 + q^2}. \qquad \blacktriangleright \tag{7.46}$$

The Fourier inversion of eqn (7.46) shows that the OZ approximation

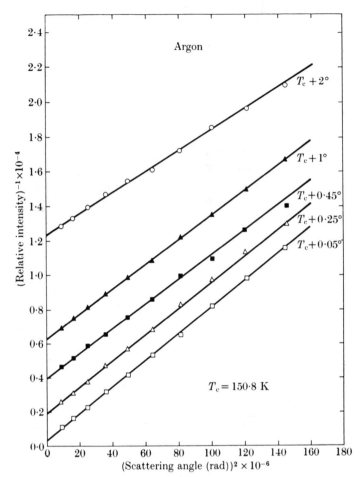

FIG. 7.6. Dependence on q^2 of the inverse scattering intensity for argon. Apparently the predictions of the Ornstein–Zernike theory are borne out by these measurements. Taken from Thomas and Schmidt (1963). Note that the slopes of the lines increase as $T \to T_c$, corresponding to an increase in the parameter R^2 defined in eqn (7.43).

to the asymptotic form for large r of the total correlation function $G(r)$ is

$$G(r) \sim \frac{1}{R^2} \frac{e^{-\kappa_1 r}}{r}. \qquad \blacktriangleright \quad (7.47)$$

From the definition (7.44), we see that $\kappa_1^2 \to 0$ as $T \to T_c$. Clearly, κ_1^{-1} has the dimensions of length and we set

$$\kappa_1 = \kappa \equiv \xi^{-1} \qquad (7.48)$$

where κ is called the inverse correlation length.

From eqns (3.8), (7.45), and (7.48) we find that as $T \to T_c$, $\kappa \sim \epsilon^{\frac{1}{2}\gamma}$. Hence from the exponent definitions in eqn (3.15), we have

$$\nu' = \tfrac{1}{2}\gamma' \tag{7.49}$$

and

$$\nu = \tfrac{1}{2}\gamma . \tag{7.50}$$

7.5. Further developments of the Ornstein–Zernike theory

7.5.1. *Failure of OZ theory for two-dimensional systems*

Suppose we consider applying the OZ approximation to a fluid system of arbitrary dimensionality d. The development of the proceeding section proceeds through eqn (7.46) essentially unchanged. To get $G(r)$ we must invert eqn (7.46) for general dimensionality d.

The d-dimensional Fourier transform,

$$G(\mathbf{r}) = \int S(\mathbf{q}) \, e^{-i\mathbf{q}\cdot\mathbf{r}} \, d\mathbf{q} \quad \text{with} \quad \mathbf{q} = (q_1, q_2, \ldots q_d) \tag{7.51}$$

has a volume element $d\mathbf{q} = q^{d-1} \, dq \, (\sin \theta_1)^{d-2} \, d\theta_1 \, d\Omega_{d-1}$, where $q = |\mathbf{q}|$, and $d\Omega_{d-1}$ is an element of solid angle in $d-1$ dimensions. Hence eqn (7.51) becomes, for an isotropic system,

$$G(r) = \int S(q) \left\{ \int_0^\pi e^{iqr\cos\theta_1} (\sin \theta_1)^{d-2} \, d\theta_1 \right\} d\Omega_{d-1} \, q^{d-1} \, dq. \tag{7.52}$$

Now the integral inside the braces in (7.52) can be evaluated in terms of Bessel functions (Arfken 1970), whence

$$G(r) \propto \int_0^\infty S(q) \, \frac{J_{d/2-1}(qr)}{(qr)^{d/2-1}} \, q^{d-1} \, dq, \tag{7.53}$$

where $J_\ell(x)$ is a Bessel function of the first kind. Note that the kernel multiplying $S(q)$ in the integrand reduces for $d = 3$ to the familiar expression $(\sin qr)/qr$.

Fisher (1962) has carried out the Fourier inversion in detail for general d, with the following results:

Case (i). Fixed $T > T_c$ (and hence fixed $\kappa > 0$) but let $r \to \infty$.

$$G(r) \propto \frac{e^{-\kappa r}}{r^{(d-1)/2}} \left\{ 1 + \mathcal{O}\left(\frac{d-3}{\kappa r}\right) \right\}. \tag{7.54}$$

Case (ii). Fixed r, but let $T \to T_c^+$ ($\kappa \to 0$).

$$G(r) \propto \begin{cases} (\ln r) \, e^{-\kappa r} \left\{ 1 + \mathcal{O}\left(\dfrac{1}{\ln \kappa r}\right) \right\} & d = 2 \\[2ex] \dfrac{e^{-\kappa r}}{r} & d = 3 \\[2ex] \dfrac{e^{-\kappa r}}{r^{d-2}} \left\{ 1 + \mathcal{O}(\kappa r) \right\} & d > 3. \end{cases} \tag{7.55}$$

For $d = 3$, we see that the higher order terms in eqn (7.54) vanish identically; thus case (i) and case (ii) both give the same answer, $G(r) \propto r^{-1} e^{-\kappa r}$.

The correlation function, for large values of r, is found from (7.55) to behave as

$$G(r)\big|_{T=T_c} \propto \begin{cases} \ln r & d = 2 & (7.56) \\ r^{-(d-2)} & d \geq 3. & (7.57) \end{cases}$$

From eqn (7.56) we see that the Ornstein–Zernike theory predicts that at the critical temperature of a two-dimensional fluid, the correlation increases with distance—this is clearly a nonsensical result! Indeed, for the two-dimensional Ising model, we can obtain the two-spin correlation function exactly (Kaufman and Onsager, 1949, Stephenson 1964) and find that

$$G(r)\big|_{T=T_c} \propto r^{-1/4} \qquad [d = 2 \text{ Ising}]. \qquad (7.58)$$

7.5.2. Fisher's Modification of Ornstein–Zernike Theory

For many systems one finds experimentally that plots of inverse scattering intensity against q^2 are not straight lines but rather that as

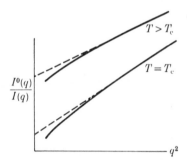

FIG. 7.7. Sketch of the normalized inverse scattering intensity as found for many systems. The curvature is smaller for many systems (as, e.g. for Ar in Fig. 7.6.) and the result $\eta = 0$ is not outside the experimental error. By eqn (7.34), $I^0(q)/I(q) = n/S(q)$.

$T \to T_c$ they become concave as sketched in Fig. 7.7. Fisher (1964) explains this downward turn of the isotherms for small q by means of the assumption that

$$S(q)\big|_{T=T_c} \sim q^{-2+\eta} \qquad [q \sim 0], \qquad (7.59)$$

where the index η is zero for the Ornstein–Zernike theory. Equivalently, on Fourier inverting (7.59), we have

$$G(r)\big|_{T=T_c} \sim r^{-(d-2+\eta)} \qquad [r \text{ large}], \qquad (7.60)$$

which is the definition of η used in eqn (3.16).

For the two-dimensional Ising model, eqn (7.60) is rigorously true and, using (7.58), we have $\eta = \frac{1}{4}$. Moreover, $\nu = 1$, and $\gamma = \frac{7}{4}$ so that the relation $\gamma = (2 - \eta)\nu$ is satisfied. For the three-dimensional Ising model, the best numerical approximations (Moore, Jasnow, and Wortis, 1969) suggest that

$$\gamma = 1.25, \tag{7.61}$$

$$\nu = 0\cdot638 \, {}^{+0\cdot002}_{-0\cdot001}, \tag{7.62}$$

and

$$\eta = 0\cdot041 \, {}^{+0\cdot006}_{-0\cdot003}. \tag{7.63}$$

The estimate of η was obtained from the estimates of γ and ν and using the relation $\gamma = (2 - \eta)\nu$.

For the spherical model (the one exactly-soluble three-dimensional magnetic model), $\gamma = 2$, $\nu = 1$, and $\eta = 0$. Most experimental measurements suggest a value of η between zero and $0\cdot1$, but the experimental error in η is frequently so large as to make excluding $\eta = 0$ impossible.

Suggested further reading
Ornstein and Zernike (1914).
Fisher (1964).
Münster (1966).
Egelstaff (1967).
Rowlinson (1969).
Stell (1969).

PART IV

MODELS OF FLUID AND MAGNETIC PHASE TRANSITIONS

8

RESULTS PROVIDED BY EXACT SOLUTION OF MODEL SYSTEMS

ALTHOUGH both the van der Waals theory of a fluid system and the molecular field theory of a magnetic system were originally introduced as purely phenomenological descriptions of interacting many-particle systems, we have seen in Chapters 5 and 6 that the predictions of both of these classical theories can be regarded as obtainable from a simple model of interparticle interactions in which one makes the (rather unrealistic) assumption that each particle interacts equally with all the other particles in the system. Unfortunately such a crude model of interparticle interactions predicts values of the critical-point exponents that are independent of the lattice dimensionality and that disagree with almost all experimental measurements.

Clearly a detailed study of the predictions of more realistic model systems would be extremely desirable in order that we may obtain some clues regarding which features of interparticle interactions are important for determining critical properties and which are not. In this chapter we introduce several specific model systems (see Table 8.1) that are thought to be fairly good approximations to the true interparticle interactions in at least a few physical systems. We shall see, however, that almost all these model systems cannot be solved exactly for a three-dimensional system! Therefore we devote this chapter to considering those cases (such as one-dimensional systems with short-range interactions) that can be solved exactly (see Table 8.2), and we defer to the following chapter a discussion of the approximation procedures that are used to obtain results for the more realistic cases that have not yet yielded to exact solution.

TABLE 8.1

Special cases of the model Hamiltonian (8.1) The parameter D is the spin dimensionality. After Stanley and Lee (1970).

D	Hamiltonian	Name	System
1	$\mathscr{H} = -J \sum_{\langle ij \rangle} S_{ix}S_{jx}$	Ising model	one-component fluid; binary alloy, mixture
2	$\mathscr{H} = -J \sum_{\langle ij \rangle} (S_{ix}S_{jx} + S_{iy}S_{jy})$	plane rotator model (Vaks-Larkin model)	λ-transition in a Bose fluid
3	$\mathscr{H} = -J \sum_{\langle ij \rangle} (S_{ix}S_{jx} + S_{iy}S_{jy} + S_{iz}S_{jz})$	classical Heisenberg model	ferromagnet; antiferromagnet
$\vdots$			
∞	$\mathscr{H} = -J \sum_{\langle ij \rangle} \left(\sum_{n=1}^{\infty} S_{in}S_{jn} \right)$	spherical model	none

TABLE 8.2

Some of the cases in which the model Hamiltonian (8.1) is exactly soluble. Here the notation n.n. stands for nearest-neighbour interactions only. A blank indicates that the system has not yet been solved. D and d denote the spin and lattice dimensionality respectively. After Stanley and Lee (1970)

D	$d = 1$	$d = 2$	$d = 3$	$d > 3$
1	all H; both n.n. and $1/r^{d+x}$	$H = 0$; n.n.	—	—
2	$H = 0$; n.n.	—	—	—
3	$H = 0$; n.n.	—	—	—
$\vdots$				
∞	all H; both n.n. and $1/r^{d+x}$	all H; both n.n. and $1/r^{d+x}$	all H; both n.n. and $1/r^{d+x}$	only critical-point exponents are known exactly

8.1. A model Hamiltonian for a classical spin system: the generalized Heisenberg model

The techniques to be discussed in this and in the following chapter can all be illustrated by applying them to a single model Hamiltonian

(Stanley 1968c, d, 1969a). We shall see that this Hamiltonian reduces for four special cases to four commonly-studied models. More complicated interactions can be taken into account by systematic modifications of this model.

The model Hamiltonian may be written as

$$\mathcal{H}^{(D)} = -J \sum_{\langle ij \rangle} \mathbf{S}_i^{(D)} \cdot \mathbf{S}_j^{(D)}, \qquad \blacktriangleright \ (8.1)$$

where the spins $\mathbf{S}_i^{(D)}$ are D-dimensional unit vectors and $-J$ is the energy of a nearest-neighbour pair $\langle ij \rangle$ of parallel spins localized on sites i and j of the lattice. Thus $(S_{i1}, S_{i2}, \ldots, S_{iD})$ are the Cartesian coordinates of $\mathbf{S}_i^{(D)}$, then $\sum_{n=1}^{D} S_{in}^2 = 1$ for $1 \leq i \leq N$, and the scalar product in eqn (8.1) denotes the quantity $\sum_{n=1}^{D} S_{in} S_{jn}$.

For example, if $D = 1$, the spins are simply one-dimensional 'sticks' capable of assuming the two discrete orientations $+1$ (up) and -1 (down), and the model Hamiltonian (8.1) reduces to the simple (spin half) Ising model. Although it was first introduced as a crude model of ferromagnetism, the Ising model has come to serve as a practical model for many systems such as a one-component fluid and a binary alloy. Perhaps better—or at least more consistent names—for the case $D = 1$ would be the Lenz model, after its inventor (Lenz 1920), or the Ising–Onsager model, after the men who solved it for one- and two-dimensional lattices, respectively (Ising 1925, Onsager 1944).

For $D = 2$, the Hamiltonian (8.1) describes a system of isotropically-interacting two-dimensional unit vectors, and is generally called the planar Heisenberg model, the plane rotator model, or the classical planar model. It has also been called the Vaks–Larkin model, because Vaks and Larkin have considered it to be a lattice model for the super-fluid transition in a Bose fluid (Vaks and Larkin 1966).

Finally, for the case of three-dimensional spins (8.1) reduces to the classical Heisenberg model. The classical Heisenberg model can also be considered as the $S \to \infty$ limit of the quantum-mechanical Heisenberg model; that is, instead of allowing the Heisenberg spins a finite number $(2S + 1)$ of discrete orientations, we assume that they can take on an entire continuum of orientations. Also, we renormalize the magnitude of each spin from $\{S(S + 1)\}^{\frac{1}{2}}$ to 1. The classical Heisenberg model was studied in the low-temperature domain as early as 1934 (Heller and Kramers 1934), but it is only within recent years that it has been realized that the classical model is an extremely realistic approximation to the quantum-mechanical case for temperatures near

T_c (Stanley and Kaplan 1966a, Wood and Rushbrooke 1966, Joyce and Bowers 1966a). It is now believed that critical point indices are either independent of spin quantum number S or they depend on S so weakly that to an extremely good approximation the spin dependence may be neglected. Hence, although the classical Heisenberg model is an unrealistic approximation to the quantum-mechanical case in the low-temperature domain, it is extremely realistic in the neighbourhood of T_c as regards critical indices.

It is important to stress, however, that even the quantum-mechanical Heisenberg model is not valid for a very wide range of real magnetic materials, as it makes the rather stringent assumptions of (a) well-localized spins and (b) complete isotropy of interaction. Most magnetic materials in nature fail in one or the other of these two stringent assumptions. For example, the $3d$ transition metals probably do not satisfy assumption (a), whereas most of the rare-earth elements do not satisfy assumption (b). Fortunately, certain materials have been discovered in recent years which satisfy both assumptions to a fair extent. Hence, when we compare the predictions of the Heisenberg model with experiments, we will have in mind materials such as the europium chalcogenides EuO and EuS.

Worth noting here is the fact that the lowest energy state of eqn (8.1) depends on the sign of the exchange parameter J. The minus sign appears in (8.1) in order to conform to the convention that if J is positive, then parallel alignment of neighbouring spins is favoured. Conversely, if J is negative, then neighbouring spins will want to be antiparallel. This is possible if the spins are situated on a loose-packed lattice, i.e. a lattice that is decomposable into two interpenetrating sublattices such that all the nearest neighbours of a given spin on one sublattice belong to the other sublattice. Interesting theoretical problems are posed by the other case—the so-called close-packed lattices—of which the plane triangular and face-centred cubic lattice are examples. A typical antiferromagnetic material that is probably well described by the Heisenberg interaction is $RbMnF_3$.

For $D > 3$, eqn (8.1) continues to describe a well-defined system—although it becomes rather difficult to draw a picture of the spins. Moreover, no physical system has been put forward that corresponds to, say, four-dimensional spins. Nevertheless, when the Hamiltonian (8.1) was first considered (Stanley 1968c) the author was led to investigate whether any simplifications occurred in the limit $D \to \infty$, that might render the system soluble for lattices of dimensionality $d > 1$. He found that the model of infinite dimensional spins was in fact quite simple to solve—even for a three-dimensional lattice (Stanley 1968d, Helfand 1969, Kac and Thompson, 1971). This was particularly exciting, in as much as only one other non-trivial interacting many-body system has yielded to solution for three-dimensional lattices—namely the Berlin–Kac spherical model (or spherical approximation to the Ising model). Surprisingly, it turns out that the expression for the partition function of $\mathscr{H}^{(D)}$ in the limit $D \to \infty$ is in fact *identical* to the spherical model partition function. This result will be discussed in § 8.4.

A study of eqn (8.1) will allow us to determine the variation of critical properties with both spin dimensionality D and lattice dimensionality d, and thereby to consider a variety of physical systems near their critical points. However, we should take care to point out explicitly that there are several assumptions that (8.1) makes for all D and d that are probably

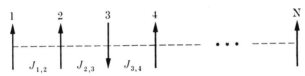

FIG. 8.1. Schematic diagram representing a chain of N Ising spins (i.e. $d = D = 1$) with a non-uniform interaction $J_{i,i+1}$ between spins situated on sites i and $i + 1$. In § 8.2 we denote $J_{i,i+1}$ by J_i for simplicity.

not realistic assumptions for most of the physical systems that we have been discussing above. These assumptions are made largely in order that the mathematical problem of obtaining critical-point predictions from a model Hamiltonian remains one of manageable complexity. Fortunately there is evidence that the assumptions that the interactions are (i) uniform in direction, (ii) nearest-neighbour in range, and (iii) isotropic in symmetry do not significantly affect the critical-point predictions. We now indicate what happens in the cases for which these assumptions do *not* hold.

(i) *Non-uniform interactions.* One simple modification of (8.1) is

$$\mathscr{H}^{(D)} = - \sum_{\langle ij \rangle} J_{ij} \mathbf{S}_i^{(D)} \cdot \mathbf{S}_j^{(D)} \tag{8.2}$$

where, as before, two spins can interact only if they are on neighbouring sites of the lattice, but the interaction strength J_{ij} depends on the sites i and j. Two soluble examples of (8.2) will be discussed in § 8.2 and in Appendix B. In the first of these, the case $D = 1$ (Ising model) and $d = 1$ (linear chain lattice) will be considered, and the exchange interactions $J_{i,i+1}$ are arbitrary positive numbers (see Fig. 8.1). We find that the critical behaviour is not affected by our choice of the exchange interactions. A second example is that of a two-dimensional square lattice with Ising spins ($d = 2$, $D = 1$) which interact with their nearest neighbours in the horizontal and vertical directions with energies J_h and J_v respectively (see Fig. 8.2). Here also, the critical predictions that can be obtained exactly are independent of the magnitudes of J_h and J_v.

(ii) *Further than nearest-neighbour interactions.* A more general (and mathematically considerably more complex) case is

$$\mathscr{H}^{(D)} = - \sum_{i,j} J_{ij}\mathbf{S}_i^{(D)} \cdot \mathbf{S}_j^{(D)},\qquad(8.3)$$

where now we allow for the (very realistic) possibility that spins other than nearest neighbours may interact with each other. In the absence

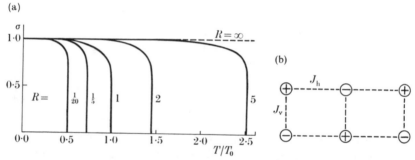

Fig. 8.2. Temperature dependence of the reduced zero-field magnetization $\sigma \equiv M(T, H = 0)/M(T = 0, H = 0)$ for the Ising model when the lattice is a two-dimensional square net, and when the exchange interaction is J_h between neighbouring spins situated in the same row and J_v between neighbours in the same column. (This lattice is sometimes called a rectangular net, reducing to the square net when $J_h = J_v$.) Here T_0 is the critical temperature when $J_h = J_v$. This figure should serve to remind us of two facts. The first of these is that there is a phase transition (i.e. $T_c > 0$) regardless how small is the ratio $R \equiv J_v/J_h$, i.e. as soon as one turns on J_v, T_c becomes different from zero (and, in fact, we see that T_c rises rather steeply with J_v/J_h for small values of this ratio). The second fact is that the critical exponent β has the value $\frac{1}{8}$ for all positive values of J_v/J_h, i.e. the exponent β is independent of the non-uniformity of exchange interaction. After Chang (1952).

of exact solutions to (8.3) we appeal to the results of numerical calculations of the sort described in Chapter 9. These have been carried out for the Ising model, when J_{ij} has the value J for spins that are either nearest neighbours, second nearest neighbours, or third nearest neighbours, and J_{ij} is zero otherwise. The results of the numerical analysis indicate that the critical indices studied probably do not change from the simple nearest-neighbour case (Dalton and Wood 1969, and references contained therein). However, when one assumes that the exchange interaction is of extremely long range, the exponents are found to change from their values for the case of nearest-neighbour interactions. For example, the particularly tractable case $J_{ij} = r^{-(d+x)}$ has been studied for both Ising and spherical models, where here $r \equiv |\mathbf{r}_i - \mathbf{r}_j|$, the distance between sites i and j (Hiley and Joyce 1965, Joyce 1966). It appears that for large values of the parameter x the critical-point exponents assume their finite range values, but for small values of x (corresponding to longer range interactions) the exponents change—in the direction of the results predicted by the mean field theory (infinite range interactions).

(iii) *Anisotropic interactions.* Returning to the case of nearest-neighbour interactions, we consider the system defined by

$$\mathscr{H}^{(D)} = - \sum_{\langle ij \rangle} \sum_{n=1}^{D} J_n S_{in} S_{jn}, \tag{8.4}$$

where S_{in} denotes the nth Cartesian coordinate of the spin $\mathbf{S}_i$. Clearly (8.4) reduces to the isotropic limit, eqn (8.1), when $J_n = J$ for $n = 1$, $2, \ldots, D$. Again the best indication of the effect of such anisotropy on the critical-point prediction is obtained from numerical studies. For example, Jasnow and Wortis (1968) have considered the anisotropic classical Heisenberg model ($D = 3$), in which case the interaction energy between neighbouring spins is of the form $J_x S_{ix} S_{jx} + J_y S_{iy} S_{jy} + J_z S_{iz} S_{jz}$. They investigated the fashion in which the $D = 3$ critical-point exponents approach the $D = 1$ exponents when J_x and J_y approach zero, and they concluded that the critical-point exponents (which they studied) jump discontinuously from their $D = 3$ values to their $D = 1$ values as soon as the slightest amount of anisotropy occurs.

With the above reservations, we proceed in the following sections to discuss some of the cases (cf. Table 8.2) in which the Hamiltonian of eqn (8.1) is exactly soluble.

8.2. Exact solution of the case $d = 1$, $D = 1$, $H = 0$: the linear chain Ising model in the absence of an external magnetic field

Probably the most easily solved interacting many-body system is the one-dimensional (or linear chain) Ising model in the absence of an external field. This system corresponds to the case $d = 1$, $D = 1$ of our model Hamiltonian (8.1). Although we shall consider (8.1) for all D in the next section, we nevertheless feel that the straightforward elegance of the solution for the special case $D = 1$ merits the special treatment given here.

The problem of the linear chain Ising model has an interesting history (Brush 1967). It was first put by Wilhelm Lenz to his student Ising in the early 1920s—several years before the advent of quantum mechanics and hence before the Heisenberg model (Heisenberg 1928). At that time the best model of magnetism was the phenomenological molecular field theory of Pierre Weiss (Weiss 1907, Langevin 1905, Van Vleck 1945). The hope was that the Ising system could provide a specific model of interparticle interactions that would display a magnetic phase transition. Ising succeeded in solving the model that is now named after him (though it is still occasionally called the Lenz–Ising model) only for the case of a one-dimensional lattice. He was extremely disappointed when the solution failed to display a phase transition. His disappointment is perhaps not surprising in the historical context, since the predictions of Weiss' mean field theory are independent of lattice dimensionality and hence the linear chain is predicted to undergo a phase transition at non-zero T_c. We often say that the Ising chain has a phase transition at a critical temperature $T_c = 0$ since the system indeed undergoes long-range ordering at absolute zero.

It is less surprising to us nowadays that the linear chain Ising model does not have a phase transition, especially in the light of the oft-repeated heuristic argument that one-dimensional systems with short-range forces of finite magnitude can not have phase transitions with $T_c > 0$. The idea of this argument is easily illustrated using the Ising chain with nearest-neighbour interactions. Suppose the contrary, namely that there exists a phase transition with $T_c > 0$. Then for some $T < T_c$, let us 'flip' half the chain to assume an opposite polarity. Clearly this process increases the enthalpy by an amount $2J$, the enthalpy required to change the relative polarity of a nearest-neighbour pair of spins. However, since it can be carried out in N ways (according to which of the N bonds is 'flipped') there will be a resultant increase in entropy of order (log N), and the Gibbs potential $G \equiv E - TS$ will decrease rather than increase! Thus the system is unstable against such 'flips', and we conclude that for an infinite system an ordered phase for $T > 0$ cannot exist.

In order to be able to obtain easily the two-spin correlation function directly from the partition function, it is useful to consider the case in which each of the nearest-neighbour interaction energies can be of arbitrary magnitude. Thus our Hamiltonian is of the form

$$\mathscr{H} = -\sum_{i=1}^{N-1} J_i s_i s_{i+1},\tag{8.5}$$

where the spins s_i are one-dimensional unit vectors assuming only the discrete values $+1$ and -1 and J_i is the interaction energy between spins situated on sites i and $i+1$. Note that J_i replaces our clumsier (but more symmetrical) notation $J_{i,i+1}$ used in § 8.1 and in Fig. 8.1.

The goal of our calculation is to obtain, in closed form, the partition function

$$Z_N \equiv Z_N(J_1, J_2, \ldots, J_{N-1}) = \sum_{s_1=-1}^{1}\sum_{s_2=-1}^{1}\cdots\sum_{s_N=-1}^{1}\exp\left(\sum_{i=1}^{N-1}\mathscr{J}_i s_i s_{i+1}\right)\tag{8.6}$$

where $\mathscr{J}_i \equiv \beta J_i \equiv J_i/kT$ is a dimensionless exchange energy and the N summations extend over the 2^N configuration states of the system. The method of solution is to calculate the effect of adding one additional spin to the end of the chain. That is, we seek to express the partition function of the 'extended chain', Z_{N+1}, in terms of the partition function for the original chain Z_N. Once we succeed in accomplishing this, we shall have a *recursion relation* which we can iterate to find the solution.

From eqn (8.6) it follows directly that

$$Z_{N+1} = \sum_{s_1=-1}^{1}\sum_{s_2=-1}^{1}\cdots\sum_{s_N=-1}^{1}\exp\left(\sum_{i=1}^{N-1}\mathscr{J}_i s_i s_{i+1}\right)\sum_{s_{N+1}=-1}^{1}$$
$$\exp\left(\mathscr{J}_N s_N s_{N+1}\right).\tag{8.7}$$

Now the last summation in (8.7) is

$$\sum_{s_{N+1}=-1}^{1}\exp\left(\mathscr{J}_N s_N s_{N+1}\right) = 2\cosh\left(\mathscr{J}_N s_N\right),\tag{8.8}$$

and we discover that, since the spin s_N can take on only the values $+1$ and -1, the right-hand side of (8.8) is independent of the orientation of s_N. Hence eqn (8.7) may be written in the form

$$Z_{N+1} = 2Z_N\cosh\mathscr{J}_N.\tag{8.9}$$

Equation (8.9) may be iterated easily, with the result

$$Z_{N+1} = Z_1 2^N\left(\cosh\mathscr{J}_1\cosh\mathscr{J}_2\cosh\mathscr{J}_3\ldots\cosh\mathscr{J}_N\right).\tag{8.10}$$

Now it is simple to see that $Z_1 = 2$, since Z_1 is just the partition function for a system consisting of a single Ising spin—so the 'sum on states' is simply the 'number of states'. Hence from (8.9) and (8.10) we have the final result for the partition function of a system of N spins with arbitrary nearest-neighbour interactions,

$$Z_N = 2^N \prod_{i=1}^{N-1} \cosh \mathscr{J}_i. \qquad \blacktriangleright \quad (8.11a)$$

In the uniform case ($J_i = J$ for all i), (8.11a) becomes

$$Z_N = 2^N \cosh^{N-1} \mathscr{J}. \qquad (8.11b)$$

In the present form the partition function (8.11) tells us nothing concerning the ordering temperature T_c. Therefore, we calculate the two-spin correlation function

$$\Gamma_k(r) \equiv \langle s_k s_{k+r} \rangle = Z_N^{-1} \sum_{\{s\}} s_k s_{k+r} \exp \left(\sum_{i=1}^{N-1} \mathscr{J}_i s_i s_{i+1} \right) \qquad (8.12)$$

where the summation symbol in (8.12) denotes the N-fold summation of eqn (8.6); i.e. a summation over all 2^N states of the system. Note that r is a measure of the distance between spins, in units of a lattice constant. Thus the nearest-neighbour correlation function is simply $\Gamma_k(1)$, which satisfies the relation

$$Z_N \Gamma_k(1) = \sum_{\{s\}} s_k s_{k+1} \exp \left(\sum_{i=1}^{N-1} \mathscr{J}_i s_i s_{i+1} \right) = \frac{\partial}{\partial \mathscr{J}_k} \sum_{\{s\}} \exp \left(\sum_{i=1}^{N-1} \mathscr{J}_i s_i s_{i+1} \right). \qquad (8.13)$$

Clearly the argument leading to eqn (8.13) can be generalized to arbitrary r, with the result

$$Z_N \Gamma_k(r) = \frac{\partial}{\partial \mathscr{J}_k} \frac{\partial}{\partial \mathscr{J}_{k+1}} \cdots \frac{\partial}{\partial \mathscr{J}_{k+r-1}} (Z_N). \qquad (8.14)$$

Hence we obtain from differentiation of eqn (8.11a)

$$\Gamma_k(1) = \tanh \mathscr{J}_k \qquad (8.15)$$

and

$$\Gamma_k(r) = (\tanh \mathscr{J}_k)(\tanh \mathscr{J}_{k+1}) \cdots (\tanh \mathscr{J}_{k+r-1}) = \prod_{i=1}^{r} \tanh \mathscr{J}_{k+i-1}, \qquad \blacktriangleright \quad (8.16)$$

which is just the product of the intermediate nearest-neighbour correlation functions. Equation (8.16) reduces, for the case of a uniform interaction, to

$$\Gamma_k(r) \equiv \langle s_k s_{k+r} \rangle = \tanh^r \mathscr{J}. \qquad (8.17)$$

which is independent of site k. For example, the correlation between two spins near the end of the chain is identical to that between two spins in the centre of the chain, providing the spins are separated by the same distance r. We shall achieve some insight into this result in Chapter 9, when we consider the allowed interaction paths between two spins in a linear chain lattice.

An instructive check on the results (8.11a) and (8.16) is to permit one of the $\mathscr{J}_i$ to approach zero so that the chain becomes simply two separated entities. Equation (8.11a) predicts that the partition function of the total system is given by the product of the partition functions of the separated chains. Equation (8.16) predicts that if the break occurs between spin s_k and spin s_{k+r} the correlation function $\Gamma_k(r)$ becomes zero, while if the break occurs elsewhere in the chain, $\Gamma_k(r)$ is unaffected.

We are now prepared to find the temperature at which long-range order sets in, i.e. the temperature at which the two-spin correlation function falls off sufficiently slowly with interspin distance r that the magnetization becomes non-zero. By analogy with eqn (7.9),

$$\sigma^2 = \lim_{r \to \infty} \Gamma_k(r), \tag{8.18}$$

where $\sigma \equiv M(T, H = 0)/M(T = 0, H = 0)$ is the normalized zero-field magnetization. From eqns (8.16) and (8.17) we see that so long as all the parameters $\mathscr{J}_i$ are finite, the values of the hyperbolic tangents are less than one and hence the products approach zero in the limit of infinite r. However, in the limit of zero temperature all the parameters $\mathscr{J}_i \equiv \beta J_i = J_i/kT$ approach infinity and the products in (8.16) and (8.17) approach 1. The linear chain therefore displays *zero* spontaneous magnetization for all *non-zero* values of the temperature, but 'suddenly' at $T = T_c = 0$ the magnetization acquires its full value. Thus we see that the Ising model, unlike the mean field theory, predicts no phase transition for a linear chain system.

The absence of a spontaneous magnetization for non-zero temperatures does not imply that all the other thermodynamic functions fail to have singularities except at $T = 0$. In fact, there exist some model systems in which the zero-field susceptibility diverges at a temperature T_{c1}, even though there is no zero-field magnetization until we reach T_{c2}, where $T_{c2} < T_{c1}$ (see Fig. 8.3). According to Ehrenfest's classification of phase transitions, the fact that the susceptibility, which is proportional to a second derivative of the Gibbs potential, diverges at T_{c1} means that the system undergoes a second-order phase transition at T_{c1} even though the spontaneous magnetization does not appear until a lower temperature is reached.

Let us then calculate the zero-field susceptibility for the one-dimensional Ising chain. The reader might think that since $\chi_T(T, H = 0) \equiv (\partial M/\partial H)_{H=0}$, we need first to calculate the magnetization in the presence of a non-zero magnetic field. This is, however, not necessary, for according to the fluctuation–dissipation relation discussed in Appendix A, the susceptibility is also given by

$$\chi_T(T, H = 0) = \frac{\bar{\mu}^2}{kT} \sum_{i=1}^{N} \sum_{j=1}^{N} \langle s_i s_j \rangle, \tag{8.19}$$

where, as in Chapter 6, $\bar{\mu} \equiv g\mu_B$. Hence it is sufficient to carry out the lattice summation (8.19), which is straightforward for a one-dimensional

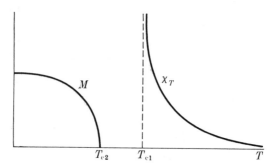

FIG. 8.3. A rare but real possibility: the susceptibility diverges at a different temperature than the onset of spontaneous magnetization.

lattice. For the uniform interaction case, $\langle s_i s_j \rangle = v^{|i-j|}$, where $v \equiv \tanh \mathscr{J}$. Hence the contributions to (8.19) are the following:

(i) N terms for which $|i - j| = 0$, each of which contributes a factor v^0,

(ii) $2(N - 1)$ terms for which $|i - j| = 1$ (corresponding to the $N - 1$ nearest neighbour pairs in the open linear chain lattice), each of which contributes a factor v^1,

(iii) $2(N - 2)$ terms for which $|i - j| = 2$ (corresponding to the $N - 2$ next nearest neighbour pairs in the lattice), each of which contributes a factor v^2, and so forth until we have the two terms for which $|i - j| = N - 1$, each of which contributes a factor v^{N-1}.

Hence we may express the double summation in (8.19) in the form

$$\chi_T(T, H = 0) = \frac{\bar{\mu}^2}{kT}\left\{ N + 2\sum_{k=1}^{N-1}(N - k)v^k \right\}. \tag{8.20}$$

The geometric series in (8.20) may be easily evaluated using the elementary relations

$$\sum_{k=0}^{N-1} v^k = (1 - v^N)/(1 - v) \tag{8.21}$$

and

$$\sum_{k=0}^{N-1} kv^k = v(\partial/\partial v)\sum_{k=0}^{N-1} v^k \tag{8.22}$$

with the result

$$\chi_T(T, H = 0) = \frac{\bar{\mu}^2}{kT}\left\{ N\left(1 + \frac{2v}{1 - v}\right) - \frac{2v(1 - v^N)}{(1 - v)^2} \right\}. \tag{8.23}$$

Equation (8.23) is an exact expression for the two-spin correlation

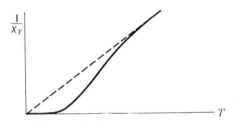

FIG. 8.4. Temperature dependence on the inverse susceptibility for the linear chain Ising model. Note that the singularity is an essential singularity, i.e. $\chi_T \sim e^{1/T}$ as $T \to 0$.

function for an open linear chain of N spins. In the thermodynamic limit, $N \to \infty$, (8.23) simplifies to

$$\chi_T(T, H = 0) = \frac{\bar{\mu}^2 N}{kT}\frac{1 + v}{1 - v} = \frac{\bar{\mu}^2 N}{kT}\frac{1 + \tanh \mathcal{J}}{1 - \tanh \mathcal{J}} = \frac{\bar{\mu}^2 N}{kT}e^{2\mathcal{J}}. \tag{8.24}$$

The result (8.24) is shown as a function of temperature in Fig 8.4; note that in the neighbourhood of the critical temperature $T_c = 0$, the zero-field susceptibility of (8.24) varies as

$$\chi_T(T, H = 0) \sim \frac{1}{T}e^{2J/kT}. \tag{8.25}$$

Thus the susceptibility has an essential singularity of the type $e^{1/x}$ where $x \to 0$ (cf. Fig. 8.4).

Although the susceptibility diverges to infinity as T approaches T_c, ($T_c = 0$), the specific heat remains finite—in fact it approaches zero (as one might expect from the third law of thermodynamics). The simplest method of obtaining the specific heat is by differentiating with respect to temperature the enthalpy, given by (Wannier 1966)

$$E(S, H = 0) = -\frac{\partial}{\partial \beta}\{\ln Z_N (T, H = 0)\} = -\sum_{i=1}^{N-1} J_i \tanh \mathscr{J}_i \quad (8.26a)$$

$$= -J(N-1)\tanh \mathscr{J}, \quad (8.26b)$$

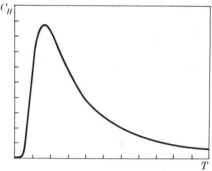

FIG. 8.5. Temperature dependence of the zero-field specific heat for the $d = 1$ Ising model. The specific heat of the paramagnet (cf. Chapter 6) has the same temperature dependence if we make the identification $J \leftrightarrow H$.

where the last equality holds for the uniform case, $J_i = J$. Hence we have, from (2.48), that

$$C_H(T, H = 0) = k\sum_{i=1}^{N-1} (\mathscr{J}_i \operatorname{sech} \mathscr{J}_i)^2, \quad (8.27a)$$

$$= k(N-1)(\mathscr{J} \operatorname{sech} \mathscr{J})^2. \quad (8.27b)$$

The specific heat for the uniform case, eqn (8.27b), is plotted in Fig. 8.5.

Finally, we consider the zero-field entropy, which is related to the Gibbs potential through eqn (2.45c), whence

$$S_N (T, H = 0) = \frac{\partial}{\partial T}\left\{\frac{1}{\beta}\ln Z_N (T, H = 0)\right\}$$

$$(8.28)$$

$$= k \ln Z_N (T, H = 0) + kT\frac{\partial}{\partial T}\{\ln Z_N (T, H = 0)\}.$$

Hence for the uniform model we obtain

$$S_N(T, H = 0) = k\{N \ln 2 + (N - 1) \ln \cosh \mathcal{J} - (N - 1) \mathcal{J} \tanh \mathcal{J}\}. \tag{8.29}$$

As a check on this expression, we notice that if we calculate the specific heat from (8.29), using the relation (2.48), the result agrees with (8.27b). Note that the entropy approaches $k \ln 2$ at low temperatures (with zero slope), and $(Nk \ln 2)$ at high temperatures—see Fig. 8.6.

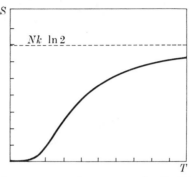

FIG. 8.6. Dependence of entropy upon temperature for the Ising chain. Notice that S approaches zero as $T \to 0$ with zero limiting slope, consistent with the fact that $C_H = T(\partial S/\partial T)_H$ approaches zero, as we see in Fig. 8.5.

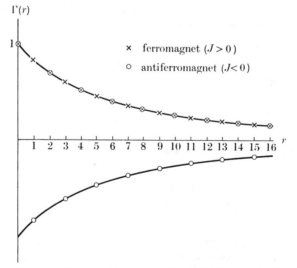

FIG. 8.7. Dependence on site separation r (in units of a lattice constant) of the two-spin correlation function $\Gamma(r) \equiv \langle s_j s_{j+r} \rangle$ for the linear chain Ising model in zero magnetic field for a uniform interaction. Both the ferromagnetic $(J > 0)$ case and the antiferromagnetic $(J < 0)$ case are shown.

In this section we have emphasized the wide range of properties that can be obtained from a knowledge of the two-spin correlation function. Before concluding, we also note that the case of an anti-ferromagnetic interaction ($J < 0$) is easily obtained by simply changing the sign of J in many of the expressions thus far derived. It is interesting to note that although the spin correlation function now alternates in sign from site to site (see Fig. 8.7), certain quantities are nevertheless changed only slightly from the ferromagnetic case. In fact, on physical grounds we expect the $H = 0$ enthalpy, specific heat, and entropy to be unchanged; our expectations are confirmed

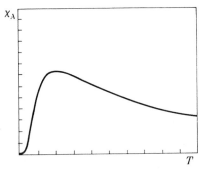

FIG. 8.8. Temperature dependence of the susceptibility χ_A for the linear chain Ising antiferromagnet ($J < 0$). Note that χ_A remains finite for all temperatures, in marked contrast (cf. Fig. 8.4) to the ferromagnetic case. This corresponds to the fact that for the antiferromagnet the spin correlation function alternates in sign from site to site (cf. Fig. 8.7).

by examination of eqns (8.26), (8.27), and (8.29). On the other hand, the susceptibility depends upon the summation of the two-spin cor-relation function over all sites in the entire lattice, and therefore be-haves quite differently in the antiferromagnetic case (see Fig. 8.8).

8.3. The linear chain of arbitrary dimensional spins in zero magnetic field

In the previous section we considered the case of the linear chain Ising model in the absence of a magnetic field, i.e. the case $D = 1$, $d = 1$, and $H = 0$ of our model Hamiltonian equation (8.1). In the remainder of this chapter we shall consider other cases in which (8.1) is exactly soluble, although we shall not carry through the solutions in most cases. In particular, in this section we consider the general-ization of the results of the previous section to arbitrary spin dimension-ality D (Stanley 1969d, Stanley, Blume, Matsuno, and Milošević, 1970).

Before undertaking discussion of yet another linear chain problem, perhaps we should comment on why one-dimensional problems merit the consideration of physicists (whose ultimate goal is, presumably, to understand a three-dimensional universe).

(i) One motivation for considering exactly soluble one-dimensional models is that their solutions frequently help us in judging the validity of approximation techniques that are used for three-dimensional systems.

(ii) A second motivation is that a one-dimensional model may serve as a reasonable approximation to some special physical system. For example, there exist materials in which the magnetic ions may be considered to form linear chains in the sense that interactions between spins within the chains are appreciably stronger than interactions between spins belonging to different chains (Skalyo, Shirane, Friedburg, and Kobayashi 1970). We can often obtain reasonable agreement with experimental results from a model in which intrachain interactions are treated exactly and the interchain interactions are approximated in some simple fashion (Stanley and Kaplan 1967b).

(iii) A third motivation is that results discovered for one-dimensional models are sometimes generalizable to higher dimensionalities. In fact, we shall find that two results of the $d = 1$ calculation in this section lead to analogous results for $d > 1$.

The partition function for the general D linear chain may be calculated in straightforward analogy to the $D = 1$ case, with the result

$$Z_N^{(D)}(T, H = 0) = (1 + \delta_{1,D})^N \prod_{i=1}^{N-1} (\mathscr{J}_i/2)^{1-D/2} \, \Gamma(D/2) I_{D/2-1}(\mathscr{J}_i), \quad (8.30)$$

where $\delta_{1,D}$ is the Kronecker delta function, $\Gamma(x)$ is the gamma function, and $I_\nu(x)$ is the modified Bessel function of the first kind of order ν. The nearest-neighbour two-spin correlation function is obtained from an expression completely analogous to eqn (8.13) above, with the result

$$\langle \mathbf{S}_k^{(D)} \cdot \mathbf{S}_{k+1}^{(D)} \rangle = I_{D/2}(\mathscr{J}_k)/I_{D/2-1}(\mathscr{J}_k), \quad (8.31)$$

while the spin correlation function for two spins a distance r apart is given by the appropriate product of nearest-neighbour spin correlation functions in analogy with (8.16). Since $I_{-\frac{1}{2}}(x) = (\frac{1}{2}\pi x)^{-\frac{1}{2}} \cosh x$ and $I_{\frac{1}{2}}(x) = (\frac{1}{2}\pi x)^{-\frac{1}{2}} \sinh x$, we see that (8.30) and (8.31) reduce to (8.11a) and (8.15) for the Ising model ($D = 1$).

For the uniform case ($J_i = J$), the magnetic enthalpy E is simply proportional to the nearest-neighbour spin correlation function (cf. (8.26b)); Fig. 8.9 shows $E/NJ \simeq -\Gamma_k(r = 1)$ plotted as a function of kT/J and of J/kT. We are struck by the fact that the spin correlation function decreases monotonically with increasing spin dimensionality for a given temperature. This result is somewhat plausible on intuitive grounds, if one regards the higher dimensionality spins as being a more 'floppy' system in the sense that there is more phase space available. (In particular, since the nearest-neighbour spin correlation function is strictly less than unity for any non-zero temperature, there is no spontaneous magnetization for any value of D except when $T = 0$.)

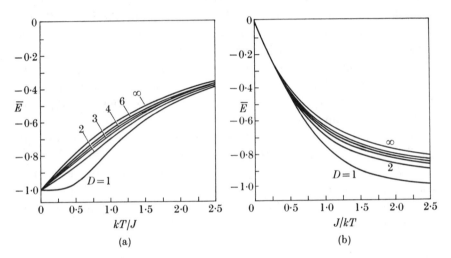

FIG. 8.9. Normalized enthalpy $\bar{E} = E/NJ \equiv -y_D$ as a function of (a) kT/J, and (b) $J/kT \equiv \mathscr{J}$ for the various spin dimensionalities D. Curves for other values of D lie in between the values shown. The correlation function between two spins located r sites apart is proportional to y_D raised to the power r—hence the correlation between *any* two given spins *decreases* monotonically with *increasing* spin dimensionality (Milošević, Matsuno, and Stanley 1970). After Stanley (1969d).

The fact that the correlation function between any two spins in the lattice is a monotonically decreasing function of spin dimensionality has not yet been proved rigorously for lattices of higher dimensionality, but this result is consistent with recent results using approximation methods of the sort to be described in the following chapter. Moreover, these same approximation methods suggest that critical-point exponents are also monotonic functions of spin dimensionality. For example, in Table 8.3 we show, for the case $d = 3$, the dependence on D of the exponents γ, α, and ν and some related quantities. These $d = 3$ values

TABLE 8.3

Estimates of critical properties for D-dimensional spins situated on a three-dimensional (fcc) lattice. The variation with D was found to be monotonic (and smooth) for D between 1 and ∞. The entries in the sixth and seventh columns are predicted to be zero by certain of the scaling relations; that the departure from zero appears to be a monotonically decreasing function of spin dimensionality should be viewed with caution—the errors in these small numbers are roughly as large as the numbers themselves. The quantity β tabulated in the eighth column is obtained by assuming the validity of the scaling relation α + 2β + γ = 2. The fact that β so defined turns out to be γ/4 (regardless of whether or not we choose α and γ to be the closest rational fraction) suggests the validity of the scaling relation γ = β(δ − 1) with δ = 5. If we assume the validity of γ = β(δ − 1) (or, equivalently, of α + β(δ + 1) = 2), we obtain the values of δ given in the last column. These exponents, plus all others which have been calculated thus far, have values for all D that are close to the 'rational fraction' given by the expression λ(D) = λ(∞) (a_λ + D)/(b_λ + D), with b_λ = 7 for all λ and a_λ = 4, −2, 4, 4, 7, and 4 for λ = γ, α, ν, β, δ, and Δ respectively. (Betts, Ditzian, Elliott, and Lee 1971; Stanley, Hankey, and Lee 1971).

D	T_c/T_M	γ	α	ν	$\eta \equiv 2 - \gamma/\nu$	$d\nu - (2 - \alpha)$	$\beta \equiv (2 - \alpha - \gamma)/2$	$\delta \equiv \dfrac{2 - \alpha + \gamma}{2 - \alpha - \gamma}$
1	0·816	1·25	0·125	0·638	0·041	0·04	0·3125(5/16)	5
2	0·804	1·32(~4/3)	0·02(~0)	0·675	0·04	0·04	0·33(~1/3)	5
3	0·793	1·38(~11/8)	−0·07(~ −1/16)	0·70	0·03	0·03	0·345(~11/32)	5
...		(~7/5)	(~ −1/10)				(~7/20)	
∞	0·7436	2	−1	1	0	0	½	5

are obtained by approximation methods to be described in the following chapter.

A second result of our linear chain calculation which has led to interesting results for systems with $d > 1$ is the fact that the partition function approaches, in the limit $D \to \infty$, the partition function for the spherical model. In the following section we shall find that this result is valid for all d.

8.4. The spherical model as the limit of infinite spin dimensionality

In this section we discuss the fact that one can solve for the partition function of our model system (8.1) in the limit $D \to \infty$, even for a three-dimensional lattice. The solution turns out to be identical with the Berlin–Kac spherical model or spherical approximation to the Ising model. This work is of particular interest because it represents essentially the only case of a many-body system that can be solved exactly for three-dimensional lattices.

The spherical model has had a very colourful history, ever since it was conceived by Professor Mark Kac in 1947. It seems that Professor George Uhlenbeck had interested Kac in the Ising model, which had recently been solved for the case of $d = 2$ by Onsager in 1944 but had not yet yielded to solution in three dimensions. To quote from Kac's own account (Kac 1964):

'With his usual clarity, Uhlenbeck explained to me what the problem was all about and discussed the remarkable implications of the Onsager solution. That afternoon, having noticed that Uhlenbeck suffered from a bad cold, I persuaded him to take a nap, and during that nap I sat down in my study to think about the Ising problem.

It soon became obvious that it was not a problem one solves on the spur of the moment, and, in the best mathematical tradition, not being able to solve the original problem, I looked around for a similar problem which I could solve. I then proceeded to replace the discrete spins by continuous ones distributed according to the Gaussian distribution. In no time I had the free energy per spin calculated, and to my amazement and pleasure the answer looked remarkably like Onsager's. I could hardly wait for Uhlenbeck to finish his nap, and only my regard for his cold kept me from waking him. When he finally got up, and I presented him with what I thought was a shattering discovery, I recall that he too was a bit startled. Our enthusiasm dampened considerably a little later when we noticed that the solution suffers from what may be called the 'low temperature catastrophe', i.e. the answer blows up below a certain temperature. Clearly the model, which became known as the Gaussian model, was not terribly good. In the ensuing weeks I tried to get rid of the catastrophe, and it finally occurred to me that one should try to assume the spins s_1, s_2, ... to satisfy the condition: $s_1^2 + s_2^2 + \cdots = N$ = number of lattice sites, and to average over the surface of the sphere which the above condition defines.'

This second model became known as the spherical model, because if we represent each of the configurations of the Ising system of N spins by the 2^N corners of an N-dimensional hypercube, then the spherical model replaces this discrete set of configurations by a con-

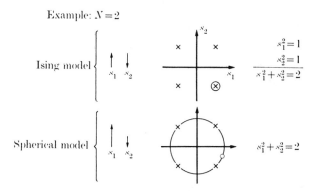

FIG. 8.10. The spherical model was originally defined by Kac as a 'continuum modification of the Ising model' in which the one-dimensional spins s_i are no longer restricted to the discrete value $+1$ and -1, but instead are allowed to assume any values whatsoever, so long as the sum of their squares is equal to the total number of spins in the system. This figure illustrates the case $N = 2$, and represents the $2^N = 4$ allowed states of the Ising system by the corners of an N-dimensional hypercube. The spherical model replaces this discrete set of allowed states with a continuum of states, by replacing the N-dimensional hypercube with an N-dimensional hypersphere. Note that the spins are still one-dimensional scalar quantities—all that we have done is to replace the strong constraints that each spin be of unit magnitude by the weak constraint that the sum of the squares of all the N spins in the system add up to N. After Stanley (1971c).

tinuum of configurations defined by the N-dimensional hypersphere that circumscribes the hypercube. An illustration for the case $N = 2$ is provided in Fig. 8.10. One might think at first glance that the spherical model would be harder to solve instead of easier, because in replacing each of the N separate constraints,

$$s_i^2 = 1, \qquad i = 1, 2, 3, \ldots N, \qquad (8.32)$$

by a single constraint,

$$\sum_{i=1}^{N} s_i^2 = N, \tag{8.33}$$

we have effectively allowed a much larger number of states to enter into our partition function. The simplifying feature of the spherical model is that instead of our summing over the discrete number of 2^N states, we can now integrate over the continuum of states described by the hypersphere. The partition function for the $d = 3$ Ising model was thus transformed by Kac from an unmanageable N-fold summation to an unmanageable N-fold integration. After some effort at performing the spherical model integral, Kac sought the help of his colleague at Cornell, Richard Feynman, who succeeded in doing the integral for a $d = 1$ lattice only. However, Feynman's method did not generalize to higher d, and it looked for a time as if the spherical model was *harder* to solve than the Ising model. Fortunately, Kac presented a lecture on the spherical model at which the late Professor Theodore H. Berlin was present. Berlin became interested in the problem, and to continue the story (Kac 1964),

'Very shortly after returning to Ithaca, I received a long letter from Ted, in which by a remarkable application of the method of steepest descent he produced a solution in one, two, and three dimensions. I had hardly heard of the method of steepest descent, and what I had heard was mainly a warning against its use because of the difficulties of rigorous justification. Here, before my eyes on about ten handwritten pages, was one of the most advanced applications of the method. Only a combination of the optimism of a physicist with the fantastic power and skill of a classical analyst could have produced the results. To say that I was immensely impressed would be a pale understatement of my reaction to that letter.'

The most exciting feature of the solution (Berlin and Kac 1952) to the spherical model is that it displays a phase transition for a three-dimensional lattice (although, unlike the Ising model, $T_c = 0$ for $d = 2$). Moreover, the nature of the singularities in the various thermodynamic functions is of the power law type, and the critical-point exponents are numbers that are of the same general order-of-magnitude as those for the Ising model (cf. Table 3.4).

The spherical model has received considerable attention during the past two decades from pure mathematicians to applied metallurgists (the Ising or 'lattice–gas' model also provides a realistic model of a binary alloy system). However, the spherical model was not very satisfying on various physical grounds; for example, the constraint

(8.33) allows as states of the system exceedingly unusual possibilities such as the length of one spin being $N^{\frac{1}{2}}$ and all other spins being of zero length! Moreover, as an approximation to the Ising model, it was essentially a single-shot approximation in the sense that there are no intermediate models that link-up the Ising model and the spherical model. For these and other reasons, it was particularly gratifying when it was discovered that the exact solution in the limit $D \rightarrow \infty$ of the Hamiltonian (8.1) reduces to the Berlin–Kac spherical model. The original derivation of this result (Stanley 1968d, 1969a) is rather lengthy to include here; instead we refer the interested reader to a very recent derivation of this same result (Kac and Thompson 1971) which has the pedagogical advantage over the original argument in that it does not use the method of steepest descent.

8.5. The transfer matrix method: application to the $d = 1$ Ising model in a magnetic field

So far in this chapter we have considered in detail the cases in which the model Hamiltonian (8.1) can be solved exactly. Unfortunately these are few in number; moreover, the exact solutions which are of greatest interest in the study of critical phenomena, the $d = 2$ Ising model ($D = 1$) and the $d = 3$ spherical model ($D = \infty$), are so complex that we can include only the former (see Appendix B). Hence we have concentrated on those elementary exact solutions that serve to illustrate methods that are used in obtaining exact solutions to more difficult problems. The transfer matrix technique led to Onsager's solution of the $d = 2$ Ising model (Onsager 1944), and to Lieb's solutions of certain two-dimensional ferroelectric and antiferroelectric models (see Appendix F). Hence it is perhaps appropriate that we conclude this introduction to exact solutions by applying the transfer matrix technique to derive an expression for the partition function $Z_N(T, H)$ for a linear chain Ising model with nearest-neighbour interactions situated in a non-zero magnetic field H.

We begin, then, with the Ising Hamiltonian

$$\mathscr{H} = -kT \mathscr{J} \sum_{i=1}^{N} s_i s_{i+1} - kTh \sum_{i=1}^{N} s_i, \tag{8.34}$$

where $\mathscr{J}$ and h are the exchange parameter and magnetic field, respectively, in dimensionless units, and where we choose periodic boundary conditions $(s_{N+1} \equiv s_1)$ for reasons that will become apparent below.

Next we define

$$U(s_i, s_{i+1}) = -kT\mathscr{J} s_i s_{i+1} - \tfrac{1}{2}(kTh)(s_i + s_{i+1}) \qquad (8.35)$$

and the transfer function

$$f(s_i, s_{i+1}) = \exp\{-\beta U(s_i, s_{i+1})\} \qquad (8.36)$$

in terms of which we can write the partition function as

$$Z_N(T, H) = \sum_{s_1=-1}^{1} \cdots \sum_{s_N=-1}^{1} f(s_1, s_2) f(s_2, s_3) \cdots f(s_N, s_1). \qquad (8.37)$$

The next step in the conventional derivation is to introduce the transfer matrix, defined by

$$\mathbf{T} \equiv \begin{pmatrix} T_{++} & T_{+-} \\ T_{-+} & T_{--} \end{pmatrix} = \begin{pmatrix} e^{\mathscr{J}+h} & e^{-\mathscr{J}} \\ e^{-\mathscr{J}} & e^{\mathscr{J}-h} \end{pmatrix}, \qquad (8.38)$$

where the matrix elements are

$$T_{\pm\pm} \equiv f(s_i = \pm 1, s_{i+1} = \pm 1) \qquad (8.39)$$

By using eqn (8.38) we can perform the summations in the partition function of eqn (8.37) over the spins $s_2, s_3, \ldots s_N$, with the result

$$Z_N(T, H) = \sum_{s_1=-1}^{1} (\mathbf{T}^N)_{s_i, s_i} = \text{trace } (\mathbf{T}^N). \qquad (8.40)$$

Thus for the closed chain we have reduced the problem of finding the partition function to that of finding the sum of the diagonal elements (or trace) of the Nth power of the transfer matrix. Now the trace of a matrix is simply the sum of its eigenvalues, and the eigenvalues of $\mathbf{T}^N$ are simply $\lambda_{\pm}^N$, where $\lambda_{\pm}$ are the eigenvalues of $\mathbf{T}$, determined by the equation

$$\begin{vmatrix} e^{\mathscr{J}+h} - \lambda & e^{-\mathscr{J}} \\ e^{-\mathscr{J}} & e^{\mathscr{J}-h} - \lambda \end{vmatrix} = 0 \qquad (8.41)$$

with solutions

$$\lambda_{\pm} = e^{\mathscr{J}} \cosh h \pm (e^{2\mathscr{J}} \sinh^2 h + e^{-2\mathscr{J}})^{\frac{1}{2}}. \qquad (8.42)$$

Hence the partition function is simply

$$Z_N(T, H) = \lambda_+^N + \lambda_-^N, \qquad (8.43)$$

where the $\lambda_{\pm}$ are given by eqn (8.42). For a fixed non-zero value of the

temperature, in the limit of large N, we may write (8.43) in the form
$Z_N(T, H) = \lambda_+^N \{1 - (\lambda_-/\lambda_+)^N\}$ and neglect the second term with respect
to the first (since $\lambda_- < \lambda_+$), with the result

$$Z_N(T, H) \sim \lambda_+^N, \quad N \text{ large.} \tag{8.44}$$

For example, if we set $H = 0$, the eigenvalues (8.42) become

$$\lambda_\pm = e^{\mathscr{J}} \pm e^{-\mathscr{J}} \tag{8.45}$$

and hence from (8.43)

$$Z_N(T, H = 0) = 2^N (\cosh^N \mathscr{J} + \sinh^N \mathscr{J}), \tag{8.46}$$

which differs from our earlier expression (8.11b) due to the effect of the
periodic boundary conditions that were assumed in deriving (8.46).
However, in the thermodynamic limit, the Gibbs potential per spin is
identical for both cases (the open and the closed chains),

$$\bar{G}(T, H = 0) = -kT \lim_{N \to \infty} \frac{1}{N} \ln Z_N(T, H = 0) = -kT \ln (2 \cosh \mathscr{J}). \tag{8.47}$$

Suggested Further Reading

Domb (1960).
Lieb and Mattis (1966).
Brush (1967).
Lebowitz (1968).
Montroll (1968).
Thompson (1971).

RESULTS OBTAINED FROM MODEL
SYSTEMS BY APPROXIMATION METHODS

IN the previous chapter we saw that the model Hamiltonian (8.1) was soluble for three-dimensional lattices only in the limit of infinite spin dimensionality $(D \to \infty)$. Moreover, the $D \to \infty$ limit or spherical model does not correspond to any known physical system. It would, of course, be extremely desirable to find another, more realistic, model Hamiltonian which is exactly soluble for three-dimensional lattices. Unfortunately, attempts along these lines have not met with any success so far. Therefore if we wish to obtain predictions of critical properties for realistic models, we must resort to numerical approximation techniques. We shall discuss in this chapter certain of these techniques which in fact have been so successful in predicting accurate values of critical temperatures and critical-point exponents that one often forgets that they are only approximations.

Of course an exact solution is always to be preferred over a numerical approximation. However, as regards comparison with experiment, the errors in the numerical procedures are often smaller than the errors in the corresponding experimental measurements. In addition to predicting precise numerical values, exact solutions frequently provide insight into qualitative features of the model, but in the case of the model systems we have been discussing, the complexity of the derivation appears to obscure much of the physical insight we might have hoped to obtain. For example, in the case of the two-dimensional Ising model, Yang's derivation (Yang 1952) provides no more understanding of why $\beta = \frac{1}{8}$ than does the numerical result $\beta \simeq 0 \cdot 125$. On the other hand, the efforts that have been invested in numerical approximation procedures have produced not only extremely accurate numbers, but in addition have led to some qualitatively new ideas concerning the critical region. As an example, it is worth mentioning that the concept of a power law singularity in the susceptibility of the Ising and Heisenberg models arose first from the numerical work of C. Domb, W. Marshall, and collaborators.

9.1. Successive approximation concept

The basic philosophy of the approximation methods to be discussed in this chapter concerns the reliability of a method of successive approximations. An analogy with perturbation theory is perhaps appropriate. In most physics work using perturbation theory, one accepts the results of a first- or second-order calculation. But suppose we wish to attach confidence limits of, say, 1 per cent to a result obtained using second-order perturbation theory. Then we should seek (i) a proof that the perturbation series converges and (ii) a method for estimating the magnitude of the remainder terms. Since (i) and (ii) in general do not exist, the next best thing might be to perform a third-order perturbation theory calculation in order to ascertain whether the second-order calculation is affected by as much as 1 per cent. It turns out that for many physical quantities the perturbation series are sufficiently slowly convergent that one needs to calculate more than two terms in order to achieve accuracies of the order of 1 per cent. In particular, for the series that we shall discuss in this chapter, many more than two terms are necessary. Hence the calculation becomes fairly complex, and ingenious diagrammatic procedures have been devised to assist in rendering this complexity manageable. However, having learned to cope with the complexity of proceeding beyond second order, it is tempting not to stop at whatever order the 1 per cent accuracy is obtained, and for some model systems predictions believed to be as accurate as 0·01 per cent have been obtained.

It is important to stress that there is no rigorous proof that the series which we shall discuss in this chapter are in fact convergent, nor do we have methods that permit us to estimate the magnitude of the remainder that we neglect. Hence a reasonable strategy might be the following: (a) calculate L terms in a series; (b) make a prediction concerning the limiting ($L \to \infty$) behaviour based on these L terms; (c) calculate the $(L + 1)$st term; (d) make a revised estimate based upon the extended series; and (e) ascribe as our confidence limits roughly the difference between the predictions made in steps (b) and (d).

We shall attempt through worked examples to illustrate this strategy. The main point is that there is no guarantee that the predictions toward which the successive approximations appear to converge are in fact the exact answers. Nevertheless, our confidence in these methods is considerably enhanced because when we carry out such procedures for exactly-soluble cases (such as $D = 1$ with $d = 1$–2, and $D = \infty$

with $d = 1$–3) we find that the predictions are extremely close to the exact answers—for example, the exactly-known critical temperature of the two-dimensional Ising model is located by these approximation methods to an accuracy of better than $0 \cdot 01$ per cent. Moreover, no examples have yet been produced in which the successive approximations appear to converge to an answer that is known *not* to be correct.

9.2. Series expansion methods

Successive approximation methods of the sort described in the preceding section are based on expanding the thermodynamic function of interest in ascending powers either of temperature (a low-temperature expansion) or of inverse temperature (a high-temperature expansion). Since the former case is restricted in applicability to the Ising interaction, we shall focus mainly on high-temperature expansions; however, much of the discussion in this chapter is equally applicable to both cases.

The basic idea of the high-temperature expansion method is the expansion of the exponential $\exp(-\beta \mathscr{H})$ as a power series in its argument,

$$\exp(-\beta \mathscr{H}) = 1 - \beta \mathscr{H} + \tfrac{1}{2}(\beta \mathscr{H})^2 + \cdots. \qquad (9.1)$$

The motivation for such an expansion is that the partition function $Z_N(T, H)$ is directly related to $\exp(-\beta \mathscr{H})$ through

$$Z_N(T, H) = \mathrm{Tr}\,\{\exp(-\beta \mathscr{H})\}. \qquad (9.2)$$

Here the symbol 'Tr' denotes a summation over all the allowed states of the system in the case of a discrete (finite) number of states (e.g. for the Ising model and the quantum-mechanical or finite-S Heisenberg model). For a system in which the constituent magnetic moments are free to assume a continuum of orientations (such as the classical or infinite-S Heisenberg model or the system of eqn (8.1) for $D > 1$), the symbol 'Tr' denotes an integral over all allowed portions of the phase space of each magnetic moment. In this chapter most specific examples will be for the Ising model so that the former definition of 'Tr' will apply.

Now if we can calculate the first L terms in an expansion of the partition function, then we can obtain, by straightforward mathematical manipulations, expressions for the first L terms in expansions of other thermodynamic functions of interest. We shall illustrate this fact by considering briefly just how one could obtain an expression for the Gibbs potential,

$$G(T, H) = -kT \ln Z_N(T, H), \tag{9.3}$$

from an expansion of the partition function—this relation is similar to that between the moments and cumulants of a distribution and is familiar from probability theory.

We begin by substituting the expansion (9.1) into (9.2), with the result

$$Z_N(T, H) = \text{Tr}\,(1)\,\{1 - \beta\,\text{tr}\,(\mathcal{H}) + \tfrac{1}{2}\beta^2\,\text{tr}\,(\mathcal{H}^2) + \cdots\}, \tag{9.4}$$

where $\text{tr}\,\mathcal{H}^\ell$ denotes the normalized trace,

$$\text{tr}\,\mathcal{H}^\ell \equiv \frac{\text{Tr}\,\mathcal{H}^\ell}{\text{Tr}\,(1)}, \tag{9.5}$$

and where the quantity $\text{Tr}\,(1)$ denotes the total number of states of the system; for example, for an Ising system of N spins, $\text{Tr}\,(1) = 2^N$, while for a quantum-mechanical Heisenberg model with spin quantum number S, $\text{Tr}\,(1) = (2S + 1)^N$. The coefficients of successive powers of β in eqn (9.4) are proportional to the moments of the Hamiltonian, $b_\ell \equiv \text{tr}\,(\mathcal{H}^\ell)$. Since the thermodynamic functions of interest are related to the Gibbs potential, (9.3), we must calculate the logarithm of the series (9.4),

$$\ln Z_N(T, H) = \ln\{\text{Tr}\,(1)\} + \ln\{1 - \beta\,\text{tr}\,(\mathcal{H}) + \tfrac{1}{2}\beta^2\,\text{tr}\,(\mathcal{H}^2) + \cdots\}. \tag{9.6}$$

It is customary to write eqn (9.6) in the form

$$\ln Z_N(T, H) = \sum_{\ell=0}^{\infty} \left(\frac{c_\ell}{\ell!}\right)\beta^\ell, \tag{9.7}$$

where the coefficients c_ℓ defined by eqn (9.7) are called the cumulants, corresponding to the moments b_ℓ. Expressions for the first few cumulants c_ℓ can be obtained easily, using the expansion $\ln(1 + x) = x - \tfrac{1}{2}x^2 + \tfrac{1}{3}x^3 + \cdots$. Thus we find

$$c_0 = \ln\{\text{Tr}\,(1)\}, \tag{9.8a}$$

$$c_1 = -b_1 = -\text{tr}\,(\mathcal{H}), \tag{9.8b}$$

$$c_2 = b_2 - b_1^2 = \text{tr}\,(\mathcal{H}^2) - \{\text{tr}\,(\mathcal{H})\}^2, \tag{9.8c}$$

and so on. An instructive example for the reader is to find a general expression for the cumulant of arbitrary order ℓ.

Thus we see that if we calculate the first few moments b_ℓ, we can obtain the first few coefficients in the expansion of the Gibbs potential and hence the first few coefficients in any of the thermodynamic functions derived from $G(T, H)$. For example, consider the magnetic specific heat C_H, which from eqn (2.48) we see is related to $G(T, H)$ through

$$C_H = -T \left(\frac{\partial^2 G}{\partial T^2} \right)_H = k\beta^2 \frac{\partial^2}{\partial \beta^2} (\ln Z_N). \qquad (9.9)$$

From eqns (9.7), (9.8c) and (9.9) it follows that the leading term in an expansion in powers of β of the specific heat involves the cumulant c_2,

$$C_H \simeq k\beta^2 c_2 = \frac{1}{kT^2} [\text{tr } (\mathcal{H}^2) - \{\text{tr } (\mathcal{H})\}^2]. \qquad (9.10)$$

Thus we see that the specific heat is predicted to fall off with temperature as $1/T^2 + \mathcal{O}(1/T^3)$ above T_c—in marked contrast to the incorrect prediction of the molecular field theory that $C_H = 0$ for $T > T_c$. We caution the reader that although eqn (9.10) looks similar to the fluctuation–dissipation result, (A.23), the latter is an exact expression in terms of the expectation value of the Hamiltonian while the former involves only the trace of the Hamiltonian (i.e. the $\beta = 0$ expectation value).

Although the leading coefficient for any of the thermodynamic functions is relatively simple to calculate, higher-order coefficients present rather more difficulty. Therefore sophisticated book-keeping techniques have been developed in order to render the resulting complexity manageable. These techniques in general involve a correspondence between a given contribution to a moment and a graph on the appropriate lattice being considered. Indeed, these graphical or diagrammatic techniques have been developed to the extent that they have come to form a fascinating branch of study in their own right.

Rather than develop this formalism in detail, we shall illustrate the basic concepts of these graphical procedures by means of a calculation of the high-temperature expansion of the partition function $Z_N(T, H)$ and the two-spin correlation function for a system of spins situated on a one-dimensional lattice and interacting through the Ising Hamiltonian —this is the case $D = d = 1$ of our model Hamiltonian (8.1).

9.3. Calculation of the coefficients in the high-temperature expansion of the partition function

In this section we illustrate the principles of the high-temperature calculation methods by calculating the coefficients in the high-temperature expansion for the Ising model partition function for the case of a one-dimensional lattice ($d = 1$). Although it is only for the case $d = 1$ that we shall be able to calculate easily the arbitrary term in the expansion and to sum the resulting series, the methods that we develop are applicable to lattices of higher dimensionality. In fact, one of the most pedagogical of the many derivations of the zero-field partition function for the $d = 2$ Ising model essentially involves summing the high-temperature expansion and, as we have stressed earlier, our most credible approximation procedures concerning the $d = 3$ Ising model are based upon extrapolations from the first few terms of the appropriate high-temperature expansion.

Consider, then, the zero-field partition function for the Ising model

$$Z_N\,(T, H = 0) = \sum_{\{s\}} \mathrm{e}^{-\beta \mathscr{H}} = \sum_{\{s\}} \prod_{<ij>}' \mathrm{e}^{\mathscr{J} s_i s_j} \qquad (9.11)$$

where, as before, $\mathscr{J} \equiv J/kT$, the summation symbol $\sum_{\{s\}}$ denotes the N-fold summation of eqn (8.6), and the product $\prod_{\langle ij \rangle}'$ is restricted to all nearest-neighbour pairs of sites $\langle ij \rangle$ in the lattice. Our analysis begins by applying the identity

$$\mathrm{e}^{A\eta} = \cosh A + \eta \sinh A = \cosh A \,(1 + \eta \tanh A) \qquad (9.12)$$

to the exponential in eqn (9.11), where η is any variable that can have only the values ± 1 (here it is $s_i s_j$). We thereby obtain

$$Z_N(T, 0) = [\cosh \mathscr{J}]^{\mathscr{P}} \sum_{\{s\}} \prod_{\langle ij \rangle}' (1 + s_i s_j v), \qquad (9.13)$$

where we have defined the variable v through

$$v \equiv \tanh \mathscr{J}, \qquad (9.14)$$

and where we have factored out the quantity $\cosh \mathscr{J}$ from the product so that $\mathscr{P}$ denotes the total number of nearest-neighbour pairs in the lattice. For a lattice with periodic boundary conditions, then,

$$\mathscr{P} = Nq/2, \qquad (9.15)$$

where q is the number of nearest-neighbour sites to a given site (the coordination number).

Next we introduce a 1:1 association between each factor in the product of (9.13) and a graph on the lattice. The motivation for this association is that only a certain topological type of graph (and hence only a certain type of factor in the product) corresponds to a non-vanishing contribution when the summation $\sum_{\{s\}}$ is finally carried out.

For simplicity, we shall first carry out this graphical analysis for the case of the $N = 3$ Ising model with periodic boundary conditions (cf. Fig. 9.1(a)); the generalization to an arbitrary lattice with arbitrary N will then be straightforward. For $N = 3$, there will be $\mathscr{P} = 3(2)/2 = 3$ factors in the product in eqn (9.13), $(1 + vs_1s_2)(1 + vs_2s_3)(1 + vs_3s_1)$,

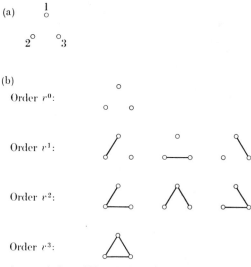

FIG. 9.1. (a) A lattice consisting of $N = 3$ sites. (b) The eight graphs that correspond to the eight terms in the Ising model partition function for the three-site lattice. Note that by virtue of eqn (9.17) only the terms corresponding to the graphs of order v^0 and of order v^3 make a non-zero contribution to (9.13); hence for this particularly simple lattice, there are only two terms in the high-temperature expansion.

so that, on expanding this product in successive powers of v, we obtain $2^{\mathscr{P}} = 8$ terms in the partition function,

$$Z_{N=3}(T, 0)/(\cosh \mathscr{J})^3 = \sum_{s_1 = -1}^{1} \sum_{s_2 = -1}^{1} \sum_{s_3 = -1}^{1}$$
$$\times \{1 + v(s_1s_2 + s_2s_3 + s_3s_1)$$
$$+ v^2(s_1s_2s_2s_3 + s_1s_2s_3s_1 + s_2s_3s_3s_1)$$
$$+ v^3(s_1s_2s_2s_3s_3s_1)\}. \tag{9.16}$$

Notice that our expansion variable for the Ising model is actually not $\beta \equiv 1/kT$ but rather $v \equiv \tanh \mathscr{J} \equiv \tanh J/kT$. This illustrates the important fact that there is nothing special about the expansion parameter β, and that any function that goes to zero when $\beta \to 0$ (i.e. $T \to \infty$) is adequate.

We now associate one graph with each of the eight terms inside the curly brackets of eqn (9.16), by drawing a bond for each pair of vertices that occurs in the product. This set of eight graphs is shown in Fig 9.1(b). Notice that since v enters into the product in (9.13) linearly with a product of the form $s_i s_j$, all graphs of order v^j will contain j bonds.

Next we utilize the topology of the graphs to enable us to carry out the multiple summation in eqn (9.16) easily. The first term of order v^0 on the right-hand side of (9.16) is simply $2^N = 8$. Consider now the three terms of order v; since $\sum_{s_j = -1}^{1} s_j = 0$, each of these terms will vanish. Similarly, each of the three terms of order v^2 contains at least one of the spin variables raised to an odd power, so that these too will vanish. In general, since

$$\sum_{s_j = -1}^{1} s_j^\ell = \begin{cases} 2 & \ell \text{ even} \\ 0 & \ell \text{ odd} \end{cases} \tag{9.17}$$

the term of order v^3 will contribute, so that

$$Z_{N=3}(T, 0) = \cosh^3 \mathscr{J}(8 + 8v^3) = 2^3(\cosh^3 \mathscr{J} + \sinh^3 \mathscr{J}). \tag{9.18}$$

Turning again to Fig 9.1(b), we observed that the only graphs that correspond to non-vanishing terms in the partition function are those graphs all of whose vertices are even (i.e. an even number of bonds emanate from each vertex); these we shall call *closed graphs*. This observation is in fact quite general, since each bond emanating from site j corresponds to a product of the form $s_i s_j$. Only if there are an even number of bonds emanating from site j will the quantity s_j raised to an even power enter into eqn (9.13) and hence, by eqn (9.17), only then will the summation $\sum_{\{s\}}$ produce a non-zero contribution to $Z_N(T, H)$. It is therefore obvious that although there will in general be 2^N graphs for a closed Ising chain of N spins, only the graph of order v^0 (with no bonds) and the graph of order v^N (with N bonds) will contribute to the partition function (cf. Fig. 9.2(a)). Hence we can immediately write down the generalization of (9.18) to all N,

$$Z_N(T, 0) = \cosh^N \mathscr{J}(2^N + 2^N v^N) = 2^N(\cosh^N \mathscr{J} + \sinh^N \mathscr{J}), \tag{9.19}$$

which agrees with our earlier expression, eqn (8.46), for the partition function of a closed Ising chain.

The difference between the open and closed chain is also easily interpreted in terms of graphs. If we do not impose periodic boundary conditions, then there will be no graphs whose vertices are all even

except the graph—with no bonds—that is of order v^0 (cf. Fig. 9.2(b)).
Hence eqn (9.13) becomes

$$Z_N(T, 0) = 2^N \cosh^{N-1} \mathscr{J} \tag{9.20}$$

which agrees with eqn (8.11b). Notice that the exponent of the hyperbolic cosine in eqn (9.20) is $N - 1$ rather than N because for an open
chain there are only $\mathscr{P} = N - 1$ pairs of lattice sites.

The foregoing graphical analysis is by no means restricted to one-

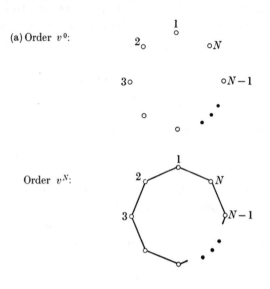

FIG. 9.2. (a) The two graphs that correspond to the two non-vanishing terms in the
high-temperature expansion of the Ising model partition function for a closed linear chain
lattice of N spins. Since these graphs contain zero and N bonds, respectively, they
correspond to terms of order v^0 and v^N. (b) The unique graph that corresponds to the
only non-vanishing term in the high-temperature expansion of the Ising model partition
function for an open linear chain lattice. This graph has no bonds, and hence corresponds
to a term in (9.21) which is of order v^0. For lattices of dimensionality $d > 1$, more complex graphs must be considered.

dimensional lattices, and therein lies its power. In general, for the Ising
model on *any* lattice eqn (9.13) may be written as

$$Z_N(T, 0) = (\cosh \mathscr{J})^{\mathscr{P}} 2^N \sum_{\ell=0}^{\mathscr{P}} g(\ell) v^\ell, \tag{9.21}$$

where $g(\ell)$ is the number of graphs that we can draw on the lattice using ℓ lines such that each vertex of the lattice is even ($\mathscr{g}(0) \equiv 1$). Thus the calculation of the partition function of the Ising model has been reduced to the problem of counting closed graphs on a lattice.

We might think that by using this approach we could obtain Onsager's result for the zero-field partition function of a two-dimensional lattice. This is indeed the case, but the analysis is so lengthy that we shall present it in Appendix B. Particularly clear treatments are given by Landau and Lifschitz (1969) and Glasser (1970); also the early work of Kac and Ward (1952), Burgoyne (1963), and Vdovichenko (1964, 1965) can be consulted.

For three-dimensional lattices, although extremely ingenious techniques have been developed for counting the graphs that contribute to (9.21) for finite values of ℓ, no one has so far seen a way of counting all the graphs; at present the coefficients $\mathscr{g}(\ell)$ in the expansion (9.21) have been calculated through order $\ell \simeq 30$ for some simple three-dimensional lattices and even through order 15 for the rather complex face-centred cubic lattice. We should also point out that the 'graphology' for the case of a non-zero magnetic field becomes somewhat more complex, and for this reason slightly fewer terms are at present known for most of the susceptibility series than for, say, the specific heat series.

In addition to requiring different series for different lattices, we require different series for different thermodynamic functions. We have concentrated so far on the high-temperature expansion of the zero-field partition function, the terms of which are used in obtaining the exponent α for the specific heat. In Table 9.1 we attempt to provide the reader with a guide to the literature, so that if he seeks a given series for a given lattice he can easily find it. Note that Table 9.1 pertains only to the three-dimensional Ising model, and that the references shown represent the longest published series (for certain series, additional coefficients have been calculated but not yet published—presumably because the all-important checking procedures are still under way).

What about models other than the Ising model? There is nothing that limits the range of applicability of the general calculational procedures that we have illustrated with the Ising model to any particular assumptions concerning the interactions among the spins. However, for systems other than the Ising model, eqn (9.17) is not valid and hence the coefficients in the high-temperature expansion are not simply the total number of closed graphs but instead they must be obtained from the consideration of graphs of somewhat more complex topology.

Perhaps more significant is the fact that the series expansion method

TABLE 9.1

A guide to the literature of high- and low-temperature expansions for the Ising model for two- and three-dimensional lattices (the one-dimensional lattice was considered in detail in the text). The references given are for the longest published series; for some functions longer series are available (but as yet unpublished). As an example, if we wish to find the critical-point exponent α' for the diamond lattice, we can find the longest published low-temperature series expansion for the specific heat in Ref. 11 (Essam and Sykes 1963). This table has been adapted from Guttmann (1969a).

	α'	α	β	γ'	γ	δ	ν
Two-dimensional lattices							
square	13	12	13	9	3 ·	5*	7
triangular		13	13	9	10	5*	7
honeycomb			13	9	3	5*	
Three-dimensional lattices							
s.c.	5*	12	9	9	1	5*	1, 7
b.c.c.	5*	12	5	5*	1	5*	1, 7
f.c.c.	5*	4	6	9	1	5*	1, 7
diamond	11	11	5	5*	11	5*	
cristobalite		2			2		

References
1. Moore, Jasnow, and Wortis (1969).
2. Betts and Ditzian (1968).
3. Sykes and Fisher (1962).
4. Sykes, Martin, and Hunter (1967).
5. Sykes, Essam, and Gaunt (1965).
5* This series may be derived from Appendix IV of Ref. 5.
6. Fisher (1967).
7. Fisher and Burford (1967).
8. Sykes (1963).
9. Essam and Fisher (1963).
10. Sykes and Zucker (1961).
11. Essam and Sykes (1963).
12. Baker Jr. (1963a).
13. Domb (1960).

is not limited to systems for which particular assumptions such as those of § 8.1 are valid. Thus, for example, calculations have been carried out for model systems in which the interactions are (i) not uniform in direction, (ii) further than nearest-neighbour in range, and (iii) not isotropic in symmetry. Not surprisingly, such calculations for more general types of interactions involve more complicated graphs, and consequently not as many terms in the high-temperature expansions have been obtained. Nevertheless, the basic idea is the same, and almost as many terms have been calculated for, say, the classical Heisenberg model as have been calculated for the Ising model.

One reason for mentioning the classical Heisenberg model is that although the critical phenomena predictions of the classical and the quantum-mechanical (finite S) Heisenberg models differ little, the calculation of the high-temperature expansions proceeds much more simply in the classical case (Stanley and Kaplan 1966a, Wood and Rushbrooke 1966, Joyce and Bowers 1966a, b) and as a result many more terms are now known in the classical series than in the general S quantum-mechanical series (Moore 1969). However, when the method of high-temperature expansions was first proposed (Kramers 1936, Opechowski 1937, Van Vleck 1937), it was applied to the quantum-mechanical Heisenberg model. The early authors succeeded in calculating only the first three or four terms in the expansion, and the twenty or so years following their initial work witnessed many additional papers—some adding a new term and others correcting the errors in the previously-calculated terms. One of the most monumental efforts was that of Rushbrooke and Wood (1958), who calculated—correctly— the first seven terms in the partition function series for general spin quantum number S. Needless to say, the coefficients become increasingly difficult to calculate as the order increases. In his Ph.D. thesis, Wood (1958) remarks that 'the labour of calculating one more term in one of the series is considerably greater than that of calculating all the previous terms'. Nevertheless, the series for some functions of especial interest have been pushed quite a long way—at the cost of a rather considerable quantity of 'labour'.

9.4. Calculation of the coefficients in the high-temperature expansion of the two-spin correlation function

In addition to the partition function there is another useful function that involves the exponential $\exp(-\beta\mathcal{H})$, and this is the two-spin correlation function. We consider in this section the analogous high-temperature expansion of the correlation function, because this function contains information not contained in the expansion of the partition function. For example, high-temperature expansions of the spin correlation function have been used (Dwight, Menyuk, and Kaplan 1965; Stanley 1967a, 1968a) to obtain predictions of the elastic paramagnetic neutron scattering cross-section, which is directly related to the Fourier transform of the spin correlation function. A second reason, and perhaps the more important reason, for considering the high-temperature expansion of the spin correlation function is that knowledge of it, together with the fluctuation–dissipation relation (A.20), permits us to obtain the zero-field susceptibility without needing to consider the partition function in a non-zero magnetic field, as would be necessary if we were to obtain the susceptibility from differentiating the Gibbs potential $G(T, H)$ with respect to magnetic field (Stanley and Kaplan 1966a; Fisher and Burford 1967; Stanley 1967a, b; Jasnow and Wortis 1968). A third motivation for presenting the series expansion of the two-spin correlation function will become apparent in § 9.5, where we shall attempt to show that the coefficients in this expansion have a rather simple and direct physical interpretation.

We shall limit our discussion to the case of the Ising interaction, and we shall see again that the linear chain lattice provides a case for which all the coefficients in the high-temperature expansion can be calculated exactly. However, the formalism that we develop is by no means limited in its applicability to the case of the one-dimensional lattice.

The two-spin correlation function for the Ising model is defined through the relation (cf. eqn (8.12))

$$\langle s_m s_n \rangle = Z_N^{-1} \sum_{\{s\}} s_m s_n \, e^{-\beta \mathcal{H}} = Z_N^{-1} \sum_{\{s\}} s_m s_n \prod_{\langle ij \rangle}' e^{\mathcal{I} s_i s_j}. \quad (9.22)$$

The high-temperature expansion of eqn (9.22) proceeds exactly as with the partition function (9.11). Thus, in analogy with eqn (9.13) we find

$$\langle s_m s_n \rangle = Z_N^{-1} (\cosh \mathcal{I})^{\mathscr{P}} \sum_{\{s\}} \prod_{\langle ij \rangle}' s_m s_n \, (1 + v s_i s_j). \quad (9.23)$$

As before, we can construct $2^{\mathscr{P}}$ graphs that stand in 1:1 correspondence with each of the $2^{\mathscr{P}}$ terms that are obtained when the product in eqn (9.23) is expanded. In fact, these graphs will be identical to the graphs for the partition function (9.13), because we shall not represent the factor $s_m s_n$ by a bond (m and n are generally not nearest neighbours). However, the rules determining which graphs make a non-zero contribution to the summation $\sum_{\{s\}}$ in (9.23) must be based on the identity (9.17) as before, and since we now have an extra factor of $s_m s_n$ we must modify the rule concerning which graphs contribute. Specifically, we must require that the vertices m and n have only an odd number of bonds emanating from them, so that, together with the factor $s_m s_n$ that is not represented in the graph, we will have each contributing spin raised to an even power. (We could, of course, have retained our former rules if we were to draw an additional line between sites m and n; however this is rather clumsy since sites m and n will not be nearest-neighbours in general.) Hence the analogue of eqn (9.21) for the zero-field partition function is

$$\langle s_m s_n \rangle = \{Z_N(T, H)\}^{-1} (\cosh \mathscr{J})^{\mathscr{P}} \, 2^N \sum_{\ell=1}^{\mathscr{P}} \eta_{mn}(\ell) v^\ell, \qquad (9.24)$$

where $\eta_{mn}(\ell)$ is the number of graphs of ℓ lines, with even vertices except at sites m and n. Equation (9.24), like its counterpart, eqn (9.21), is valid for lattices of any dimensionality—the topology of the lattice enters through the factor $\eta_{mn}(\ell)$.

As an example of eqn (9.24), we consider the linear chain Ising model, without periodic boundary conditions. It is obvious from Fig. 9.3 that the only graph with a non-vanishing contribution for this lattice is

Order $v^{|n-m|}$:

$$\begin{matrix} \underset{1}{\circ} & \underset{2}{\circ} & \underset{3}{\circ} & \bullet\bullet\bullet & \underset{m}{\circ}\!-\!\circ\!-\!\circ\!-\!\circ\!-\!\circ\!-\!\underset{n}{\circ} & \bullet\bullet\bullet & \underset{N-1}{\circ} & \underset{N}{\circ} \end{matrix}$$

Fig. 9.3. The single graph that corresponds to the single non-vanishing term in the high-temperature expansion (9.24) for the two-spin correlation function $\langle s_m s_n \rangle$ for the linear chain Ising model without periodic boundary conditions. This graph is a chain of $(n\text{–}m)$ bonds between sites m and n. For higher dimensional lattices more complicated graphs will contribute, but the dominant contribution is found to be the simple chain of bonds (called a *self-avoiding walk* in the research literature).

the graph consisting of nearest-neighbour bonds joining all the lattice sites between sites m and n. Hence there will be only one term that enters into the summation (9.24), and this is the term of order $v^{|n-m|}$, whence

$$\langle s_m s_n \rangle = (2^N \cosh^{N-1} \mathscr{J})^{-1} (\cosh^{N-1} \mathscr{J}) \, 2^N \, v^{|n-m|} = v^{|n-m|}, \qquad (9.25)$$

which agrees with eqn (8.17).

Having obtained the zero-field spin correlation function, we are in a position to use the fluctuation–dissipation relation (A.20) to calculate the zero-field susceptibility. It is somewhat more conventional in series expansion calculations to consider the reduced susceptibility, defined by the relation

$$\bar{\chi}_T(T, H = 0) \equiv \chi_T(T, H = 0)/\chi_{\text{Curie}}(T, H = 0), \qquad (9.26)$$

where χ_{Curie} is the susceptibility in the non-interacting limit (for the

Ising model, $\chi_{\text{Curie}} = N\bar{\mu}^2/kT$). The normalization adopted in (9.26) is chosen such that in the high-temperature limit, $v \equiv \tanh J/kT \to 0$, $\bar{\chi}_T(T, H = 0) \to 1$, and hence the leading term in the high-temperature expansion of the reduced susceptibility must be unity. Hence on using (9.25) and (8.19), we obtain, for the large N limit, the simple series expansion $\bar{\chi}_T = 1 + 2v + 2v^2 + 2v^3 + \cdots$. This series may be summed exactly by first writing $\bar{\chi}_T = 1 + 2v(1 + v + v^2 + v^3 + \cdots)$, which is recognized as being identical to the expression of eqn (8.24),

$$\bar{\chi}_T = 1 + \frac{2v}{1 - v} = \frac{1 + v}{1 - v}. \tag{9.27}$$

If we wish to express the susceptibility in terms of the natural variable $\mathscr{J} \equiv J/kT$ instead of the variable $v \equiv \tanh \mathscr{J}$, then we simply substitute the expansion for v in terms of $\mathscr{J}$ into the 'v-expansion' of eqn (9.26), with the result

$$\bar{\chi}_T(T, H = 0) = \sum_{\ell=0}^{\infty} a_\ell \mathscr{J}^\ell, \tag{9.28}$$

where the coefficients a_ℓ are given, in general, by the expression

$$a_\ell = \frac{2^\ell}{\ell!}. \tag{9.29}$$

9.5. Physical interpretation of the terms in the high-temperature expansion of the correlation function

We shall now consider briefly the nature of the graphs that make a non-vanishing contribution to the two-spin correlation function $\langle s_m s_n \rangle$ for a lattice of dimensionality greater than 1. We shall see that each of these graphs can be interpreted as corresponding to a unique path in the lattice whereby the direct interaction of strength J (which only extends between nearest-neighbour sites in the lattice) is propagated from site m—via intervening pairs of lattice sites—to site n. If the graph involves ℓ bonds, then the 'interaction path' will involve ℓ pairs of intervening lattice sites, and it will enter first in the contribution of the order $(J/kT)^\ell$. We present this argument because it should help to clarify the mechanism whereby nearest-neighbour interactions produce a significant effect between spins that are situated a long distance apart, and this is the essence of what gives rise to critical phenomena and phase transitions in the first place.

Although for a one-dimensional lattice, the simple chain of nearest-neighbour bonds joining the sites m and n is the *only* graph to contribute to the correlation function $\langle s_m s_n \rangle$, for lattices of higher dimensionality there will be much more complicated graphs as well. For example, we show in Fig 9.4 the graphs that make a non-vanishing contribution to the nearest-neighbour correlation function (i.e. sites n and m are nearest neighbours of each other) for orders v^1–v^5 for the simple quadratic

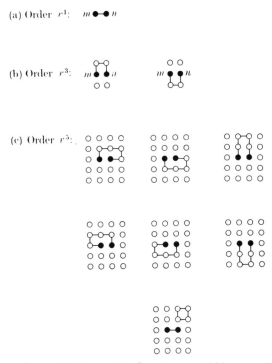

FIG. 9.4. (a) The only graph that corresponds to a non-vanishing contribution of order v (and, since $v \equiv \tanh J/kT$, the only contribution of order J/kT) in the high-temperature expansion of the two-spin correlation function $\langle s_m s_n \rangle$ of a square lattice for the case when the lattice sites m and n are nearest neighbours of each other. (b) The two graphs that make a contribution of order v^3. For the square lattice there is no contribution of order v^2 because there is no way in which the correlation can be propagated from site m to site n by means of two intervening interactions. (c) The seven graphs that make contributions of order v^5. All but one of these consists of a self-avoiding walk between sites m and n. The approximation in which one neglects all but the self-avoiding walks in calculating the correlation function $\langle s_m s_n \rangle$ leads to predictions which, not surprisingly, do not agree with exact solutions but are considerably more realistic than the classical theories (such as the molecular field approximation) in that, for example, they depend on lattice dimensionality.

lattice. We see up to order v^3 the only non-vanishing graphs are those that represent a simple chain of bonds, and even though polygons begin to enter in the order v^5 contribution (for example, one graph consists of a single bond between sites m and n plus an isolated square elsewhere in the lattice), the dominant contribution would appear to arise from the six different possibilities of drawing a simple chain of five bonds between sites m and n.

The reason for emphasizing the graphs of Fig. 9.4 is that we can see therein precisely how the correlation between two nearest-neighbour spins arises: there is a contribution, of order v (and hence of order J), which is due to the direct interaction (of strength J) between the nearest neighbour sites. The remaining contributions to the nearest-neighbour correlation function arise from the indirect interaction of the moment on sites m and n through the other moments in the lattice. For example, the contribution of order v^3 begins to contribute only in order J^3, since $v \equiv \tanh J/kT \simeq J/kT + \cdots$ when it becomes possible for sites n and m to interact through *three* intervening interactions. This argument is appealing to the intuition, for it implies that the contribution to the nearest-neighbour two-spin correlation function—which is the probability that if a spin at site m is 'up' a spin at site n will also be 'up'—is made up of two sorts of effects: (1) the one direct interaction, which is of order J, and (2) a set of indirect interactions, each of which occurs as the result of ℓ other direct interactions, and each of which therefore contributes first in order J^ℓ. Since our high-temperature expansion is really an expansion in the variable J/kT (i.e. J enters linearly with $1/T$), what we are really doing when we perform a high-temperature expansion of the correlation function is performing a J expansion. This involves considering all possible paths by which two spins can interact, i.e. we are literally examining what the system actually does in adding up these interaction paths.

For example, if we calculate the first ten terms in the high-temperature expansion of the nearest-neighbour correlation function, we are in effect taking into account, exactly, all possible paths by which the spin at site n can communicate with its nearest neighbour at site m and which involve first one interaction (the direct interaction), then two interactions, and so on through ten intervening interactions. Of course, if we stop our calculation with ten terms, we are clearly neglecting indirect interactions that proceed by, say, eleven intervening direct interactions. However, the extrapolation method used in obtaining information from high-temperature expansions involves the crucial assumption that we do not truncate our series at, say, ten terms, but rather we seek to extrapolate the regular behaviour of the first ten coefficients and thereby guess the behaviour of the eleventh. Such extrapolation procedures will be explained in § 9.6.

We have seen in this section that the topology of the lattice inter-connections plays a very substantial role in determining the value of the correlation function between two spins, for it is this that determines the nature of the allowed interaction paths between the spins. Therefore it would seem that we are in a position now to appreciate the observation that critical behaviour depends very strongly upon lattice dimensionality. A somewhat stronger statement is often made—that critical phenomena depend much more strongly on lattice dimensionality than they do on most features of the interaction Hamiltonian. For example, it was once believed that the critical-point exponent γ for the Heisenberg model had different values for $S = \frac{1}{2}$ and for $S = \infty$ (Stanley and Kaplan 1967b, Baker, Gilbert, Eve, and Rushbrooke 1967; Stanley 1967b, Wood and Rushbrooke 1966, Joyce and Bowers 1966b). More recent evidence would appear to support the conjecture that γ is independent of S (Bowers and Woolf 1969, Stanley 1969c, Lee and Stanley 1971), so that, in particular, the $S = \frac{1}{2}$ Heisenberg model has the same critical-point exponents as the classical (or $S = \infty$) Heisenberg model. Although the critical-point exponents are now believed to be independent of S, the critical temperature is not. For example, the critical temperature of three-dimensional lattices obtained from extrapolations of high-temperature expansions may be fitted closely by the mnemonic formula (Rushbrooke and Wood 1958),

$$\frac{kT_c}{J} = \tfrac{5}{96}(q - 1)\{11S(S + 1) - 1\}. \tag{9.30}$$

Equation (9.30) is to be compared with the analogous result of the mean field approximation, eqn (6.51), which is predicted to be valid for *all* lattice dimensionalities.

Before concluding this section, we remark that the molecular field approximation does not consider the *indirect* interactions between spins on sites n and m and, for this reason, it fails to predict the correct expressions for the thermodynamic functions. For example, the molecular field approximation predicts that $C_H = 0$ for $T > T_c$, when in fact we need only to consider the second coefficient in the high-temperature expansion to obtain the rigorous result that $C_H \sim 1/T^2 + \mathcal{O}(1/T^3)$.

We shall not dwell any longer on either the derivation or the meaning of the coefficients in the high-temperature series expansions, but we shall instead turn directly to the question of how we attempt to 'guess' the limiting behaviour of thermodynamic functions—given only a finite (and sometimes rather small) number of coefficients in an infinite series.

9.6. Extrapolation procedures for estimating the limiting behaviour of a power series from the behaviour of its first few terms

Our goal in this section is to discuss methods of obtaining information about the limiting behaviour of a thermodynamic function, given only the knowledge of the first L coefficients in the high-temperature expansion of that function. This is in fact a problem in applied mathematics that has engaged the best of minds for over a century, and we shall here discuss only those aspects of this fascinating area that have been applied to the specific problem of inferring the behaviour of a singular thermodynamic function in the critical region.

We begin by considering why it is necessary to extrapolate in order to obtain a reliable estimate of the singular behaviour. The answer is that if we do not extrapolate, there will be no singular behaviour of any sort predicted. Suppose we have calculated ten coefficients in a series representation for the reduced susceptibility $\bar{\chi}_T$ of the form of (9.28), and we wish to find the critical temperature T_c. The salient feature of the susceptibility is that it diverges to infinity as $T \to T_c$. However, the successive approximations to $\bar{\chi}_T$ obtained by truncating the expansion (9.28) (at say, 5 terms, 6 terms, . . ., and 10 terms) will all be simple polynomials and hence will not diverge at any value of $\mathscr{J} \equiv \beta J$ except $\mathscr{J} = \infty$, corresponding to $T = 0$. Hence the consideration of successive truncations of the series (9.28) would appear to be fruitless.

A more useful approach stems from the realization that what we are seeking is essentially the radius of convergence of the power series, since $\bar{\chi}_T$ is finite for all values of $J/kT < J/kT_c \equiv \mathscr{J}_c$, and it diverges when $\mathscr{J} \to \mathscr{J}_c$ from below. Actually the above statement is not quite true, for there might exist another non-physical singularity elsewhere in the complex plane which lies closer to the origin than the physical singularity that we are seeking to determine, and hence determining the radius of convergence would yield no useful information. In applying the ratio methods to be described in this section we are therefore assuming that the closest singularity is the physical singularity, $\mathscr{J} = \mathscr{J}_c$; when this assumption is not valid, other approaches will become necessary (cf. § 9.7). It should be remarked that if the coefficients of a series are *not* all positive, then we might suspect that the radius of convergence is *not* determined by the physical singularity, since a general property of power series is that if the nearest singularity is on the positive real axis, then all the coefficients in the series eventually become positive.

When we seek the radius of convergence of a power series, the first thing that frequently comes to mind is the ratio test, according to which the reciprocal of the radius of convergence is given by

$$r_c^{-1} = \lim_{\ell \to \infty} \frac{a_\ell}{a_{\ell-1}}, \qquad (9.31)$$

assuming that the limit exists. We shall see in this section that techniques based upon this simple approach are extremely powerful and that they predict not only the radius of convergence (or critical temperature) but also the fashion in which the function diverges at the critical temperature, i.e. the critical-point exponent.

Our detailed discussion begins by assuming that χ_T has a simple power law singularity, of the form

$$\chi_T = \mathscr{C} \, \epsilon^{-\gamma} (1 + A \epsilon^{+x} + \cdots). \qquad (9.32)$$

The correction terms may be of a sort other than that indicated in eqn (9.32), without affecting the general validity of the argument that we shall present. It is important, however, to realize that there will in general be correction terms of some sort.

For example, suppose there were no correction terms whatsoever, that is, suppose

$$\chi_T = \mathscr{C} \left(\frac{T - T_c}{T_c} \right)^{-\gamma} = \mathscr{C} \left(\frac{T}{T_c} \right)^{-\gamma} \left(1 - \frac{T_c}{T} \right)^{-\gamma} \qquad (9.33)$$

or, equivalently

$$\chi_T = \mathscr{C} (\tilde{T})^{-\gamma} (1 - \tilde{T}^{-1})^{-\gamma} = \mathscr{C} \tilde{\beta}^{\gamma} (1 - \tilde{\beta})^{-\gamma}, \qquad (9.34)$$

where $\tilde{T} \equiv T/T_c$ and $\tilde{\beta} \equiv \beta/\beta_c$ as before. Then we can expand (9.34) for $T > T_c$ (or $\beta < \beta_c$) by using the binomial theorem, with the result

$$\chi = \mathscr{C} \tilde{\beta}^{\gamma} \left\{ 1 + \gamma \tilde{\beta} + \frac{\gamma(\gamma + 1)}{2!} \tilde{\beta}^2 + \cdots + \frac{\gamma(\gamma + 1) \ldots (\gamma + \ell - 1)}{\ell!} \tilde{\beta}^{\ell} + \right.$$
$$\left. + \cdots \right\}. \qquad (9.35)$$

(Note that all the coefficients in this expansion are unity if $\gamma = 1$, as in the molecular field approximation.) Now there are only three unknown quantities in eqn (9.35): γ, $\tilde{\beta}$, and $\mathscr{C}$. Hence it should be possible to obtain all three unknown quantities by calculating the first three terms in the high-temperature expansion.

Now let us examine some expansions for actual model systems and seek to determine if indeed our assumption of no correction terms is valid. To this end, it is useful to consider the ratio of successive terms in the series of eqn (9.28). Since the expansion variable in (9.28) is $\mathscr{J} \equiv \beta J$ and not $\tilde{\beta} \equiv \beta/\beta_c$, the ratio of successive terms $a_\ell/a_{\ell-1}$ in (9.28) will be

$$\left\{\frac{\gamma(\gamma+1)\ldots(\gamma+\ell-1)}{\mathscr{J}_c^\ell\,\ell!}\right\}\bigg/\left\{\frac{\gamma(\gamma+1)\cdots(\gamma+\ell-2)}{\mathscr{J}_c^{\ell-1}(\ell-1)!}\right\}$$

$$=\frac{\gamma+\ell-1}{\ell\mathscr{J}_c}=\mathscr{J}_c^{-1}\left(1+\frac{\gamma-1}{\ell}\right), \quad (9.36)$$

where $\mathscr{J}_c \equiv \beta_c J \equiv J/kT_c$ as before. Hence we see that a plot of $a_\ell/a_{\ell-1}$ against $1/\ell$ should be a straight line which, when $1/\ell = 0$ (corresponding to $\ell = \infty$), will intercept the ordinate at the value $\mathscr{J}_c^{-1} \equiv kT_c/J$; moreover, the slope of this straight line should be given by $\mathscr{J}_c^{-1}(\gamma-1)$.

Let us see if this is the case by examining the Ising model. Fig 9.5 shows plots of the ratios of successive coefficients in the Ising model series for the linear chain lattice ($d = 1$), for the simple quadratic or square lattice ($d = 2$), for the simple cubic lattice ($d = 3$), and for the molecular field approximation. We have divided the ratios by an appropriate factor in order that the ordinate will be T_c/T_M, where here T_M denotes the critical temperature predicted by the molecular field approximation to the Ising model and is given by the expression $kT_M = qJ$. From examination of Fig. 9.5 we see that the linear chain Ising model and the molecular field approximation both (correctly) predict there are *no* correction terms, while the fact that for $d = 2$ and $d = 3$ all the ratios do *not* lie on precisely a straight line means that for the square and simple cubic lattices there are evidently correction terms. However, the fact that the observed oscillation of the ratios about the (conjectured) asymptotic behaviour seems to get smaller as ℓ increases is suggestive of the fact that the correction terms are not interfering strongly with the dominant singularity.

To summarize our findings so far, we have seen that although there apparently exist correction terms for the case of the $d = 2$ and $d = 3$ lattices, the ratios of successive coefficients appear to settle down to an apparent linear behaviour which is presumably dominated by a simple power law singularity. Moreover, the limiting value (as $\ell \to \infty$ or $1/\ell \to 0$) of the ratios of successive coefficients approaches the exactly known value for $d = 1$, for $d = 2$, and for the molecular field approximation. Therefore it is rather tempting to extrapolate the behaviour of the $d = 3$ ratios, to identify the intercept with the critical temperature of a simple cubic lattice, and to utilize the slope of the asymptotic straight line to calculate the critical-point exponent γ.

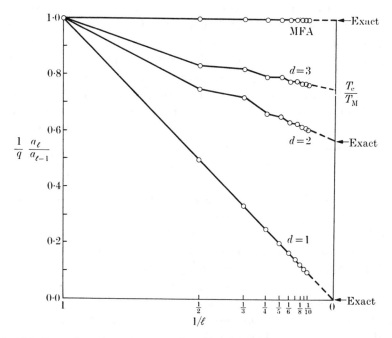

FIG. 9.5. Dependence upon inverse order $1/\ell$ of the ratios of successive coefficients a_ℓ in the reduced susceptibility series, (9.28), for the Ising model for the linear chain lattice ($d = 1$), for the simple quadratic or square lattice ($d = 2$), and for the simple cubic lattice ($d = 3$). Only the first ten ratios are shown, even though a larger number have been calculated. For purposes of comparison, we also show the first ten ratios for the molecular field approximation. For the Ising model, the ratios have been divided by the coordination number q in order that in their limiting ($\ell \to \infty$) values will be T_c/T_M, where $kT_M = qJ$. The exactly known values of T_c/T_M are indicated for the $d = 1$ and $d = 2$ cases. By courtesy of D. L. Njus.

The agreement with the exact results for $d = 1$ and $d = 2$ is so encouraging that some workers have come to refer to predictions based upon extrapolations such as this as representing exact calculations. This is certainly not the case. All that is exact is the calculation of the successive terms in the series, and the crucial step of extrapolation that is needed to obtain specific predictions (such as the values of T_c and γ) is far from rigorous. In fact there does not exist a proof that the series even converges, let alone that its asymptotic behaviour is represented by the first dozen or so coefficients in its high-temperature expansion. Nevertheless, in so far as approximation procedures near the critical point are concerned, extrapolations from coefficients of high-temperature expansions have proved to be quite reliable in the sense that there are no known cases in which they have failed to predict an exactly known answer. The evidence supporting their success is not restricted to two-dimensional lattices, for if it were the procedure might be suspect since we have seen that critical behaviour can change markedly with lattice dimensionality. For example, for the spherical model over eighty terms in the high-temperature expansion have been calculated for three-dimensional lattices (Dalton and Wood 1968, Milošević and Stanley 1971) and the agreement between predictions based upon a finite number of coefficients and the exact result is as good as it is for the $d = 2$ Ising model.

There is, however, one 'fly in the ointment'. This is the case of the two-dimensional lattice with $D > 1$ (e.g. the classical Heisenberg model, $D = 3$). It has been shown rigorously (Mermin and Wagner 1966) that there can exist no spontaneous magnetization at any temperature greater than absolute zero, yet the ratios of successive coefficients in the high-temperature expansion appear to be as regular as for the three-dimensional counterpart and they extrapolate to a value of T_c that is appreciably larger than zero (Stanley and Kaplan 1966b, 1967a, Stanley 1967c, 1968b, c, 1969b, Moore 1969, Betts, Elliott, and Ditzian 1971). This can only mean either that the ratio method fails if $d = 2$ and $D > 1$ or else that there is in fact a non-zero temperature T_c at which the susceptibility diverges but below which there is no spontaneous magnetization (cf. Fig. 8.3). This latter possibility has received limited support from recent theoretical calculations (Jancovici 1967, Reatto 1968, Lines 1971, Mubayi and Lange 1969, Watson, Vineyard, and Blume 1970, Wegner 1967), but the question is far from settled (Birgeneau, Skalyo, and Shirane, 1970).

However, when the ratios behave smoothly with order ℓ, it is possible to obtain a high degree of confidence in the extrapolations made therefrom. For example, we show in Fig. 9.6 the behaviour of the ratios of successive coefficients in the susceptibility series and in the specific heat series for the three-dimensional Ising model. Not surprisingly, both sets of ratios appear to be extrapolating to the same intercept as

$1/\ell \to 0$. Moreover, we see that one can use the intercept determined from the susceptibility series to help determine the exponent for the specific heat series. Thus the dotted lines labelled $\alpha = 0$ and $\alpha = \frac{1}{8}$ correspond to the limiting slopes that one would expect for these two choices of the exponent α. Evidently the coefficients would seem to

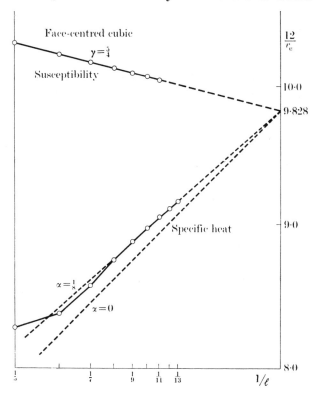

FIG. 9.6. Dependence upon $1/\ell$ of the ratios of successive coefficients in the v-expansions of the susceptibility and specific heat of the Ising model for the fcc lattice. The dashed lines are obtained by assuming the value of $q/v_c \equiv q/\tanh (J/kT_c) \simeq 9.828$ as obtained from extrapolations based upon the susceptibility series, and then assuming that $\alpha = \frac{1}{8}$ (upper curve) and $\alpha = 0$ (lower curve). It is evident that the specific heat series strongly favours $\alpha = \frac{1}{8}$ over $\alpha = 0$. This procedure is typical of methods that permit confidence limits to be placed on numerical estimates obtained by extrapolation from high-temperature expansions. Based upon examination of this data, we would place confidence limits at least as small as ± 0.1 on the estimate $\alpha = 0.125$. This graph was privately communicated to the author by M. F. Sykes.

favour the choice $\alpha = \frac{1}{8}$ over the choice $\alpha = 0$, so that the specific heat of the three-dimensional Ising model would appear not to have the same sort of logarithmic divergence as the two-dimensional case—as was originally believed (cf. § 4.1).

We conclude this section by remarking that we can systematically carry out these expansions for all lattice dimensionalities as well as all spin dimensionalities, and we find that the critical properties so obtained appear to vary monotonically with both spin and lattice dimensionality, as shown in Fig. 9.7 and Table 8.3. Just as the limit $D \to \infty$ corresponds to the spherical model, so the limit $d \to \infty$ corresponds to the molecular field theory. Although this latter result has not yet been proved, it is at least intuitively plausible, for as the lattice dimensionality increases without limit, so also does the lattice coordination number and hence the effective interaction range (for a hypercubical lattice, $q = 2d$). One somewhat intriguing result that has arisen from the analysis of lattices with $d > 3$ is the following: rather than the anticipated mean field behaviour setting in gradually as $d \to \infty$, the mean field critical-point exponents appear to be obtained for all values of $d \geq 4$ (Milošević and Stanley 1971).

9.7. Padé approximants and transformation methods

9.7.1. *When the ratios are not smooth*

We have seen that the ratio method provides an extremely simple and straightforward method for estimating both T_c and the critical-point exponent for a thermodynamic function, given that the first few terms of its high-temperature expansions are sufficiently regular with order ℓ. Moreover, there exist no examples of thermodynamic functions for which the ratios of successive coefficients appear to be regular, yet the extrapolated estimates are incorrect. This is not to say, however, that it is difficult to construct a function for which the ratios are not a smooth function of $1/\ell$, and in fact there exist some thermodynamic functions for which the ratios are exceedingly irregular. The most notable of these examples is the low-temperature series for the zero-field magnetization of the Ising model for three-dimensional lattices. For example, for the fcc lattice, the first twenty-five terms are (Fisher 1965)

$$M(T, H = 0) = 1 - 2u^6 - 24u^{11} + 26u^{12} - 48u^{15} - 252u^{16} + 720u^{17}$$
$$- 438u^{18} - 192u^{19} - 984u^{20} - 1008u^{21} + 12924u^{22}$$
$$- 19536u^{23} + 3062u^{24} - 8280u^{25} + \cdots , \qquad (9.37)$$

where here

$$u \equiv e^{-4J/kT} \qquad (9.38)$$

is the low-temperature expansion variable ($u \to 0$ as $T \to 0$). We see

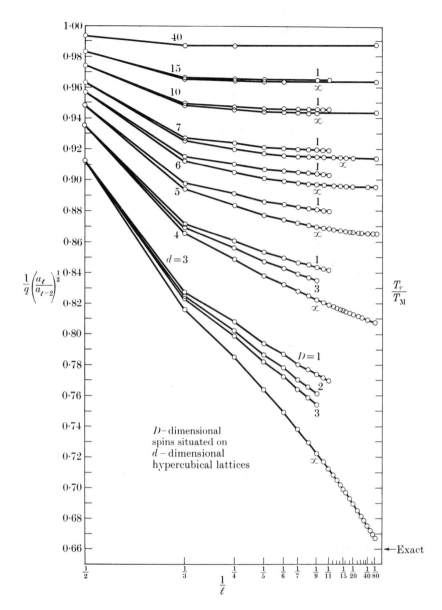

Fɪɢ. 9.7. Dependence upon $1/\ell$ of the square root of the ratio of alternate coefficients. According to eqn (9.40), these plots should also approach T_c/T_M (they have been normalized in analogy with Fig. 9.5). Also, we can show by arguments analogous to those leading to eqn. (9.36) that the limiting slope should again be proportional to $\gamma - 1$. The subject of critical behaviour for lattices of dimensionality $d > 3$ is of considerable theoretical interest, as the critical-point exponents appear to have their mean field values even though the critical temperatures do not. After Stanley (1969b).

that the ratios of successive coefficients are oscillating wildly in sign, as well as in magnitude, and clearly the ratio test will not be applicable.

There are many possible reasons for irregular behaviour such as in eqn (9.37). One possible reason is simply that the number of terms in the expansion that have been calculated is not sufficiently large for us to obtain information about the singularity. This reason is always an explanation for erratic behaviour, no matter how many terms in the series may have been calculated. As many authors have pointed out, a knowledge of the first L terms in the series

$$\bar{\chi} \equiv \sum_{\ell=0}^{L} a_\ell \beta^\ell + \beta^{L+1}(\beta_c - \beta)^{-\gamma} \tag{9.39}$$

tells us nothing whatsoever concerning the singularity at $\beta = \beta_c$.

A second possible way to obtain information from a series, the ratios of whose coefficients are somewhat irregular, is to consider other means of determining the radius of convergence. For example, plots against $1/\ell$ of $(a_\ell/a_{\ell-2})^{\frac{1}{2}}$, $(a_\ell)^{1/\ell}$, and similar functions are frequently found to be smoother than the simple ratios $a_\ell/a_{\ell-1}$. However, in the case of the low-temperature magnetization series it is evident from inspection of (9.37) that none of these functions will be smoothly varying with $1/\ell$. This by no means implies that the radius of convergence is undefined, for tests such as the ratio test are all of the form that the radius of convergence r_c is given by the relation

$$r_c^{-1} = \begin{cases} \lim_{\ell \to \infty} & a_\ell/a_{\ell-1} \\ \lim_{\ell \to \infty} & (a_\ell/a_{\ell-2})^{\frac{1}{2}} \\ \vdots & \\ \lim_{\ell \to \infty} & (a_\ell)^{1/\ell}, \end{cases} \tag{9.40}$$

providing that the limit exists, and it is not difficult to construct power series with well-defined singularities for which the $\ell \to \infty$ limits in eqn (9.40) do not exist.

A third possible explanation for the erratic behaviour of the magnetization series (9.37) is that the singularity closest to the origin in the complex u plane is not the physical singularity $u_c \equiv \exp(-4J/kT_c)$ about which we are seeking information. This possibility is supported by the fact that the signs of the known coefficients in (9.37) are by no means all positive, while if the radius of convergence of a power series is determined by a singularity on the positive real axis, the terms in the series must (eventually) become all positive. We shall see that the singularities of the series (9.37) are distributed throughout the complex u-plane in a fashion roughly as shown in Fig. 9.8(a), and that the radius of the circle of convergence is not determined by the position of the physical singularity (i.e. the nearest singularity that is on the positive

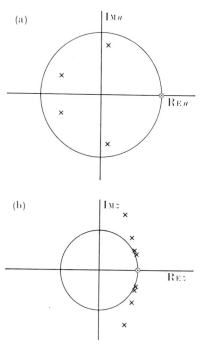

FIG. 9.8. (a) Distribution of singularities in the complex u-plane, where $u \equiv \exp(-4J/kT)$, as estimated by extrapolations from the low-temperature series (9.37) for the zero-field magnetization of the fcc Ising model. The principal feature of this figure is that the radius of convergence is determined by singularities which are *not* on the positive real axis. (b) Distribution of singularities in the complex z plane, where z is related to the original variable u through the transformation of eqn (9.48). Notice that the transformation has shifted the relative location of the physical and non-physical poles so that in the z-plane the radius of convergence is determined by the physical singularity on the positive real axis. Not surprisingly, the coefficients in the z-expansion, unlike those in the u-expansion (9.37), are of uniform sign. After Guttmann (1969a).

real axis) but rather by a complex conjugate pair of 'non-physical' singularities that are considerably closer to the origin. The elucidation of the behaviour shown in Fig. 9.8(a) can be facilitated by means of the method of Padé approximants, which we shall now discuss.

9.7.2. *Analytic continuation beyond the non-physical singularity: the method of Padé approximants*

Although the method of Padé approximants has a colourful history dating back almost a century, it is only within the last decade that this procedure has become a part of the 'bag of tricks' of the working physicist (Baker, 1961, Gammel, Marshall, and Morgan 1963, Baker 1965). The Padé approximant $\mathscr{P}_D^N(x)$ to the function $F(x)$ is simply the ratio of two polynomials, of orders N and D respectively,

$$\mathscr{P}_D^N(x) = \frac{n_0 + n_1 x + \cdots + n_N x^N}{d_0 + d_1 x + \cdots + d_D x^D}. \qquad \blacktriangleright \quad (9.41)$$

Without loss of generality, we can choose $d_0 = 1$, so that there are only $N + D + 1$ undetermined coefficients on the right-hand side of (9.41). Suppose, now, that we have obtained the first $L + 1$ coefficients in the expansion of $F(x)$,

$$F(x) = \sum_{\ell=0}^{\infty} f_\ell x^\ell. \qquad (9.42)$$

Next we equate the series of (9.42) truncated at $\ell = L$ to the function $\mathscr{P}_D^N(x)$ defined in (9.41); it is clear that since there are $N + D + 1$ unknown coefficients in the function $\mathscr{P}_D^N(x)$ and $L + 1$ known coefficients in the truncated series of (9.42), the truncated series can uniquely determine the function $\mathscr{P}_D^N(x)$ only if

$$N + D \leq L. \qquad (9.43)$$

Thus it is possible to form all Padé approximants $\mathscr{P}_D^N(x)$ to the function $F(x)$ for which $N + D \leq L$. The actual process of determining the coefficients n_ℓ and d_ℓ involves solving the set of L simultaneous, homogeneous equations obtained when one combines (9.41) and (9.42) in the form

$$(f_0 + f_1 x + \cdots + f_L x^L) \times (1 + d_1 x + \cdots + d_D x^D)$$
$$= n_0 + n_1 x + \cdots + n_N x^N, \quad (9.44)$$

expands the cross product on the left-hand side, and equates coefficients of this new polynomial term-by-term with the corresponding coefficients of the polynomial on the right-hand side of (9.44). Thus we have essentially 'fitted' the function $F(x)$ of (9.42) to the ratio of two polynomials!

The motivation for this procedure becomes clear when we apply the Padé approximant method to a thermodynamic function, such as the susceptibility, which has a power law singularity,

$$\bar{\chi}_T = \sum_{\ell=0}^{\infty} a_\ell \mathscr{J}^\ell \sim (\mathscr{J}_c - \mathscr{J})^{-\gamma}. \tag{9.45}$$

Note that if we have calculated $L + 2$ coefficients $a_\ell(\ell = 0, 1, \ldots L + 1)$ in the series for $\bar{\chi}_T$, we can obtain, by straightforward differentiation, $L + 1$ coefficients $(\ell = 0, 1, \ldots, L)$ in the series for logarithmic derivative of $\bar{\chi}_T$. Choosing the function $F(x)$ of eqn (9.42) to be given by this logarithmic derivative series, we obtain

$$F(\mathscr{J}) = \sum_{\ell=0}^{\infty} f_\ell \mathscr{J}^\ell \sim \frac{\gamma}{\mathscr{J}_c - \mathscr{J}}. \tag{9.46}$$

We see that this logarithmic derivative series is predicted to have a simple pole at $\mathscr{J} = \mathscr{J}_c$ with a residue equal to γ. Thus if we go on to form the Padé approximants to the *logarithmic derivative* series— that is if we 'fit' the function $F(\mathscr{J})$ of eqn (9.46) to the ratio of two polynomials—then we might expect to find that among the roots of the denominator polynomial is the root $\mathscr{J} = \mathscr{J}_c$ and that the residue corresponding to this root is precisely the critical-point exponent γ. Our expectations are borne out in the sense that for each Padé approximant constructed there is usually—but not always—*one* root that lies somewhat near the value $\mathscr{J}_c$ that we are seeking. Hence we must essentially take an average of these approximate values of $\mathscr{J}_c$ in order to obtain a reliable estimate for the true $\mathscr{J}_c$.

Perhaps the most dramatic success of the Padé approximant method in obtaining critical-point exponents concerns the series (9.37) for the zero-field magnetization of the Ising model, for which the nearest singularity to the origin is evidently not the physical singularity. What the Padé approximants do, essentially, is analytically continue the power series (9.37) beyond the radius of convergence (which is determined by the closer non-physical singularity). In this fashion the Padé approximant method has succeeded in locating the critical temperature as well as obtaining the critical exponent ($\beta \simeq 5/16$) for the series of eqn (9.37).

9.7.3. *Transformation methods that separate the physical and the non-physical singularities*

A second method of obtaining critical-point predictions when the nearest singularity is not the physical singularity—and hence the ratios are not smooth, as in eqn (9.37)—involves changing the expansion variable. We saw earlier that we could consider high-temperature expansions in terms of the variable $\mathcal{J}$ or in terms of the variable $v \equiv \tanh \mathcal{J}$—in short, there is nothing special about the expansion variable, and one variable might well result in a smoother series than another (Stanley 1967c, d). For example, Guttmann, Thompson, and Ninham (1970) have analysed the series (9.37) for the zero-field magnetization of the Ising model by substituting for u the variable

$$z \equiv 1 - \tanh (J/kT) = 1 - v. \tag{9.47}$$

The new variable z is simply related to the old variable $u \equiv \exp (-4J/kT)$ through

$$u = \frac{z^2}{(2 - z)^2}. \tag{9.48}$$

Guttmann *et al.* find that the series (9.37), when expanded in terms of the variable z, has coefficients that are of uniform sign. Since there is a $2:1$ correspondence between each point in the u-plane and each point in the z-plane, it would seem reasonable to conjecture that the effect of the transformation (9.48) has been to shift the relative location of the physical and the unphysical singularities in precisely such a fashion that the radius of convergence is now determined by the physical singularity. This is indeed what has happened, as we see in Fig. 9.8(b), where the locations of the singularities in the z-plane are shown.

The general idea of finding an appropriate transformation method such that the singularities near the physical singularity are moved further away has in fact led to substantial progress in analysing series for which other singularities interfere with the physical singularity (Guttmann 1969a, Betts, Elliott, and Ditzian 1971, Lee and Stanley 1971, Stanley, Hankey, and Lee 1971; see also Danielian and Stevens 1957).

9.8. Conclusions

In summary, then, we have seen that series expansion methods afford us the opportunity of obtaining fairly reliable information concerning a thermodynamic function. High-temperature expansions permit us to learn about functions for $T \gtrsim T_c$, while low-temperature expansions give information for $T \lesssim T_c$. Although low-temperature expansions are practical only for the Ising model, they have been sufficiently successful for this case that it is possible to plot essentially all of the thermodynamic functions of the Ising model with an accuracy

that is believed to be exceedingly high. For example, we show in Fig. 9.9
the zero-field magnetization $M(T, H = 0)$ for the Ising model as obtained by exact calculation for two-dimensional lattices and as obtained
by extrapolation from low-temperature series for three-dimensional
lattices. It is perhaps appropriate to conclude our discussion with this
figure in mind, as it demonstrates the several points that we have tried
to develop in this chapter.

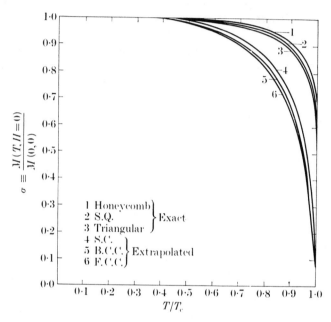

FIG. 9.9. Dependence upon T/T_c of the normalized zero-field magnetization. The curves
labelled 1–3 are obtained from exact solution of the Ising model, while the curves
labelled 4–6 are obtained from extrapolations based upon low-temperature series
expansions of the form (9.37). After Burley (1960).

(i) Expansion methods enable us to obtain numerical predictions
of model systems for three-dimensional lattices, while exact solutions
(with the exception of the spherical model) are not known.

(ii) Expansion methods are not rigorous, but they produce predictions that are thought to be, in many cases, as accurate as the experimental data with which one might wish to compare them and hence
they afford the possibility of gauging the validity of a microscopic
model by comparison with experiment.

(iii) Expansion methods provide us with a direct physical picture
of how one spin is correlated with another through a sequence of

distinct interaction paths. In particular, we achieve some insight into the apparent fact that cooperative phenomena in general (and critical phenomena in particular) depend very sensitively on lattice structure and especially upon the lattice dimensionality. For example, we see in Fig. 9.9 that the exact curves for the three two-dimensional lattices differ only slightly from one another, while they differ a good deal from the curves that are obtained by extrapolation from low-temperature expansions for three-dimensional systems. In particular, the magnetization critical-point exponents are identical for all three two-dimensional lattices ($\beta = \frac{1}{8}$) and the extrapolations strongly suggest that β is the same ($\beta \simeq \frac{5}{16} = 0.3125$) for the three three-dimensional lattices shown.

Suggested further reading
Rushbrooke and Wood (1958).
Domb (1960).
Gammel, Marshall, and Morgan (1963).
Fisher (1965).
Baker (1965).
Fisher (1967).
Domb (1970).
Stanley, Hankey, and Lee (1971).

PART V

PHENOMENOLOGICAL THEORIES OF PHASE TRANSITIONS

10

LANDAU'S CLASSIC THEORY OF EXPONENTS

WE have seen in Chapters 5–7 that the equations of state predicted by the classical theories fail to predict experimentally-observed behaviour near the critical point. In this and in the following chapter we describe some attempts to provide a more realistic equation of state. Our efforts will take the direction of making assumptions concerning the form of a thermodynamic potential near the critical point. Of course once we have guessed a thermodynamic potential—any thermodynamic potential—then we can obtain all of the thermodynamic properties using the methods described in § 2.2.

10.1. Expansions about the critical point

One of the simplest and most elegant speculations concerning the possible general form of a thermodynamic potential near the critical point is due to Landau (Landau 1937a, b, c, d; 1965). Since particularly readable accounts of Landau's original argument have appeared in other books (Landau and Lifshitz (1969, Chapter 14) and Landau and Lifshitz (1960, Chapter 5)), we shall concentrate in this chapter on the specific predictions of the Landau theory for critical-point exponents. We shall find that the values of the critical-point exponents predicted by the Landau theory are identical to those predicted by the classical theories presented in Chapters 5–7, and hence they disagree drastically with the results of most experiments in the immediate neighbourhood of the critical point. The source of this discrepancy can be traced to an unrealistic assumption which rests at the foundation of the Landau approach, namely, that one can expand the thermodynamic potential in a power series about the critical point. Since in Chapter 9 we expanded thermodynamic potentials in power series about $\mathscr{J} \equiv J/kT = 0$ (with no proof that the resulting power series were convergent) it might seem consistent to do likewise for the critical point. However, we know for certain that an expansion about the critical point cannot be convergent. For example, suppose we wish to expand the Gibbs potential of a magnetic system, $G(T, H)$ about the critical point, $T = T_c, H = 0$. The coefficients in this expansion will involve the partial derivatives of $G(T, H)$ with respect to T and H, evaluated at $T = T_c$ and $H = 0$. However from eqn (2.49) we see that

$$- (\partial^2 G/\partial H^2)_{T=T_c, H=0} = \chi_T(T = T_c, H = 0). \qquad (10.1)$$

Since we expect the isothermal susceptibility of a magnetic system to be infinite at the critical point, such an expansion of the Gibbs potential cannot be convergent. If we try to expand instead the Helmholtz potential, $A(T, M)$, we find that although the derivative

$$(\partial^2 A/\partial M^2)_{T=T_c, M=0} = \chi_T^{-1}(T = T_c, M = 0) \qquad (10.2)$$

is finite (it is in fact zero), the derivative

$$- T(\partial^2 A/\partial T^2)_{T=T_c, M=0} = C_M(T = T_c, M = 0) \qquad (10.3)$$

is probably infinite. Thus it would appear that attempts to generate a convergent power series expansion about the critical point are fruitless.

It would be, however, quite unfair to Landau to ascribe to him the notion of a convergent expansion about the critical point. In fact he rather carefully qualifies the matter, pointing out that there will be singularities in the higher-order coefficients of the power-series expansion. He supposes that the singular coefficients are of higher order than that of the terms of the expansion used in the calculation, and proceeds on this basis to examine the lower-order coefficients in order to obtain certain predictions regarding the critical region. It is in this spirit that we shall proceed with the Landau argument.

10.2. Assumptions of the Landau theory

We begin, then, by assuming that we can expand the Helmholtz potential about $T = T_c$ and $M = 0$ in a standard Taylor series form for functions of two variables,

$$A(T, M) = \sum_{j=0}^{\infty} L_j(T) M^j = L_0(T) + L_2(T) M^2 + L_4(T) M^4 + \cdots,$$

$$\blacktriangleright(10.4)$$

where the coefficients $L_j(T)$ themselves can be expanded about $T = T_c$,

$$L_j(T) = \sum_{k=0}^{\infty} \ell_{jk} (T - T_c)^k = \ell_{j0} + \ell_{j1} (T - T_c) + \cdots. \quad (10.5)$$

Notice that we have set $L_j(T) = 0$ for j odd, because $A(T, M)$ is an even function of M. From eqns (2.45d) and (10.4) it follows that the equation of state in the Landau theory is

$$H = H(M, T) = \sum_{j=1}^{\infty} j L_j(T) M^{j-1} = 2 L_2(T) M + 4 L_4(T) M^3 + \cdots.$$
$$\blacktriangleright(10.6)$$

Note that on differentiating the Helmholtz potential (10.4) once more with respect to M, we find that the inverse isothermal susceptibility $\chi_T^{-1} = (\partial^2 A/\partial M^2)_T = (\partial H/\partial M)_T$ is given by

$$\chi_T^{-1} = \sum_{j=2}^{\infty} j(j - 1) L_j(T) M^{j-2} = 2 L_2(T) + 12 L_4(T) M^2 + \cdots. \quad (10.7)$$

Next we place some restrictions on the nature of the coefficients $L_j(T)$, based upon what we know about the magnetic system. For example, in the limit as $T \to T_c^+$, we expect that the zero-field susceptibility approaches infinity; from (10.5) and (10.7) we have

$$\chi_T^{-1}(T, 0) = 2 L_2(T) = 2\{\ell_{20} + \ell_{21} (T - T_c) + \ell_{22}(T - T_c)^2 + \cdots\},$$
$$(10.8)$$

so that we choose $\ell_{20} = 0$.

Another restriction is that the Helmholtz potential be a convex function of the magnetization at fixed temperature (cf. § 2.10). For $T > T_c$, $A(T, M)$ will be a convex function of M if we choose all the coefficients ℓ_{jk} to be non-negative, $\ell_{jk} \geq 0$ for all j, k. However this choice would violate the requirement that $A(T, M)$ be a concave function of temperature, and for $T < T_c$ it is not even sufficient to guarantee convexity with respect to M. For example, suppose that $\ell_{21} \geq 0$ in eqn (10.8); then as T approaches T_c from below, the isothermal susceptibility will become negative. One simple modification of (10.4) and (10.5) in order to restore the convexity of $A(T, M)$ with respect to M below T_c is to replace $(T - T_c)$ in (10.5) by $|T - T_c|$. This modification would, however, lead to other complications. For example, the presence of the quantity $|T - T_c|$ in (10.5) would mean that for any fixed value of M (not just $M = 0$), there will occur a mathematical singularity in $A(T, M)$ at $T = T_c$—whereas we want $A(T, M)$ to be an analytic function of T for all $M \neq 0$. Some authors restore convexity below T_c instead by means of the 'double tangent construction' shown in Fig. 10.1, which is in some respects similar to the Maxwell construction used in Chapter 5 to restore convexity to $A(T, V)$ in the van der Waals theory.

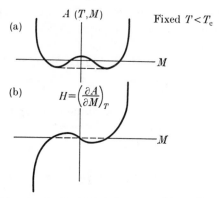

FIG. 10.1. (a) The solid curve is the Helmholtz potential predicted by eqn (10.4); the convexity catastrophe occurring for small M is repaired by the double tangent construction (dashed curve) (cf. Figs. 2.9(b) and 5.2(b)). (b) H–M isotherms obtained from part (a) (cf. Figs 2.9(d) and (5.2(a)). Note that the choice of a horizontal double tangent in (a) corresponds to the property of a zero value of H in (b).

10.3. Critical-point predictions of the Landau theory

10.3.1. *Magnetization exponent* β

From the equation of state (10.6) it follows that for $H = 0$ and M small,

$$0 = \{\ell_{21}(T - T_c) + \cdots\} + 2M^2\{\ell_{40} + \ell_{41}(T - T_c) + \cdots\} + \ldots, \quad (10.9)$$

so that

$$M = (\ell_{21}/2\ell_{40})^{\frac{1}{2}} (T_c - T)^{\frac{1}{2}}. \quad (10.10)$$

Thus $\beta = \frac{1}{2}$ in the Landau theory.

10.3.2. *Susceptibility exponents* γ *and* γ′

From (10.7) we have

$$\chi_T^{-1}(T, M) = 2\{\ell_{21}(T - T_c) + \cdots\} + 12 M^2\{\ell_{40} + \ell_{41}(T - T_c) + \cdots\}. \quad (10.11)$$

For $T > T_c$, $M = 0$ when $H = 0$ and the zero-field susceptibility is

$$\chi_T^{-1}(T, 0) = 2\ell_{21}(T - T_c) + \cdots, \quad (10.12)$$

so that $\gamma = 1$. For $T < T_c$, the zero-field magnetization is of course non-zero; on substituting M^2 from eqn (10.10) into (10.11), we obtain

$$\chi_T^{-1}(T, 0) = 4\ell_{21}(T_c - T) + \cdots. \tag{10.13}$$

Thus $\gamma' = 1$ also, and $\gamma = \gamma'$, yet the inverse isothermal susceptibility rises twice as fast from zero for $T < T_c$ as for $T > T_c$ (see Fig. 10.2) Thus the Landau theory not only predicts the same exponents as for the van der Waals and mean field theories, but it also predicts the same relation $\mathscr{C} = 2\mathscr{C}'$ among the coefficients defined in eqn (3.8).

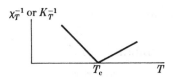

FIG. 10.2. The inverse zero-field isothermal susceptibility for a magnet—or the inverse isothermal compressibility for a fluid—as predicted by the Landau theory (cf. Fig. 5.4). Notice that $\mathscr{C} = 2\mathscr{C}'$, where the coefficients $\mathscr{C}$ and $\mathscr{C}'$ are defined in eqn (3.8).

10.3.3. Critical isotherm exponent δ

To obtain the curvature of the $M - H$ isotherm for $T = T_c$, we set $T = T_c$ in eqn (10.6). Thus

$$H(T_c, M) = 4\ell_{40} M^3 + \cdots, \tag{10.14}$$

and $\delta = 3$.

10.3.4 Specific heat exponents α and α'

The exponents α and α' describe the behaviour of $C_H(T, H = 0) = -\Gamma(\partial^2 G/\partial T^2)_H$, whereas our Landau expansion (10.4) is for $A(T, M)$. Perhaps the best approach is to first calculate $C_M = -(T(\partial^2 A/\partial T^2)_M$ and then to calculate C_H from the identity (2.54).

For $T > T_c$, $H = 0$ implies $M = 0$, and we have from (10.4) and (2.47) that

$$C_H = C_M = -T\{2\ell_{02} + 6\ell_{03}(T - T_c) + \mathcal{O}([T - T_c]^2)\}. \tag{10.15}$$

For $T < T_c$, on using (10.10) we obtain

$$C_M = -T[2\ell_{02} + \{6\ell_{03} - (\ell_{21}\ell_{22}/\ell_{40})\}(T - T_c) + \cdots]. \tag{10.16}$$

To calculate C_H from (2.54), we first use (10.6) to find $\alpha_M \equiv (\partial H/\partial T)_M$ and then we use (10.13) to find χ_T. Thus we obtain

$$C_H - C_M = T(\ell_{21}^2/2\ell_{40})\{1 + \mathcal{O}(T - T_c)\}. \tag{10.17}$$

We see from (10.16) and (10.17) that C_H undergoes a simple jump discontinuity at $T = T_c$ of magnitude

$$\Delta C_H = \{(\ell_{21}^2/2\ell_{40}) - 2\ell_{20}\}T_c, \qquad (10.18)$$

and of course $\alpha = \alpha' = 0$.

To check this result, observe that for the mean field theory we can calculate the Landau coefficients ℓ_{jk} directly from (6.29), and we find, using (10.18), that

$$\Delta C_H = \frac{3}{2} Nk, \qquad (10.19)$$

which agrees with the result of (6.31).

10.4. Critique of the Landau theory

The Landau approach fails to provide a general theory of critical phenomena in the following two respects.

(i) The existence of the expansion (10.4) assumes that all of the partial derivatives of $A(T, M)$ with respect to its arguments exist and are finite. In particular, the expansion (10.4) is not valid for a system whose specific heat diverges to infinity. For example, the Helmholtz potential in zero magnetic field (which is identical to the Gibbs potential) for the $d = 2$ Ising model can be written for $T > T_c$ as (Fisher 1967)

$$A(T, 0) = A(T_c, 0) + a(T - T_c) + b(T - T_c)^2 \ln (T - T_c) + \cdots. \qquad (10.20)$$

Hence the specific heat, given by the second derivative with respect to temperature, has a logarithmic divergence of the form

$$C_M = - 2bT \ln (T - T_c) - 3bT + \cdots. \qquad (10.21)$$

Of course it is quite possible that many real fluid and magnetic systems have specific heats that are finite at T_c. There is some recent experimental evidence to suggest this and, in any case, one can certainly never prove experimentally that any quantity is truly infinite. Measurements on the systems that may have $C(T = T_c)$ finite strongly suggest, however, that $C(T)$ has a cusp-like singularity at $T = T_c$. Thus the Landau expansion would still fail, though at one higher order. It is perhaps worth remarking that the Landau theory provides a much better description of systems such as ferroelectrics, superconductors, and liquid crystals in which the specific heat may not exhibit a divergence at the transition temperature.

(ii) The Landau theory predicts values of the critical-point exponents that disagree with most experiments. In a sense, however, the particular values obtained correspond to the assumptions we made in § 10.2. The question naturally arises whether or not we might make different assumptions which will lead to the experimentally-observed exponents, while still being consistent with the constraints of thermodynamic stability and the like.

For example, we notice that it was necessary to require the coefficient ℓ_{40} to be non-zero in deriving the results $\beta = \frac{1}{2}$ and $\delta = 3$ from eqns (10.10) and (10.14). If, on the other hand, $\ell_{40} = 0$ but also $\ell_{60} > 0$, then the arguments of § 10.3 imply that $\beta = \frac{1}{4}$ and $\delta = 5$, with $\gamma' = 1$ and $\alpha' = 0$ as before. Since experimentally β is generally smaller than $\frac{1}{2}$ and δ is generally larger than 3, this might appear to be an improvement on the original Landau approach. However, the astute reader will have noted that the values $\alpha' = 0$, $\beta = \frac{1}{4}$, and $\gamma' = 1$ fail to satisfy the Rushbrooke inequality, and hence are inadmissible on thermodynamic grounds.

Suppose that, since the prediction $\gamma' = 1$ followed from our assumption (10.5), we allow for an arbitrary value of γ' by replacing (10.5) with

$$L_2(T) \equiv \ell_{21} \left| T - T_c \right|^{\gamma'}, \tag{10.22}$$

with $\ell_{21} > 0$, $\ell_{j0} \geq 0$ for $j \geq 4$, and where we have used the absolute value sign in order to avoid the occurrence of complex numbers. Although eqn (10.22) leads to an unwanted singularity for any non-zero value of M, it is nevertheless interesting to observe that we would obtain

$$\chi_T^{-1} = 2 \ell_{21} \left| T - T_c \right|^{\gamma'} \tag{10.23}$$

instead of (10.11); similarly (10.10) becomes

$$M = (\ell_{21}/2 \, \ell_{40})^{\frac{1}{2}} \left| T - T_c \right|^{\gamma'/2} \tag{10.24}$$

[i.e. $\beta = \frac{1}{2} \gamma'$] and $\delta = 3$ as before.

In general, if $L_j(T) = 0$ for $2 < j \leq I$ (where I is some integer), then $\delta = I + 1$ and

$$\beta = \gamma'/I = \gamma'/(\delta - 1) \tag{10.25}$$

so that

$$\gamma' = \beta(\delta - 1). \qquad \blacktriangleright \tag{10.26}$$

The relation (10.26) was first conjectured by Widom (1964), and we shall see in Chapter 11 that it arises from certain rather more general assumptions which, in particular, do not require that δ be an odd integer.

It is interesting, however, that for many models δ appears to be an odd integer and, moreover, (10.25) appears to be obeyed. For example, the exponents predicted for the spherical model (or non-interacting Bose gas) are $\delta = 5$, $\beta = \frac{1}{2}$, and $\gamma = \frac{1}{2}(5 - 1) = 2$ (Gunton and Buckingham 1968). Similarly, for the two-dimensional Ising model δ is probably 15, $\beta = \frac{1}{8}$, and $\gamma = (15 - 1)/8 = \frac{7}{4}$. For the three-dimensional Ising model and the Heisenberg model, no exact answers are known, but the most recent numerical estimates are consistent with (10.25) and $\delta = 5$.

Suggested further reading
Landau and Lifshitz (1960).
Landau (1965).
Fisher (1967).
Kadanoff et al. (1967).
Kittel (1969).
Kuramoto (1969).
Landau and Lifshitz (1969).

11

SCALING LAW HYPOTHESIS FOR
THERMODYNAMIC FUNCTIONS

THUS far we have seen that much experimental and theoretical work
on critical-point phenomena appears to support the exponent inequalities discussed in Chapter 4, and, moreover, these inequalities are in
many cases satisfied as equalities. We would clearly like some explanation of this fact—if indeed the inequalities *are* equalities. As yet no
one has proved rigorously that any of the inequalities are equalities,
but an alternate approach has been rather intriguingly successful. This
approach, which has come to be called the *static scaling law* or *homogeneous function approach*, involves making a simple assumption
concerning the basic form of a thermodynamic potential. Although
there exist heuristic arguments of various sorts that serve to make somewhat plausible the basic scaling hypothesis no rigorous justification
has as yet been given. Hence the results of this chapter, like those of the
previous chapter on the Landau theory, rest on an unproved assumption. However, the predictions of the static scaling hypothesis are
somewhat less specific than the predictions of the Landau theory. For
example, unlike the Landau theory, which predicts that all exponents
assume their mean field values ($\beta = \frac{1}{2}$, $\delta = 3$, etc.), the scaling approach
does not lead to specific numerical values for any of the critical-point
exponents. But, because the scaling hypothesis does give rise to functional relations among the critical-point exponents (the exponent
equalities referred to above), the number of independent critical-point
exponents is restricted in number. In addition to predicting relations
among the critical-point exponents, the scaling hypothesis makes
specific predictions concerning the form of the equation of state. Both
these sets of predictions from static scaling appear to be supported by
most experimental work.

The first clear and coherent mathematical exposition of the scaling
hypothesis is due to Widom (1965*a*, *b*) and Domb and Hunter (1965);
references to other work appear at the end of this chapter. We shall
describe the heuristic arguments due to Kadanoff (1966) in Chapter 12,

where we extend the scaling concept from thermodynamic functions to the correlation functions.

11.1. Homogeneous functions of one or more variables

11.1.1. *Homogeneous functions of one variable*

This section is intended as a short review of homogeneous functions and may be omitted by those readers who are already familiar with scaling concepts.

A function $f(r)$ is by definition homogeneous if for all values of the parameter λ,

$$f(\lambda r) = g(\lambda)f(r), \qquad \blacktriangleright \quad (11.1)$$

where $g(\lambda)$ is an unspecified function. A simple example of a homogeneous function is the parabola,

$$f(r) = Br^2, \qquad (11.2)$$

for which $f(\lambda r) = B(\lambda r)^2 = \lambda^2 Br^2 = \lambda^2 f(r)$, whence $g(\lambda) = \lambda^2$.

A homogeneous function $f(r)$ has the property that if we know its value at one point, $r = r_0$, and we know the functional form of $g(\lambda)$, then we know the function everywhere. This follows because every value of r can be written in the form λr_0, and

$$f(\lambda r_0) = g(\lambda)f(r_0). \qquad (11.3)$$

Equation (11.3) says that the value of the function $f(r)$ at any point is related to the value of $f(r)$ at a reference point $r = r_0$ by a simple change of scale. Of course, this change of scale is, in general, not linear (unless $g(\lambda)$ is linear in λ).

The scaling function $g(\lambda)$ of eqn (11.1) is not arbitrary—in fact it must be of the form

$$g(\lambda) = \lambda^p, \qquad (11.4)$$

where the parameter p is generally called the *degree of homogeneity*. To prove (11.4), we need to examine the functional requirements on $g(\lambda)$ implied by the relation (11.1). Suppose we make two changes of scale, first by μ and then by λ. Equation (11.1) implies

$$f\{\lambda(\mu r)\} = g(\lambda)f(\mu r) = g(\lambda)g(\mu)f(r). \qquad (11.5)$$

An identical result could, however, be obtained by a single change of scale,

$$f\{(\lambda\mu)r\} = g(\lambda\mu)f(r). \qquad (11.6)$$

On comparing (11.5) and (11.6), we obtain the requirement on the function $g(\lambda)$,

$$g(\lambda\mu) = g(\lambda)g(\mu). \tag{11.7}$$

Now any continuous function that satisfies the functional equation (11.7) must be either identically zero or else a simple power of the form (11.4). The complete proof of this statement, though straightforward, is rather lengthy (Aczèl 1966, 1969). However, if we also assume that $g(\lambda)$ is differentiable, the proof is quite short. Differentiating both sides of eqn (11.7) with respect to μ, we obtain

$$\frac{\partial}{\partial\mu} g(\lambda\mu) = \lambda\, g'(\lambda\mu) = g(\lambda)g'(\mu). \tag{11.8}$$

Now let $\mu = 1$, and set $g'(\mu = 1) \equiv p$. Then eqn (11.8) may be written as

$$\frac{g'(\lambda)}{g(\lambda)} = \frac{\mathrm{d}}{\mathrm{d}\lambda} \ln g(\lambda) = \frac{p}{\lambda}, \tag{11.9}$$

whence

$$\ln g(\lambda) = p \ln \lambda + c \tag{11.10}$$

or

$$g(\lambda) = \mathrm{e}^c \lambda^p. \tag{11.11}$$

Now from (11.11) we have $g'(\lambda) = p\, \mathrm{e}^c \lambda^{p-1}$ and the definition $p \equiv g'(1)$ implies that the integration constant c has the value zero. Therefore (11.11) reduces to (11.4) and the proof is complete.

11.1.2. *Homogeneous functions of an arbitrary number of variables*

The concept of a homogeneous function is by no means restricted to functions of a single variable. For a function of n variables, $f(x_1, x_2, \ldots, x_n)$, eqn (11.1) becomes

$$f(\lambda x_1, \lambda x_2, \ldots, \lambda x_n) = g(\lambda) f(x_1, x_2, \ldots, x_n). \tag{11.12}$$

If we represent the n-tuple $(x_1, x_2, \ldots x_n)$ by the n-dimensional vector $\mathbf{r}$, then (11.12) may be written as simply

$$f(\lambda\mathbf{r}) = g(\lambda) f(\mathbf{r}) \tag{11.13}$$

and, by arguments similar to those presented in the case of a single variable, we can show that $g(\lambda)$ must be a simple power of λ, as in (11.4).

Since our applications of homogeneous functions in this chapter deal with functions of two variables, it is worth considering such an example in detail. In Fig. 11.1 we show the surface

$$f(x_1, x_2) = x_1^2 + x_2^2 \tag{11.14}$$

which, by eqns (11.12) and (11.4), is evidently homogeneous of degree two (i.e. $g(\lambda) = \lambda^2$). A simple geometrical interpretation of homogeneity in two dimensions will prove useful in our subsequent discussion. If we re-label the x_1 and x_2 axes such that $x_1 \to \lambda x_1$ and $x_2 \to \lambda x_2$, then the surface of the function $f(\lambda x_1, \lambda x_2)$ will be obtained if we simply

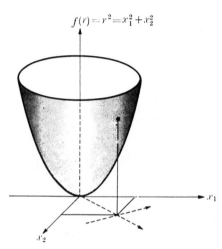

FIG. 11.1. The homogeneous function of two variables considered in eqn (11.14), $f(x_1, x_2) = x_1^2 + x_2^2$, represented in a three-dimensional graph. After Reitz and Milford (1960).

stretch the vertical coordinate by an amount $g(\lambda)$. This is shown in Fig. 11.2. Notice, incidentally, that knowledge of a homogeneous function $f(x_1, x_2)$ along any path encircling the origin ($x_1 = x_2 = 0$) suffices to determine the function $f(x_1, x_2)$ everywhere.

11.1.3. Generalized homogeneous functions

We shall see in the next section, §11.2, that the static scaling hypothesis states not that thermodynamic potentials are homogeneous of the form

$$f(\lambda x, \lambda y) = \lambda^p f(x, y), \tag{11.15}$$

but rather that they are of the somewhat more general form

$$f(\lambda^a x, \lambda^b y) = \lambda f(x, y), \qquad \blacktriangleright \tag{11.16}$$

where a and b are arbitrary numbers. Functions $f(x, y)$ that satisfy equations of the form of eqn (11.16) are sometimes called *generalized homogeneous functions*.

It is worth noting that (11.16) cannot be further generalized to an equation of the form

$$f(\lambda^a x, \lambda^b y) = \lambda^p f(x, y), \tag{11.17}$$

because without loss of generality we can choose $p = 1$ in eqn (11.17); i.e., a function $f(x, y)$ that satisfies (11.17) also satisfies the equation

$$f(\lambda^{a/p} x, \lambda^{b/p} y) = \lambda f(x, y), \tag{11.18}$$

and the converse statement is also valid. Since (11.18) is of the form of eqn (11.16), we conclude that eqn (11.17) is no more general than eqn (11.16).

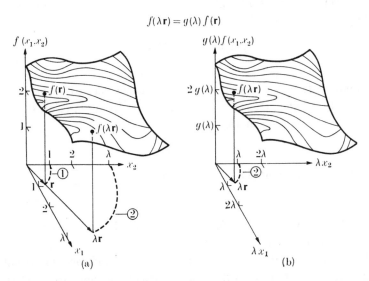

FIG. 11.2. Undergoing the scale transformation (11.13) is equivalent to relabelling the axes as shown in (b). Notice that knowledge of the function $f(\mathbf{r})$ on path 1 suffices to determine values of $f(\mathbf{r})$ on path 2. Since the parameter λ is an arbitrary number, knowledge of the function $f(\mathbf{r})$ along any path encircling the origin $\mathbf{r} = 0$ suffices to determine the function $f(\mathbf{r})$ everywhere. By courtesy of A. diSessa.

Other equivalent forms of (11.16) that frequently appear in the literature on scaling laws are

$$f(\lambda x, \lambda^b y) = \lambda^p f(x, y) \tag{11.19}$$

and

$$f(\lambda^a x, \lambda y) = \lambda^p f(x, y). \tag{11.20}$$

The main point is that there are at most two undetermined parameters for a function $f(x, y)$ of two variables.

11.1.4. *The mathematical form of a homogeneous function*

We have seen in the foregoing discussion that the assumption that a function is homogeneous places rather a severe restriction upon the nature of the function. We shall now display one particular feature of a homogeneous function that will be useful in our later discussion. For the sake of clarity we shall carry out our argument for the case of a homogeneous function of two variables,

$$f(\lambda x, \lambda y) = \lambda^p f(x, y). \qquad \blacktriangleright \quad (11.21)$$

This argument is, however, easily generalized to a homogeneous function of n variables, or to a generalized homogeneous function. Since eqn (11.21) is valid for all values of the parameter λ, it must certainly hold for the particular choice $\lambda = 1/y$,

$$f(x/y, 1) = y^{-p} f(x, y). \qquad (11.22)$$

Now the function $f(x/y, 1)$ appearing on the left-hand side of eqn (11.22) is formally a function of two variables, but the second variable is fixed at the value unity. Hence we can denote it by a function of a single variable, defining the function

$$F(z) \equiv f(z, 1). \qquad (11.23)$$

Combining (11.23) and (11.22), we obtain

$$f(x, y) = y^p F(x/y). \qquad \blacktriangleright \quad (11.24)$$

Thus (11.24) says that if a function $f(x, y)$ is homogeneous, it may be written as y^p times a function of x/y.

The converse statement is also true: if a function $f(x, y)$ satisfies (11.24), then it satisfies (11.21), that is, it is homogeneous. The proof of the converse is straightforward. From (11.24) we have

$$\begin{aligned} f(\lambda x, \lambda y) &= (\lambda y)^p F(\lambda x/\lambda y) \\ &= \lambda^p y^p F(x/y), \end{aligned} \qquad (11.25)$$

which reduces, on applying (11.24), to

$$f(\lambda x, \lambda y) = \lambda^p f(x, y), \qquad (11.26)$$

which is precisely eqn (11.21). This completes the proof that all functions $f(x, y)$ that satisfy an equation of the form of (11.21) can be written in the form of (11.24), where $F(z)$ is defined in (11.23), while all functions that satisfy an equation of the form of (11.24)—with $F(x/y)$ an arbitrary function—also satisfy (11.21).

Another form for the homogeneous function $f(x, y)$ of (11.21) is obtained by setting $\lambda = 1/x$, whence

$$f(1, y/x) = x^{-p} f(x, y). \qquad (11.27)$$

On defining

$$\mathscr{F}(z) \equiv f(1, z) \qquad (11.28)$$

we obtain

$$f(x, y) = x^p \mathscr{F}(y/x). \qquad (11.29)$$

As before, the converse statement is again true—functions $f(x, y)$ satisfying (11.29) for any function $\mathscr{F}(z)$ are homogeneous.

In summary, then, we have proved that functions of the form (11.24) —or, equivalently, of the form (11.29)—and only these functions, are homogeneous functions.

11.2. Static scaling hypothesis

In this section we present the static scaling hypothesis in the form of an *ad hoc* assumption. In Appendix C we shall discuss a geometrical interpretation of this assumption.

We shall state the static scaling hypothesis for the Gibbs potential $G(T, H)$ of a magnetic system. We shall write $G(T, H) \to G(\epsilon, H)$, where $\epsilon \equiv (T - T_c)/T_c$ is the reduced temperature, and we shall assume that any non-singular terms in the $G(\epsilon, H)$ function have been subtracted off. *The static scaling hypothesis asserts that the Gibbs potential $G(\epsilon, H)$ is a generalized homogeneous function.* Thus from the general definition of eqn (11.16), this statement is equivalent to the requirement that there exist two parameters, which we shall call a_ϵ and a_H, such that

$$G(\lambda^{a_\epsilon} \epsilon, \lambda^{a_H} H) = \lambda G(\epsilon, H) \qquad \blacktriangleright (11.30)$$

for any value of the number λ. It is important to stress that the scaling hypothesis does not specify the parameters a_ϵ and a_H. We shall see below that all of the critical-point exponents can be simply expressed in terms of a_ϵ and a_H (so that if we could specify a_ϵ and a_H then we would immediately determine all of the critical-point exponents). The fact that we cannot specify a_ϵ and a_H corresponds to the fact that the homogeneous function or scaling hypothesis does not determine the values of the critical-point exponents. On the other hand, the fact that all the critical-point exponents can be expressed in terms of only two 'scaling parameters' a_ϵ and a_H means that if two critical-point exponents are specified, all others can be determined.

We shall see in Appendix C that the assumption that $G(\epsilon, H)$ is a generalized homogeneous function implies that the other three thermodynamic potentials $A(\epsilon, M)$, $U(S, M)$, and $E(S, H)$ are generalized homogeneous functions. Therefore we could as well have made the static scaling hypothesis for, say, the Helmholtz potential, as is often done in the research literature. We chose to use the Gibbs potential here because the critical-point exponents we seek are most easily obtained from the Gibbs potential.

11.3. Predicted relations among the critical-point exponents

In this section we apply the static scaling hypothesis (11.30) to relate the various critical-point exponents to the two scaling parameters a_ϵ and a_H. Our treatment, which follows Hankey and Stanley (1971), begins by differentiating both sides of (11.30) with respect to H

$$\lambda^{a_H} \, \partial\{G(\lambda^a \, \epsilon, \lambda^{a_H} H)\}/\partial(\lambda^{a_H} H) = \lambda \, \partial \, G(\epsilon, H)/\partial H. \qquad (11.31)$$

Since the field derivative of the Gibbs potential is the negative of the magnetization (cf. eqn (2.45c)), (11.31) is equivalent to

$$\lambda^{a_H} M(\lambda^{a_\epsilon} \epsilon, \lambda^{a_H} H) = \lambda M(\epsilon, H). \qquad (11.32)$$

There are two critical-point exponents associated with the behaviour of the magnetization near the critical point. The exponent β refers to behaviour when $H = 0$ and $\epsilon \to 0$, while the exponent $1/\delta$ refers to behaviour when $\epsilon = 0$ and $H \to 0$. Consider first the case $H = 0$, whence eqn (11.32) becomes

$$M(\epsilon, 0) = \lambda^{a_H - 1} M(\lambda^{a_\epsilon} \epsilon, 0). \qquad (11.33)$$

The argument now proceeds along lines similar to those developed in § 11.1.4. Since eqn (11.33) is valid for all values of number λ, it must certainly hold for the particular choice $\lambda = (-1/\epsilon)^{1/a_\epsilon}$ whence

$$M(\epsilon, 0) = (-\epsilon)^{(1 - a_H)/a_\epsilon} M(-1, 0). \qquad (11.34)$$

But when $\epsilon \to 0^-$, we have from (3.7) that $M(\epsilon, 0) \sim (-\epsilon)^\beta$, so that evidently

$$\beta = \frac{1 - a_H}{a_\epsilon}. \qquad (11.35)$$

Equation (11.35) expresses the critical-point exponent β in terms of the unknown scaling parameters a_ϵ and a_H. The exponent δ can also be expressed in terms of the scaling parameters by setting $\epsilon = 0$ in (11.32) and letting $H \rightarrow 0$,

$$M(0, H) = \lambda^{a_H - 1} M(0, \lambda^{a_H} H). \tag{11.36}$$

If we set $\lambda = H^{-1/a_H}$, eqn (11.36) becomes

$$M(0, H) = H^{(1 - a_H)/a_H} M(0, 1). \tag{11.37}$$

When $H \rightarrow 0$, we have from eqn (3.11) that $M(0, H) \sim H^{1/\delta}$, whence

$$\delta = \frac{a_H}{1 - a_H}. \tag{11.38}$$

Equations (11.35) and (11.38) can be solved simultaneously for the scaling parameters a_ϵ and a_H, with the results

$$a_\epsilon = \frac{1}{\beta} \frac{1}{\delta + 1}, \qquad \blacktriangleright \quad (11.39)$$

and

$$a_H = \delta \frac{1}{\delta + 1}. \qquad \blacktriangleright \quad (11.40)$$

We can obtain additional exponents by forming additional partial derivatives of the Gibbs potential. For example, on differentiating twice with respect to H, we obtain

$$\lambda^{2a_H} \chi_T(\lambda^{a_\epsilon} \epsilon, \lambda^{a_H} H) = \lambda \chi_T(\epsilon, H), \tag{11.41}$$

where we have used eqn (2.49) to relate the isothermal susceptibility χ_T to the Gibbs potential. If we consider $H = 0$ and choose $\lambda = (-\epsilon)^{-1/a_\epsilon}$, eqn (11.41) becomes

$$\chi_T(\epsilon, 0) = (-\epsilon)^{-(2a_H - 1)/a_\epsilon} \chi_T(-1, 0). \tag{11.42}$$

If $\epsilon \rightarrow 0^-$, $\chi_T(\epsilon, 0) \sim (-\epsilon)^{-\gamma'}$ from eqn (3.9), and we obtain

$$\gamma' = \frac{2a_H - 1}{a_\epsilon}. \tag{11.43}$$

Since there are only two unknown scaling parameters, we would expect that the value of the exponent γ' is not independent of the values of β and δ. Indeed, if we substitute (11.39) and (11.40) into (11.43), we obtain the simple relation

$$\gamma' = \beta(\delta - 1), \tag{11.44}$$

which is recognized as the *Widom equality* of eqn (10.26). Moreover, we recall from line 5 of Table 4.1 that the relation $\gamma' \geq \beta(\delta - 1)$ is rigorously an inequality while we see from (11.44) that the scaling hypothesis predicts that the inequality holds as an equality. This is a hallmark of the predictions of the static scaling hypothesis.

A second hallmark of static scaling is the equality of the primed and unprimed critical-point exponents. To see this, we choose $\lambda = \epsilon^{-1/a_\epsilon}$ in eqn (11.41), with the result

$$\gamma = \frac{2a_H - 1}{a_\epsilon}. \tag{11.45}$$

Combining (11.43) and (11.45), we obtain

$$\gamma' = \gamma. \qquad \blacktriangleright \ (11.46)$$

Next we demonstrate that the gap exponents, Δ_ℓ, are all equal. It is evident that if we differentiate both sides of the fundamental scaling relation (11.30) ℓ times with respect to H, we obtain

$$G^{(\ell)}(\epsilon, H) = \lambda^{\ell a_H - 1} G^{(\ell)}(\lambda^{a_\epsilon} \epsilon, \lambda^{a_H} H); \tag{11.47}$$

indeed, special cases of (11.47) are (11.32) and (11.41). Hence it follows that

$$\frac{G^{(\ell)}(\epsilon, H)}{G^{(\ell-1)}(\epsilon, H)} = \lambda^{a_H} \frac{G^{(\ell)}(\lambda^{a_\epsilon} \epsilon, \lambda^{a_H} H)}{G^{(\ell-1)}(\lambda^{a_\epsilon} \epsilon, \lambda^{a_H} H)}. \tag{11.48}$$

Now the definition equation (3.19) of the gap exponent Δ'_ℓ implies that when $H = 0$, the left-hand side of (11.48) varies as $\epsilon^{-\Delta'_\ell}$. Therefore on choosing $\lambda = \epsilon^{-1/a_\epsilon}$, we obtain $\Delta'_\ell \equiv \Delta$ independent of order ℓ, where

$$\Delta = \frac{a_H}{a_\epsilon} = \beta\delta = \beta + \gamma'. \qquad \blacktriangleright \ (11.49)$$

Incidentally, we can define a different sort of gap exponent by setting $\epsilon = 0$ and letting $H \to 0$, and these also turn out to be independent of ℓ.

We can of course also differentiate the Gibbs potential with respect to temperature (and, incidentally, define still further sets of gap-like indices). In particular, from the second temperature derivative it follows from eqn (2.48) that

$$\lambda^{2a_\epsilon} C_H(\lambda^{a_\epsilon} \epsilon, \lambda^{a_H} H) = \lambda C_H(\epsilon, H). \tag{11.50}$$

On setting $H = 0$ and $\lambda = (-\epsilon)^{-1/a_\epsilon}$, we obtain

$$\alpha' = 2 - \frac{1}{a_\epsilon}. \tag{11.51}$$

Combining (11.51) and (11.39), we obtain the Griffiths inequality satisfied as an equality,

$$\alpha' + \beta(\delta + 1) = 2, \qquad (11.52)$$

while if we combine (11.52) with (11.44), we obtain the Rushbrooke inequality in the form of an equality,

$$\alpha' + 2\beta + \gamma' = 2. \qquad (11.53)$$

It should now be clear how to obtain all of the critical-point exponents in terms of the scaling parameters a_ϵ and a_H, and how to further eliminate these two parameters and obtain a whole host of equalities among the exponents. In Table 11.1 we list a few of the more frequently referenced exponent equalities. The reader will recognize many of the exponent combinations that appear in Table 11.1 from our discussion in Chapter 4 of rigorous inequalities (cf., especially, Table 4.1).

TABLE 11.1

Relations among the critical-point exponents predicted by the scaling hypothesis. These relations are not all independent of one another and in fact knowledge of two exponents suffices to determine the remaining ones. Note that many of the inequalities listed in Table 4.1 are predicted by the scaling hypothesis to be equalities.

1.	$\alpha + 2\beta + \gamma = 2$
2.	$\alpha + \beta(\delta + 1) = 2$
3.	$(2 - \alpha)\zeta + 1 = (1 - \alpha)\delta$
4.	$\gamma(\delta + 1) = (2 - \alpha)(\delta - 1)$
5.	$\gamma = \beta(\delta - 1)$ (or $\beta\delta = \beta + \gamma$)
6.	$\varphi + 2\psi - 1/\delta = 1$
7.	$\varphi\beta\delta = \alpha$
8.	$\psi\beta\delta = 1 - \alpha$
9.	$\delta = \dfrac{2 - \alpha + \gamma}{2 - \alpha - \gamma}$
10.	$\Delta = \beta + \gamma = \beta\delta$
11.	$\Delta = 1 + \frac{1}{2}(\gamma - \alpha)$
12.	$\alpha = \alpha'$
13.	$\gamma = \gamma'$
14.	$\Delta = \Delta'_\zeta = \Delta_{2\zeta}$

11.4. Magnetic equation of state: scaled magnetization and scaled magnetic field

In addition to predicting the relations among the critical-point exponents discussed in the previous section, the scaling hypothesis makes specific predictions concerning the form of the magnetic equation of state which are supported by recent experimental work on both insulating and metallic ferromagnetic systems. These predictions are obtained by arguments similar to those utilized in § 11.1.4.

The magnetic equation of state is a relation among the variables M, H, and T. Hence we begin by writing eqn (11.32) in the form

$$M(\epsilon, H) = \lambda^{a_H - 1} M(\lambda^{a_\epsilon} \epsilon, \lambda^{a_H} H). \qquad (11.54)$$

Next we set $\lambda = |\epsilon|^{-1/a_\epsilon}$, whence (11.54) becomes

$$M(\epsilon, H) = |\epsilon|^{(1 - a_H)/a_\epsilon} M(\epsilon/|\epsilon|, H/|\epsilon|^{a_H/a_\epsilon}). \qquad (11.55)$$

Using eqns (11.35) and (11.49) to eliminate the scaling parameters in favour of the critical-point exponents β and $\Delta = \beta\delta$, eqn (11.55) becomes

$$\frac{M(\epsilon, H)}{|\epsilon|^\beta} = M\left(\frac{\epsilon}{|\epsilon|}, \frac{H}{|\epsilon|^{\beta\delta}}\right). \qquad \blacktriangleright (11.56)$$

Next we define the variables

$$m \equiv |\epsilon|^{-\beta} M(\epsilon, H), \qquad (11.57)$$

called the *scaled magnetization*, and

$$h \equiv |\epsilon|^{-\beta\delta} H(\epsilon, M), \qquad (11.58)$$

called the *scaled magnetic field*. Also we observe that the function on the right-hand side of (11.56) is a function only of h and the sign of $T - T_c$ ($\epsilon/|\epsilon| = \pm 1$), motivating the definition

$$F_\pm(h) \equiv M(\pm 1, h). \qquad (11.59)$$

Hence eqn (11.56) may be written in terms of the reduced variables as simply

$$m = F_\pm(h). \qquad (11.60)$$

It is quite common to write (11.60) in the equivalent form

$$h = f_\pm(m), \qquad \blacktriangleright (11.61)$$

where the function $f_\pm(m)$ is the inverse of the function $F_\pm(h)$. Equations (11.60) and (11.61) predict that if we scale M by dividing by $|\epsilon|^\beta$ and if we scale H by dividing by $|\epsilon|^{\beta\delta}$, then plots of m vs. h should be the same for all values of temperature, in contrast to the case for plots of M vs. H, in which case the data fall on distinct isotherms as in Fig. 11.3. Recently Ho and Litster (1969) have measured M as a function of H for the insulating ferromagnet CrBr$_3$ along thirty different isotherms in the temperature range $-0.03 \leq \epsilon \leq 0.2$ (i.e. $31.9K \leq T \leq 39.5K$, where $T_c = 32.844K$), and their plot of scaled magnetization m vs. scaled magnetic field h is reproduced in Fig. 11.4.

The measurements of Ho and Litster were made on the insulating ferromagnet CrBr$_3$; similar data have been obtained (Weiss and Forrer 1926, Kouvel and Rodbell 1967, Arrott and Noakes 1967, Kouvel and Comly 1968) for the metallic ferromagnet nickel, and a plot analogous to Fig. 11.4 is shown for Ni in Fig. 11.5. Note that here the ordinate is m^2 and the abscissa is h/m. Again the data fall on two separate curves, one for $T < T_c$ and the other for $T > T_c$.

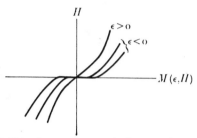

FIG. 11.3. A series of M–H isotherms for a typical magnetic system. The scaling hypothesis (11.30) permits us, in a sense, to superpose all isotherms near the critical point providing we scale M and H appropriately (cf. Figs. 11.4 and 1.2(b)).

We observe from Fig. 11.4 that h appears to be linear in m for small values of m, and from Fig. 11.5 that h/m seems to be linear in m^2. Therefore one might expect $f_\pm(m)$ to be of the form

$$h = f_\pm(m) = b_1 m + b_3 m^3 + b_5 m^5 + \cdots \qquad (11.62)$$

or, on using the definitions (11.57) and (11.58) together with $\beta\delta = \beta + \gamma$,

$$H = b_1 M |\epsilon|^\gamma + b_3 M^3 |\epsilon|^{\gamma - 2\beta} + b_5 M^5 |\epsilon|^{\gamma - 4\beta} + \cdots. \qquad (11.63)$$

The sets of critical-point exponents found experimentally are

$$\beta = 0.368 \pm 0.005, \gamma = 1.215 \pm 0.015, \text{ and } \delta = 4.28 \pm 0.1 \qquad (11.64)$$

for the insulator CrBr$_3$, and

$$\beta = 0.378 \pm 0.004, \gamma = 1.34 \pm 0.01, \text{ and } \delta = 4.58 \pm 0.05. \qquad (11.65)$$

for the metal nickel. Observe that both sets of exponents are quite different from each other. We should not be surprised by this result, since the physical interactions leading to cooperative phenomena in insulators and metals are probably quite different. Both sets of exponents are nevertheless consistent with the predictions of the scaling hypothesis, as summarized in Table 11.1.

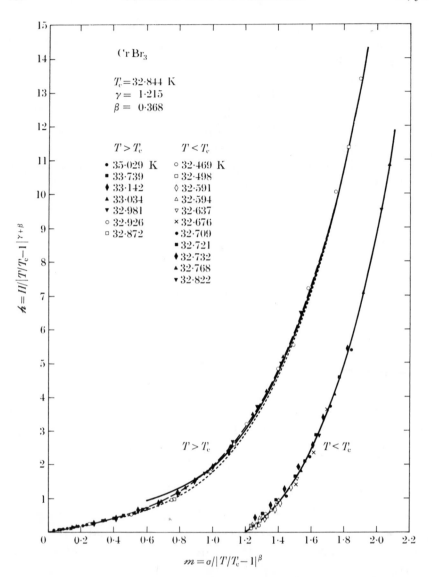

Fig. 11.4. Scaled magnetic field h is plotted against scaled magnetization m for the insulating ferromagnet CrBr$_3$, using data from seven supercritical ($T > T_c$) and from eleven subcritical ($T < T_c$) isotherms. Here $\sigma \equiv M/M_0$. After Ho and Litster (1969).

Note that the determination of the values of two of the exponents is not sufficient to check the validity of the scaling predictions; we need at least three exponents. Of course, if we assume the validity of the scaling hypothesis, (11.30), then determination of two exponents

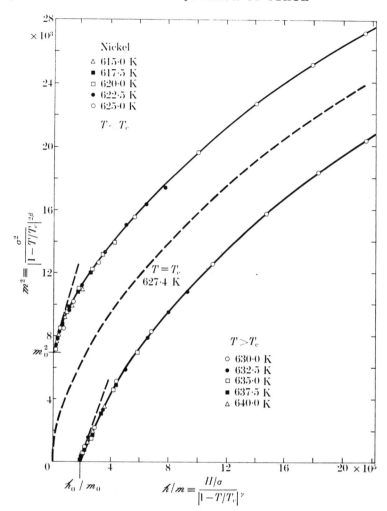

Fig. 11.5. A plot of m^2 against h/m, where m is the scaled magnetization and h is the scaled magnetic field. The data are from measurements on the metallic ferromagnet nickel. Here $\sigma \equiv M/M_0$. After Kouvel and Comly (1968).

suffices to fix the values of all the remaining exponents. For example, the reader can easily verify from Table 11.1 that for CrBr$_3$ the data of (11.64) together with the scaling assumption imply that

$$\left.\begin{aligned} \alpha &\simeq 0{\cdot}05, \\ \Delta &\simeq 1{\cdot}6, \\ \varphi &\simeq 0{\cdot}03, \\ \psi &\simeq 0{\cdot}60. \end{aligned}\right\} \qquad (11.66)$$

Similarly, for nickel the data of (11.65) lead to

$$\left.\begin{array}{l} \alpha \simeq -0.1, \\ \Delta \simeq 1.72, \\ \varphi \simeq -0.06, \\ \psi \simeq 0.64. \end{array}\right\} \tag{11.67}$$

Suggested further reading
Widom (1965*a*, *b*).
Domb and Hunter (1965).
Aczél (1966, 1969).
Patashinskii and Pokrovskii (1966).
Griffiths (1967*b*).
Pokrovskii (1968).
Josephson (1969).
Schofield (1969).
Vincentini-Missoni, Levelt-Sengers, and Green (1969).
Hankey and Stanley (1971).

SCALING OF THE STATIC
CORRELATION FUNCTIONS

KADANOFF (1966) has presented a heuristic argument that provides intuitive support for the scaling hypothesis considered in Chapter 11. His work predicts not only the relations concerning exponents for the thermodynamic functions but also certain relations among the correlation function exponents. Whereas the former are consistent with essentially all known experiments and calculations, the latter appear to be in possible conflict with certain calculations—specifically, the correlation function scaling laws agree with the exact Ising model results for two-dimensional lattices but they are just barely outside the range of values predicted by the most reliable numerical approximations for three-dimensional lattices (Ferer, Moore, and Wortis 1969).

12.1. The Kadanoff construction

Consider a system of N Ising spins situated on a d-dimensional lattice, and interacting with nearest-neighbour interactions of magnitude J, and let the assembly be in a magnetic field H. Thus the Hamiltonian is given, in units of kT, by the expression

$$\beta \mathscr{H} = -\mathscr{J} \sum_{\langle ij \rangle} s_i s_j - h \sum_{i=1}^{N} s_i, \qquad (12.1)$$

where the energy parameters $\mathscr{J} \equiv J/kT$ and $h \equiv \bar{\mu}H/kT$ are dimensionless quantities and where $\beta \equiv 1/kT$.

Now partition the lattice into 'cells' of side La, where a is the lattice constant and L is an arbitrary number but is much greater than unity. Thus there are $n \equiv N/L^d$ cells, each of which contains L^d spins. A $d = 2$ example with $L = 5$ is shown in Fig. 12.1. Next consider the domain of temperatures that are sufficiently near T_c that the correlation length ξ is much larger than the length of a cell, i.e. $\xi \gg La$. It is within this temperature region only that the Kadanoff argument is designed to apply.

So far all we have done is to construct cells mentally, each consisting of L^d spins, where $1 \ll L \ll \xi/a$, so that within a cluster of correlated spins (cf., e.g., Fig. 1.5) there are a large number of cells. Next we shall make some physical assumptions about these cells. First of all, we associate with each cell $\alpha(\alpha = 1, 2, \cdots, n)$ a magnetic moment $\tilde{s}_\alpha$. We shall assume that each of these cell moments $\tilde{s}_\alpha$ behaves in some sense like a site moment s_i in that it is either 'up' or 'down'. This assumption should be valid providing our cell is entirely inside one of the correlated clusters of Fig. 1.5. Since we have assumed that we are sufficiently close to T_c that $\xi \gg La$, there will be a large number of cells in each correlated cluster. Therefore the surface-to-volume ratio should be so small that

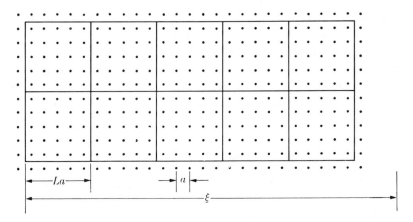

FIG. 12.1. Division of the Ising model into cells of side La, where a is a nearest-neighbour distance. In general we choose L such that $L \gg 1$ yet $La \ll \xi$. From Kadanoff et al. 1967.

the overwhelming majority of the cells will be inside a correlated cluster and hence all of the constituent magnetic moments of each cell should be pointing in the same direction.

Thus this assumption that the cells themselves behave after a fashion as Ising spins (i.e. the net moment of each cell is virtually either plus or minus L^d) would seem to be valid sufficiently close to T_c. Therefore, if we write the Hamiltonian in terms of the cell moments $\tilde{s}_\alpha$ (rather than in terms of the site moments s_i) we might expect that it will be similar in form to the Hamiltonian (12.1) for the site model, except of course that the parameters J and h may be different. Therefore we associate analogous parameters $\tilde{J}$ and $\tilde{h}$ with our cell model. Actually, since the critical temperature is a direct measure of the strength of the interaction J, we shall use $\tilde{\epsilon}$ rather than $\tilde{J}$.

Now we may wonder why we focus attention on the Hamiltonian for the cell model, since we cannot solve the Hamiltonian even for the site model (except for $d = 1$). The reason is that we wish to argue that since the Hamiltonians have the same structure for both models (only the parameters J and h are different), we may expect that the thermodynamic functions for the two models are also similar (with only the parameters being different). Therefore we assume that, for example, the Gibbs potential $G_{\text{cell}}(\tilde{\epsilon}, \tilde{h})$ for the cell model is the same analytic function of $\tilde{\epsilon}$ and $\tilde{h}$ as the Gibbs potential $G_{\text{site}}(\epsilon, h)$ of the site model. Hence we write

$$\bar{G}(\tilde{\epsilon}, \tilde{h}) = L^d \bar{G}(\epsilon, h), \qquad \blacktriangleright \quad (12.2)$$

where $\bar{G}(\tilde{\epsilon}, \tilde{h})$ is the Gibbs potential per cell and $\bar{G}(\epsilon, h)$ is the Gibbs potential per site.

To take advantage of eqn (12.2) we must relate $\tilde{h}$ to h and $\tilde{\epsilon}$ to ϵ. The constant that multiplies the field part must certainly be linear in the field, and therefore we might expect $\tilde{h}$ to be proportional to h. We expect that the proportionality constant might in general depend upon the parameter L, so we write

$$\tilde{h} = H(L)h. \qquad (12.3)$$

We can argue that $\tilde{\epsilon}$ should be proportional to ϵ in a similar fashion. The fact that the cell representation is just another formulation of what is basically the same physical system—the Ising spin system—means that the cell formalism might be expected to 'go critical' when the spin system does; that is, we assume

$$\tilde{\epsilon} = T(L)\,\epsilon. \qquad (12.4)$$

It is possible to proceed with the argument without making further assumptions (Cooper 1968). However we shall follow Kadanoff's original discussion and assume that the functions $H(L)$ and $T(L)$ are of the form

$$H(L) = L^x \qquad (12.5)$$

and

$$T(L) = L^y, \qquad (12.6)$$

where x and y are arbitrary numbers. Substituting (12.3) and (12.4)—together with the Kadanoff choices (12.5) and (12.6)—into eqn (12.2), we obtain

$$\bar{G}(L^y \epsilon, L^x h) = L^d \bar{G}(\epsilon, h). \qquad (12.7)$$

Equation (12.7) is identical in form to the generalized homogeneous function of eqn (11.17), except that the parameter λ in (11.17) is an arbitrary number, while the parameter L in eqn (12.7) is not entirely arbitrary ($1 \ll L \ll \xi/a$). If we assume that (12.7) is valid for all values of L, then eqn (12.7) essentially states that the Gibbs potential per spin is a generalized homogeneous function. That is, any function $\bar{G}(\epsilon, h)$ that satisfies (12.7) also satisfies the equation

$$\bar{G}(L^{y/d}\epsilon, L^{x/d}h) = L\,\bar{G}(\epsilon, h), \qquad \blacktriangleright \quad (12.8)$$

and the converse statement is also valid (cf. eqn (11.18)). But eqn (12.8) is precisely the static scaling hypothesis (11.30), with

$$a_\epsilon = \frac{y}{d} \qquad (12.9)$$

and

$$a_H = \frac{x}{d}. \qquad (12.10)$$

Thus the Kadanoff construction—together with a generous number of assumptions—leads to the static scaling hypothesis, and hence to all the predictions of Chapter 11 that followed from (11.30).

12.2. Application to the pair correlation function

In this section we present Kadanoff's line of reasoning which supports the assumption that the pair correlation function is a generalized homogeneous function. To this end, we introduce the pair correlation functions for the site model,

$$\Gamma(r, \epsilon) \equiv \langle (s_i - \langle s \rangle)(s_j - \langle s \rangle) \rangle, \qquad (12.11)$$

and for the cell model,

$$\Gamma(\tilde{r}, \tilde{\epsilon}) \equiv \langle (\tilde{s}_\alpha - \langle \tilde{s} \rangle)\,(\tilde{s}_\beta - \langle \tilde{s} \rangle) \rangle, \qquad (12.12)$$

where $r = |\mathbf{r}_i - \mathbf{r}_j|$ is the magnitude of the site separation vector, and $\tilde{r} = |\mathbf{r}_\alpha - \mathbf{r}_\beta| = r/L$ is the magnitude of the cell separation vector.

We next introduce a parameter $\mathscr{L}$ through the relation

$$L^{-d} \sum_{i \in \alpha} s_i = \mathscr{L}\tilde{s}_\alpha, \qquad (12.13)$$

where $\tilde{s}_\alpha = \pm 1$ by definition, and where the summation is over the L^d sites in cell α. Therefore it follows that

$$\Gamma(r, \epsilon) = \mathscr{L}^2\, \Gamma(\tilde{r}, \tilde{\epsilon}). \qquad (12.14)$$

Next we consider the field-dependent term in the Hamiltonian (12.1). For the cell model this term is

$$\beta \mathscr{H}_{\text{cell}} = -\tilde{h} \sum_{\alpha=1}^{n} \tilde{s}_{\alpha}, \tag{12.15}$$

while for the site model it is

$$\beta \mathscr{H}_{\text{site}} = -h \sum_{\alpha=1}^{n} \sum_{i \in \alpha} s_i$$

$$= -h \sum_{\alpha=1}^{n} \tilde{s}_{\alpha} L^d \mathscr{L}; \tag{12.16}$$

the second equality in (12.16) follows from (12.13). Hence we make the identification

$$\tilde{h} = L^d \mathscr{L} h. \tag{12.17}$$

Now if we assume that $\mathscr{L}$ is independent of h and ϵ but varies with L as L^z, then we find on combining (12.17) with (12.5) and (12.3) that

$$z = x - d. \tag{12.18}$$

Combining (12.18) and (12.14), we have

$$\Gamma(r, \epsilon) = L^{2(x-d)} \Gamma(\tilde{r}, \tilde{\epsilon}). \tag{12.19}$$

On substituting $\tilde{\epsilon} = L^y \epsilon$ from eqn (12.6) and $\tilde{r} = r/L$ into (12.19), we obtain

$$\Gamma(r, \epsilon) = L^{2(x-d)} \Gamma(L^{-1}r, L^y \epsilon). \qquad \blacktriangleright \tag{12.20}$$

If we assume, in the same spirit adopted in § 12.1, that eqn (12.20) is valid for all values of the parameter L, then eqn (12.20) implies that the pair correlation function $\Gamma(r, \epsilon)$ is a generalized homogeneous function of its two arguments, r and ϵ. Therefore there should be two scaling parameters, and we see that the parameters x and y in (12.20) are known from our discussion above for the thermodynamic potential $\bar{G}(\epsilon, h)$. Hence we might expect to find that (12.20) predicts relations between the critical-point exponents ν and η (which describe the behaviour of $\Gamma(r, \epsilon)$) and the critical-point exponents $\alpha, \beta, \gamma, \cdots$ which describe the behaviour of the derivatives of the thermodynamic potentials.

Our expectations are in fact borne out. In § 11.1.4 we saw that the expressions (11.24) and (11.29) are mathematically equivalent expressions for the homogeneous function (11.21). In a similar fashion, we can obtain expressions equivalent to eqn (12.20). If we choose the parameter L in eqn (12.20) to have the value $L = (1/\epsilon)^{1/y}$, then we obtain (in analogy with (11.24))

$$\Gamma(r, \epsilon) = \epsilon^{2(d-x)/y} f(r \epsilon^{1/y}), \tag{12.21}$$

where the function $f(t)$ is defined through

$$f(t) \equiv \Gamma(t, 1). \tag{12.22}$$

Similarly, if we choose $L = (1/r)^{-1} = r$, we obtain (in analogy with (11.29))

$$\Gamma(r, \epsilon) = r^{2(x-d)}g(\epsilon r^y), \tag{12.23}$$

where

$$g(t) \equiv \Gamma(1, t). \tag{12.24}$$

Therefore since $\xi \sim \epsilon^{-\nu}$ and $\Gamma(r, \epsilon = 0) \sim r^{-(d-2+\eta)}$ we have

$$\nu = \frac{1}{y} \tag{12.25}$$

and

$$-(d - 2 + \eta) = 2(x - d). \tag{12.26}$$

To relate ν and η to the exponents of Chapter 11, we recall from (12.9) that $y = d\, a_\varepsilon$, and from (11.51) we have $a_\varepsilon = (2 - \alpha')^{-1}$. Using the same argument which led to (11.46), we find $\alpha' = \alpha$, whence

$$d\,\nu = 2 - \alpha. \qquad \blacktriangleright \tag{12.27}$$

Similarly, from (12.10) we have $x = d\, a_H$, and from (11.40) $a_H = \delta/(\delta + 1)$. Hence (12.26) becomes

$$d - 2 + \eta = \frac{2d}{\delta + 1} = \frac{2d\beta}{2 - \alpha} = \frac{2\beta}{\nu}, \tag{12.28}$$

where we have used eqns (11.52) and (12.27). Equation (12.28) relates η to ν and β. It is somewhat more customary to instead relate η and ν and γ whence, on using (11.53), (11.46), and (12.27), we have

$$(2 - \eta)\nu = \gamma. \qquad \blacktriangleright \tag{12.29}$$

From eqns (12.27) and (12.29), respectively, we see that the Josephson inequality (4.55) and the Fisher inequality (4.50) are predicted by the homogeneity relation (12.20) to hold as equalities. Finally, (12.27)–(12.29) can be combined to obtain

$$d\,\frac{\delta - 1}{\delta + 1} = \frac{d\gamma'}{2\beta + \gamma'} = 2 - \eta \qquad \blacktriangleright \tag{12.30}$$

and thus we see that the Buckingham–Gunton inequalities of eqns (4.44) and (4.45) are also satisfied as equalities.

12.3. Alternative methods of obtaining the correlation function scaling relations

We have seen that a large number of theoretical predictions can be obtained by assuming that (i) a thermodynamic potential—such as $G(\epsilon, H)$—is a generalized homogeneous function, and (ii) the pair correlation function is a generalized homogeneous function. Moreover, we have seen that the Kadanoff line of reasoning, by leading us to eqns (12.7) and (12.20), serves to render these assumptions plausible on physical grounds. However the Kadanoff approach is far from rigorous, and in the years subsequent to the appearance of the Kadanoff argument several workers have put forward alternate lines of reasoning that support the homogeneity assumption. None of these alternate approaches is rigorous, however, and therefore we shall not enter into an exhaustive discussion of any of them here.

We shall instead discuss in this section the line of reasoning presented by Halperin and Hohenberg (1967, 1969), because their approach will be generalized to the case of dynamic critical phenomena in Chapter 15. We begin by replacing the independent variable ϵ by the independent variable ξ in the correlation function,

$$\Gamma(\epsilon, r) \to \Gamma(\xi, r), \tag{12.31}$$

and by $1/\xi \equiv \kappa$ in the structure factor,

$$S(\epsilon, q) \to S(\kappa, q), \tag{12.32}$$

where $S(\epsilon, q)$ denotes the spatial Fourier transform of $\Gamma(\epsilon, r)$ (cf. eqn (7.34)). (To be mathematically precise, one should of course not use the same functional notation when the independent variable is changed.) We now assume that there exists some function $f(u)$ such that the pair correlation function $\Gamma(\xi, r)$ may be written as

$$\Gamma(\xi, r) = \xi^u f(r/\xi). \tag{12.33}$$

From the argument of eqns (11.25) and (11.26), we see that this assumption is fully equivalent to the assumption that $\Gamma(\xi, r)$ is a generalized homogeneous function, i.e. that for all the values of λ,

$$\Gamma(\lambda\xi, \lambda r) \equiv \lambda^u \Gamma(\xi, r). \qquad \blacktriangleright \quad (12.34)$$

The appeal of eqn (12.33) is that it states clearly the fact that the homogeneity assumption is equivalent to the assumption that apart from a scale change, $\Gamma(r, \epsilon)$ depends on the site separation distance r only through a single function of the ratio of r to the coherence length ξ. This idea that there is only one characteristic length in the problem will be seen to have its analogue in the dynamic scaling hypothesis, for which—in addition to assuming a characteristic length—we shall assume the existence of a characteristic inverse time ('frequency').

It is also convenient to consider the analogous homogeneity assumption for the structure factor $S(\kappa, q)$,

$$S(\kappa, q) = q^v \mathscr{F}(q/\kappa). \tag{12.35}$$

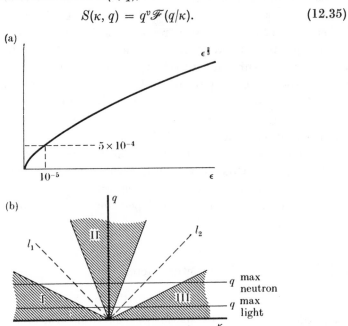

Fig. 12.2. (a) Approach to zero of the inverse correlation length as $T \to T_c^+$. In order to convey a qualitative feel for the magnitude, we have chosen the form $\kappa = \kappa_0 \epsilon^v$ with $\kappa_0 \simeq 1\,\text{Å}^{-1} = 10^8\,\text{cm}^{-1}$ and $v \simeq \frac{2}{3}$. Thus in order for the inverse correlation length to become equal to a typical value (such as $10^5\,\text{cm}^{-1}$) of the magnitude of the momentum transfer vector $q = (4\pi/\lambda)\sin\frac{1}{2}\,\theta$, we need to reach temperatures so close to T_c that $\epsilon \simeq 10^{-5}$. (b) is a graph from Halperin and Hohenberg (1967) showing the wave number q as ordinate and the inverse correlation length κ as abscissa. The three asymptotic regions shown are : I, the low-temperature hydrodynamic region ($q \ll \kappa$); II, the critical region ($q \gg \kappa$); and III, the high-temperature hydrodynamic region ($q \ll \kappa$). In light scattering experiments, the momentum transfer vectors q_{light} are sufficiently small that it is extremely difficult (though by no means impossible) to get close enough to T_c to probe region II, the critical region. On the other hand, in neutron-scattering experiments, the wavelength is about three orders of magnitude smaller so that $q_{\text{neutron}} \gg q_{\text{light}}$ and it is possible to reach region II.

Equation (12.35) says that apart from a change of scale, the Fourier transform of the correlation function depends on q only through a single function of the ratio q/κ. It is therefore convenient to consider the graph shown in Fig. 12.2, in which the abscissa is κ and the ordinate q, and the origin corresponds to the critical point $q = \epsilon = 0$. Halperin and Hohenberg distinguish three asymptotic regions in the κ–q plane, in which we might expect to find qualitatively different behaviour of the correlation function $S(\kappa, q)$. The regions marked I and III in Fig. 12.2 correspond to long-wavelength or *hydrodynamic* regions for which $q \ll \kappa$ (or $q\xi \ll 1$), whereas the region marked II is called the *critical* region and is characterized by $q \gg \kappa$ (or $q\xi \gg 1$). We assume that a single function describes the correlations over the entire κ–q plane, with a characteristic dependence upon the parameter q/κ given by eqn (12.35). Thus the correlation function $S(\kappa, q)$ is essentially determined by its limiting behaviour in the three asymptotic regions of Fig. 12.2. This means that if two forms valid in regions I and II are extrapolated to the line ℓ_1 ($q/\kappa = 1$, $T < T_c$), they must differ at most by some factor of order unity.

As an illustration of the Halperin–Hohenberg approach, we derive the scaling relation $\gamma = (2 - \eta)\nu$ of eqn (12.29). Begin by recalling from eqn (7.59) that the exponent η is equivalently defined by means of the relation

$$S(\kappa = 0, q) \sim q^{-2+\eta}. \tag{12.36}$$

Since the relation (12.36) is valid in region II, ($q/\kappa \gg 1$), we expand eqn (12.35) in the form

$$S(\kappa, q) = q^v\{1 + \mathcal{O}(\kappa/q) + \cdots\}. \tag{12.37}$$

On comparing leading powers of q in (12.36) and (12.37), we obtain

$$v = -2 + \eta \tag{12.38}$$

We can similarly write for the susceptibility, (cf. (7.23) or (A.20)),

$$\chi_T \sim S(\kappa, q = 0) \sim \epsilon^{-\gamma} \sim \kappa^{-\gamma/\nu} \tag{12.39}$$

and the analogue of (12.37) in region III is

$$S(\kappa, q) = \kappa^v\{1 + \mathcal{O}(q/\kappa) + \cdots\}. \tag{12.40}$$

Comparing (12.39) and (12.40), we have

$$v = -\gamma/\nu \tag{12.41}$$

and we obtain $\gamma = (2 - \eta)\nu$ by combining (12.38) and (12.41).

12.4. Comparison with model calculations

Two-dimensional Ising model. The $d = 2$ Ising model provides the most striking support for the scaling assumptions. Table 12.1 presents

TABLE 12.1

Comparison of scaling relations with model calculations. All entries are predicted to have the value d/y. Numerical values for the exponents are taken from Table 3.4. The scaling relations all appear to be satisfied by the two-dimensional Ising model, whereas the relations involving the lattice dimensionality d appear not to hold for the other models shown. There is, as yet, no satisfactory explanation why the relation $d\nu = 2 - \alpha = 2 - \alpha' = d\nu'$ appears to fail for the three-dimensional Ising model. It is also quite possible that $\gamma' = \frac{21}{16} > \gamma \; (=\frac{5}{4})$ and that $\alpha' = \frac{1}{16} < \alpha \; (=\frac{1}{8})$, but this lack of symmetry between the high-temperature and low-temperature exponents can be circumvented (see Stell (1968b))

Quantity	$d = 2$ Ising	$d = 3$ Ising	Classical Heisenberg $(d = 3)$	Spherical model $(d = 3)$	Mean field theory $(d = 3)$
$2 - \alpha$	2	$1\cdot875 \pm 0\cdot01$	$2\cdot1 \pm 0\cdot1$	3	2
$2 - \alpha'$	2	$1\cdot875 \pm 0\cdot06$	not available	not defined	2
$\gamma + 2\beta$	2	$1\cdot875 \pm 0\cdot02$	$2\cdot1 \pm 0\cdot1$	3	2
$\gamma' + 2\beta$	2	$1\cdot875 \pm 0\cdot06$	—	not defined	2
$\beta(\delta + 1)$	2	$1\cdot875 \pm 0\cdot05$	not yet available	3	2
$d\nu$	2	$1\cdot914 \begin{smallmatrix} +0\cdot006 \\ -0\cdot003 \end{smallmatrix}$	$2\cdot1 \pm 0\cdot03$	3	$1\cdot5$
$d\nu'$	2	—	—	3	$1\cdot5$
$d\gamma/(2 - \eta)$	2	$1\cdot914 \begin{smallmatrix} +0\cdot006 \\ -0\cdot003 \end{smallmatrix}$	$2\cdot1 \pm 0\cdot1$	3	$1\cdot5$

numerical values of the various exponent combinations, all of which are predicted to have the value d/y by the scaling hypothesis. We see that all eight quantities have the value 2 (so that $y = 1$). Also, the correlation function is predicted by (12.21) to be of the form

$$\Gamma(r, \epsilon) = \epsilon^{1/4} f(r\epsilon), \tag{12.42}$$

since $y = 1$ and $x = \frac{15}{8}$ for the $d = 2$ Ising model. Recently the asymptotic form of the pair correlation function has been calculated exactly in zero magnetic field, and the form (12.42) predicted by the scaling approach has been corroborated (Wu 1966).

Three-dimensional Ising model. The $d = 3$ Ising model exponents (cf. Table 3.4) are all obtained by extrapolations from truncated series expansions. The numbers shown in the second column of Table 12.1 are obtained from the most reliable approximations so far available. We notice that the quantities are nearly but not quite identical. Observe that all quantities would have the value $\frac{15}{8} = 1\cdot875$ providing ν were changed from $0.638^{+0.002}_{-0.001}$ to 0.625 ($=\frac{5}{8}$) and η were changed from $0.041^{+0.006}_{-0.003}$ to zero. However these values are both well outside the numerical uncertainties. Also it is quite possible that $\gamma' = \frac{21}{16} = 1.3125$ although at present we cannot exclude $\gamma' = \gamma = \frac{5}{4} = 1.25$.

(iii) *Classical* ($S = \infty$) *Heisenberg model.* The results for the classical Heisenberg model, based upon numerical approximations, are shown in the third column. We have used the estimate $\beta \simeq 0.38$ (Stephenson and Wood 1970), obtained by a method that utilizes high-temperature expansions to calculate low-temperature critical-point exponents (Baker, Eve, and Rushbrooke 1970).

(iv) *Three-dimensional spherical model.* The spherical model is exactly soluble for a three-dimensional lattice (cf. Chapter 8) and hence it provides another anchor point on which to test the correlation function scaling relations. We see from the fourth column of Table 12.1 that all the tabulated quantities have the same value, so that the scaling relations would appear to be satisfied for this three-dimensional system.

(v) *Molecular field approximation.* The predictions of this classical theory for a three-dimensional lattice are shown in the fifth column of Table 12.1, and it is clear that the scaling relations involving the lattice dimensionality d fail. Indeed, the same exponents are found for any lattice dimensionality, so the scaling relations involving d could hold for at most one value of d. In Chapter 6 we saw that the molecular field theory corresponds to a model in which each spin interacts equally with all the other spins in the entire system. Hence the Kadanoff construction of partitioning the lattice into cells of dimension much larger than the force range would appear to be impossible, at least within a molecular field framework.

(vi) *Lattices of dimensionality $d > 3$.* The scaling relations involving d are found to fail rather drastically for lattices with $d > 3$. The source of these discrepancies has been discussed in some detail by Stell (1968a) and by Domb (1968).

Suggested further reading
Kadanoff (1966)
Kadanoff *et al.* (1967).
Halperin and Hohenberg (1969).

PART VI

DYNAMIC ASPECTS OF CRITICAL PHENOMENA

13

INTRODUCTION TO DYNAMIC CRITICAL PHENOMENA IN FLUID SYSTEMS

In this, the concluding section of our introduction to phase transitions and critical phenomena, we discuss some of the recent progress that has occurred in the direction of understanding non-equilibrium behaviour near the critical point. This is the area that has come to be called *dynamic* critical phenomena, in contrast to the domain of *static* critical phenomena to which our discussion thus far has been restricted. Much of the formalism developed in our study of time-independent aspects of critical phenomena can be generalized to the time-dependent domain.

We shall focus on the time-dependent pair correlation function $\mathscr{G}(\mathbf{r}, t)$ rather than the time-independent correlation function introduced in Chapter 7. We shall see that the dynamic structure factor $\mathscr{S}(\mathbf{q}, \omega)$ (which is the Fourier transform of $\mathscr{G}(\mathbf{r}, t)$) is directly related to the inelastic scattering cross-section, so that measurement of the energy spectrum of scattered radiation provides us with a great deal of useful information about the time-dependent aspects on a microscopic level. We shall develop this formalism in the present chapter, together with the predictions of the classical theory of hydrodynamics for the behaviour of the dynamic structure factor in the long wavelength domain.

In the following chapter we shall consider recent experimental measurements of the dynamic structure factor for one-component fluid systems. We shall see the limitations of hydrodynamics in providing a description of the experimental data, and hence we shall be motivated to consider, in Chapter 15, theoretical developments that serve to extend the predictions of hydrodynamics. In particular, we shall find that a simple generalization of the static scaling hypothesis appears to be consistent with a great deal of recent experimental data. A somewhat more microscopic approach, called the *mode–mode coupling theory*, will also be seen to be useful in interpreting these recent data. Finally, we will consider in Appendix E the appropriate generalization to non-equilibrium situations of the model system that has proved so successful in describing equilibrium behaviour, the Ising model.

13.1. Critical-point exponents for transport coefficients

The hydrodynamic behaviour of a simple fluid system can be described by specifying the three transport coefficients for the system, the thermal conductivity Λ, the shear viscosity η, and the bulk viscosity ζ. In the neighbourhood of the critical temperature T_c, experimental work seems to indicate that certain of these coefficients diverge with a simple power law behaviour. We therefore introduce, in Table 13.1, the three new critical-point exponents, a, b, and c, to characterize the behaviour of Λ, η, ζ respectively. We have introduced minus signs in the definitions of a, b, and c to ensure that positive values of these exponents will correspond to divergences in the transport coefficients.

TABLE 13.1

Critical-point exponents for transport coefficients. Until recently there was little experimental or theoretical evidence concerning the values of the exponents a, b, and c describing the singularities (if any) in the transport coefficients Λ, η, and ζ. Currently it is believed that both Λ and ζ can diverge at T_c, while η remains non-singular through the critical region (i.e. a, $c \gtrsim 0$ while $b \simeq 0$). Some workers denote Λ by λ or κ, and a by ψ.

	$T < T_c$	$T > T_c$
Λ (thermal conductivity)	$(-\epsilon)^{-a'}$	ϵ^{-a}
η (shear viscosity)	$(-\epsilon)^{-b'}$	ϵ^{-b}
ζ (bulk viscosity)	$(-\epsilon)^{-c'}$	ϵ^{-c}

Until fairly recently the qualitative behaviour of Λ, η, and ζ in the critical region was not known. Although at one time some workers argued that all three transport coefficients should remain finite near the critical point, it is now widely believed that Λ and ζ diverge for many systems. However, the existence of outstanding counterexamples prevents us from assuming that any of the transport coefficients diverge for all systems. The dramatic behaviour of the thermal conductivity of sulphur hexafluoride (SF_6) illustrates this point well. While the thermal conductivity Λ diverges to infinity when the critical point is approached from below along the coexistence curve Λ appears to behave quite differently when the critical point is approached from above along the critical isochore (Benedek 1968, 1969).

This example is indicative of an important difference between the behaviour of equilibrium quantities and transport coefficients in the region near the critical point. Whereas for equilibrium quantities, systems that differ from each other in their microscopic interactions are found to behave quite similarly near the critical point, this universality is not always observed for the critical behaviour of the transport coefficients.

13.2. Time-dependent correlation functions and the dynamic structure factor $\mathscr{S}(\mathbf{q}, \omega)$

We shall find that the use of correlation functions proves to be extremely useful in the study of time-dependent cooperative phenomena. We first introduce the space- and time-dependent density–density correlation function

$$\mathscr{G}_{nn}(\mathbf{r}, t) \equiv \langle \delta n(\mathbf{r}, t)\, \delta n(\mathbf{0}, 0)\rangle. \qquad (13.1)$$

The angular brackets denote an equilibrium ensemble average, and

$$\delta n(\mathbf{r}, t) \equiv n(\mathbf{r}, t) - \langle n \rangle \qquad (13.2)$$

is the local deviation of the number density $n(\mathbf{r}, t)$ from the equilibrium value of the number density $\langle n \rangle$. By substituting eqn (13.2) into (13.1), we obtain an equivalent expression for $\mathscr{G}_{nn}(\mathbf{r}, t)$,

$$\mathscr{G}_{nn}(\mathbf{r}, t) = \langle n(\mathbf{r}, t)n(\mathbf{0}, 0)\rangle - \langle n\rangle\langle n\rangle. \qquad \blacktriangleright \ (13.3a)$$

For magnetic systems we introduce an analogous correlation function for the magnetic order parameter

$$\mathscr{G}_{\mathbf{ss}}(\mathbf{r}_{ij}, t) \equiv \langle \mathbf{S}_i(t) \cdot \mathbf{S}_j(0)\rangle - \langle \mathbf{S}_i(t)\rangle\langle \mathbf{S}_j(0)\rangle, \qquad (13.3b)$$

where $\mathbf{r}_{ij} \equiv \mathbf{r}_i - \mathbf{r}_j$ and $\mathbf{S}_i$ denotes a spin situated on a site at position $\mathbf{r}_i$.

A more useful quantity than the correlation function itself is the dynamic structure factor $\mathscr{S}_{nn}(\mathbf{q}, \omega)$, defined as the Fourier transform in space and time of $\mathscr{G}_{nn}(\mathbf{r}, t)$,

$$\mathscr{S}_{nn}(\mathbf{q}, \omega) \equiv \int d\mathbf{r} \int_{-\infty}^{\infty} dt\, \mathscr{G}_{nn}(\mathbf{r}, t) e^{-i(\mathbf{q}\cdot\mathbf{r} - \omega t)}. \qquad \blacktriangleright \quad (13.4)$$

In concert with $\mathscr{S}_{nn}(\mathbf{q}, \omega)$ we shall define the static structure factor $S_{nn}(\mathbf{q})$, where

$$S_{nn}(\mathbf{q}) \equiv \int_{-\infty}^{\infty} \frac{d\omega}{2\pi} \mathscr{S}_{nn}(\mathbf{q}, \omega). \qquad \blacktriangleright \quad (13.5)$$

In what follows we will omit the subscripts nn on correlation functions except where confusion might arise.

13.3. Relation between the dynamic structure factor and light-scattering experiments

The dynamic structure factor is directly measured by certain scattering experiments, and in this section we shall derive an explicit relationship between the structure factor and the intensity of light scattered through a wave vector $\mathbf{q}$ and with a frequency shift ω. This relationship is of fundamental significance in that it provides the link whereby theoretical calculations of the time-dependent correlation function can be compared with experimental measurements.

The connection between scattering and the correlation function is more general than the specific example of light scattering that we shall discuss. For example, it can be shown that in the scattering of neutrons from magnetic materials, the scattering cross-section is directly related to the spin correlation function of eqn (13.3b) (Marshall and Lowde 1968).

We shall consider a plane, monochromatic light wave incident upon a macroscopically small volume V of fluid (cf. Fig. 13.1). To simplify the discussion we treat the electric field as a scalar quantity. Thus, we have

$$E_0(\mathbf{r}, t) = \mathscr{E}_0 e^{i\mathbf{k}_0 \cdot \mathbf{r} - i\omega_0 t}. \qquad (13.6)$$

When a molecule of the fluid is subjected to this plane electromagnetic wave, its positive and negative charges are set into oscillation with respect to each other. This oscillating electric dipole now emits a spherical scattered wave. To calculate the electric field at a distant observation point $\mathbf{R}$, we must consider each of the individual particles in the volume to be the source of such a scattered wave.

For a homogeneous substance, we have a cancellation of the scattered waves in all but the forward direction. In the forward direction the scattered waves collaborate to replace the incident radiation with a new wave of different wavelength, giving rise to an altered velocity of magnitude

$$v = \frac{c}{n},$$ (13.7)

where n is the index of refraction of the medium.

In an inhomogeneous medium, the scattered waves no longer cancel completely in all directions other than the forward direction. Fluctuations in the density of the fluid provide the inhomogeneities necessary for scattering, so long as these fluctuations vary over a distance comparable to the wavelength of the incident radiation.

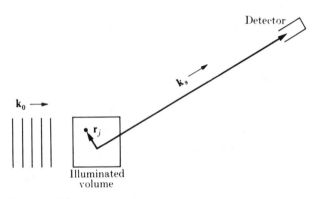

FIG. 13.1. Diagrammatic representation of scattering geometry used to obtain the relation (13.27) between the dynamic structure factor $\mathscr{S}(\mathbf{q}, \omega)/S(\mathbf{q})$ and observed intensity $\mathscr{I}(\mathbf{q}, \omega)/I(\mathbf{q})$ at point $\mathbf{R}$.

The *net* field incident upon a given molecule of the fluid is composed of the incident field plus the sum of all previously-scattered fields. Since the incident field modified by the effect of the index of refraction is so much stronger than the scattered field, we may neglect the latter. Making this assumption is essentially equivalent to making the first Born approximation. Since the amplitude of the scattered field from a particular scattering centre is proportional to the incident field, we may write the contribution to the scattered field arising from the jth particle as

$$E_j(\mathbf{R}, t) = \{\mathscr{E}_0 e^{i\mathbf{k}_0 \cdot \mathbf{r}_j(t_r) - i\omega_0 t_r}\} \left\{ \frac{\mathscr{A}}{|\mathbf{R} - \mathbf{r}_j(t_r)|} e^{ik_0 \cdot |\mathbf{R} - \mathbf{r}_j(t_r)| - i\omega_0(t - t_r)} \right\}.$$ (13.8)

In eqn (13.8) $\mathscr{A}$ is the constant of proportionality between the amplitude of the incident and scattered waves, $\mathbf{r}_j$ is the position of the jth particle, and

$$t_r \equiv t - \frac{\mathbf{R} - \mathbf{r}_j(t_r)}{c/n}. \tag{13.9}$$

The retarded time t_r is used because the field observed at $\mathbf{R}$ at time t from the jth particle is due to the scattering that took place at a different location and therefore earlier in time.

In an actual experiment, the dimensions of the illuminated volume V are small compared to the distance $R \equiv |\mathbf{R}|$ to the observation point. Hence $|\mathbf{r}_j(t_r)| \ll |\mathbf{R}|$ and we can set

$$|\mathbf{R} - \mathbf{r}_j(t_r)| \simeq |\mathbf{R}| \tag{13.10}$$

in the denominator of eqn (13.8), and

$$|\mathbf{R} - \mathbf{r}_j(t_r)| \simeq |\mathbf{R}| - \hat{\mathbf{R}} \cdot \mathbf{r}_j(t_r) \tag{13.11}$$

in the argument of the exponentials, where in (13.11) $\hat{\mathbf{R}} \equiv \mathbf{R}/|\mathbf{R}|$. Thus (13.8) becomes

$$E_j(\mathbf{R}, t) = \mathscr{B}e^{ik_0 R}e^{i(\mathbf{k}_0 - k_0\hat{\mathbf{R}})\cdot\mathbf{r}_j(t_r)}e^{-i\omega_0 t}, \tag{13.12}$$

where $k_0 \equiv |\mathbf{k}_0|$, $R \equiv |\mathbf{R}|$, and $\mathscr{B} \equiv \mathscr{E}_0\mathscr{A}/R$.

To relate $E_j(\mathbf{R}, t)$ to the density–density correlation function $\mathscr{G}(\mathbf{r}, t)$ of eqn (13.1), we introduce the particle density by the relation (7.3),

$$n(\mathbf{r}, t) \equiv \sum_{j=1}^{N} \delta\{\mathbf{r} - \mathbf{r}_j(t)\}, \tag{13.13}$$

and we introduce into (13.12) the Dirac delta function $\delta\{\mathbf{r} - \mathbf{r}_j(t_r)\}$,

$$E_j(\mathbf{R}, t) = \mathscr{B}e^{i(k_0 R - \omega_0 t)} \int d\mathbf{r}\, \delta\{\mathbf{r} - \mathbf{r}_j(t_r)\}e^{i(\mathbf{k}_0 - k_0\hat{\mathbf{R}})\cdot\mathbf{r}}, \tag{13.14}$$

where the spatial integration is over the volume V. Hence we obtain from (13.13) and (13.14) for the total scattered field at point R, due to scattering from all the particles in V,

$$E(\mathbf{R}, t) = \sum_{j=1}^{N} E_j(\mathbf{R}, t), \tag{13.15}$$

the expression

$$E(\mathbf{R}, t) = \mathscr{B}e^{i(k_0 R - \omega_0 t)} \int d\mathbf{r}\, n\left(\mathbf{r}, t - \frac{|\mathbf{R} - \mathbf{r}|}{v}\right) e^{i(\mathbf{k}_0 - k_0\hat{\mathbf{R}})\cdot\mathbf{r}}. \tag{13.16}$$

Now the power spectral density of the scattered light of frequency ω

observed at $\mathbf{R}$ is the ensemble average of the square of the temporal Fourier transform of $E(\mathbf{R}, t)$,

$$I(\mathbf{R}, \omega) = \lim_{T \to \infty} \frac{1}{T} \left\langle \left| \int_{-T/2}^{T/2} dt\, E(\mathbf{R}, t)\, e^{i\omega t} \right|^2 \right\rangle, \qquad (13.17)$$

where the angular brackets denote the equilibrium ensemble average. Substitution of (13.16) into (13.17) leads to the expression

$$I(\mathbf{R}, \omega) = \lim_{T \to \infty} |\mathscr{B}|^2 \frac{1}{T} \iint_{-T/2}^{T/2} dt dt' \iint d\mathbf{r} d\mathbf{r}'\, e^{i(\mathbf{k}_0 - k_0 \hat{\mathbf{R}})\cdot(\mathbf{r} - \mathbf{r}')}\, e^{i(\omega - \omega_0)(t - t')}$$

$$\times \left\langle n\!\left(\mathbf{r}, t - \frac{|\mathbf{R} - \mathbf{r}|}{v}\right) n\!\left(\mathbf{r}', t' - \frac{|\mathbf{R} - \mathbf{r}'|}{v}\right) \right\rangle. \quad (13.18)$$

In order to simplify the arguments of the number density operators in (13.18) we change variables through the substitutions $t \to t + |\mathbf{R} - \mathbf{r}|/v$ and $t' \to t' + |\mathbf{R} - \mathbf{r}'|/v$. Hence

$$I(\mathbf{R}, \omega) = \lim_{T \to \infty} |\mathscr{B}|^2 \frac{1}{T} \iint_{-T/2}^{T/2} dt dt' \iint d\mathbf{r} d\mathbf{r}'\, e^{i(\mathbf{k}_0 - k_0 \hat{\mathbf{R}})\cdot(\mathbf{r} - \mathbf{r}')}$$

$$\times e^{i((\omega - \omega_0)/v)(\mathbf{r} - \mathbf{r}')\cdot\hat{\mathbf{R}}}\, e^{i(\omega - \omega_0)(t - t')} \langle n(\mathbf{r}, t) n(\mathbf{r}', t') \rangle, \quad (13.19)$$

where we have not changed the limits ($\pm T/2$) on the time integrals because at the end of the calculation we shall let $T \to \infty$, and we have again used the approximation $|\mathbf{R} - \mathbf{r}| \simeq R - \hat{\mathbf{R}} \cdot \mathbf{r}$ of eqn (13.11).

The result (13.19) can be simplified by introducing the momentum transfer vector $\mathbf{q} \equiv \mathbf{k}_s - \mathbf{k}_0$ where $\mathbf{k}_0$ is the incident wave vector and $\mathbf{k}_s$ is the scattered wave vector.

$$\mathbf{k}_s = \left(k_0 - \frac{\omega - \omega_0}{c/n}\right) \hat{\mathbf{R}}. \qquad (13.20)$$

Hence eqn (13.19) becomes

$$I(\mathbf{R}, \omega) = \lim_{T \to \infty} \frac{1}{T} |\mathscr{B}|^2 \iint_{-T/2}^{T/2} dt dt' \iint d\mathbf{r} d\mathbf{r}'\, e^{-i\mathbf{q}\cdot(\mathbf{r} - \mathbf{r}')}\, e^{i(\omega - \omega_0)(t - t')}$$

$$\times \langle n(\mathbf{r}, t) n(\mathbf{r}', t') \rangle. \quad (13.21)$$

Still further simplification arises on observing that

$$\langle n(\mathbf{r}, t) n(\mathbf{r}', t') \rangle = \langle n(\mathbf{r} - \mathbf{r}', t - t') n(0, 0) \rangle, \qquad (13.22)$$

which follows from the invariance of the equilibrium ensemble to translations in both time and space. If we substitute eqn (13.22) into (13.21) and change integration variables through the substitutions $t - t' \to t$ and $\mathbf{r} - \mathbf{r}' \to \mathbf{r}$, we can perform the t' and $\mathbf{r}'$ integrations.

We thereby obtain, in the thermodynamic limit ($V \to \infty$), the expression

$$\mathscr{I}(\mathbf{q}, \omega) = |\mathscr{B}|^2 \int_{-\infty}^{\infty} dt \int d\mathbf{r} \, e^{-i\mathbf{q}\cdot\mathbf{r} + i(\omega - \omega_0)t} \langle n(\mathbf{r}, t)n(\mathbf{0}, 0)\rangle. \qquad (13.23)$$

Here we have made the notational change $I(\mathbf{R}, \omega)/V \to \mathscr{I}(\mathbf{q}, \omega)$, i.e. we describe the position of the detector by $\mathbf{q}$ rather than by $\mathbf{R}$.

Equation (13.23) is essentially the final result—that the scattered intensity is proportional to the Fourier transform of the density–density correlation function. A more conventional form is obtained by introducing the density fluctuation $\delta n(\mathbf{r}, t)$, defined in eqn (13.2), whereupon eqn (13.23) becomes

$$\mathscr{I}(\mathbf{q}, \omega) = |\mathscr{B}|^2 \int_{-\infty}^{\infty} dt \int d\mathbf{r} \, e^{-i\mathbf{q}\cdot\mathbf{r} + i(\omega - \omega_0)t} \langle \delta n(\mathbf{r}, t)\delta n(\mathbf{0}, 0)\rangle$$
$$+ |\mathscr{B}|^2 (2\pi)^4 \langle n \rangle^2 \delta(\mathbf{q})\delta(\omega - \omega_0). \qquad (13.24)$$

The first term in the integrand of eqn (13.24) is due to fluctuations in density, while the second term contributes only for scattering that is in the forward direction ($q = 0$). In practice forward scattering cannot be measured (and our approximations are not valid for $q = 0$); hence it is customary to neglect the second term in (13.24) and simply write

$$\mathscr{I}(\mathbf{q}, \omega) = |\mathscr{B}|^2 \mathscr{S}(\mathbf{q}, \omega - \omega_0), \qquad \blacktriangleright \quad (13.25)$$

where the structure factor $\mathscr{S}(\mathbf{q}, \omega)$ is, as defined in eqn (13.4), the Fourier transform in space and in time of the density–density correlation function $\mathscr{G}(\mathbf{r}, t)$.

We next integrate both sides of (13.25) over all frequencies, obtaining

$$I(\mathbf{q}) \equiv \int_{-\infty}^{\infty} \frac{d\omega}{2\pi} \mathscr{I}(\mathbf{q}, \omega) = |\mathscr{B}|^2 \int_{-\infty}^{\infty} \frac{d\omega}{2\pi} \mathscr{S}(\mathbf{q}, \omega - \omega_0) = |\mathscr{B}|^2 S(\mathbf{q}),$$
$$(13.26)$$

where the first equality in (13.26) serves to define the function $I(\mathbf{q})$ in analogy with the definition in eqn (13.5) of the function $S(\mathbf{q})$. If we now divide both sides of eqn (13.25) by $I(\mathbf{q})$, we eliminate the proportionality factor $|\mathscr{B}|^2$ and obtain the remarkably simple result

$$\frac{\mathscr{I}(\mathbf{q}, \omega)}{I(\mathbf{q})} = \frac{\mathscr{S}(\mathbf{q}, \omega - \omega_0)}{S(\mathbf{q})}. \qquad \blacktriangleright \quad (13.27)$$

This normalization of the intensity $\mathscr{I}(\mathbf{q}, \omega)$ by the function $I(\mathbf{q})$ is particularly convenient; in the quasielastic approximation, $I(\mathbf{q})$ is the total intensity at a given $\mathbf{q}$ (scattering angle) when no frequency analysis of the scattered radiation is carried out.

Thus eqn (13.27) provides a direct link between an experimentally-measurable quantity $\mathscr{I}(\mathbf{q}, \omega)$ and a microscopic property of a system, $\mathscr{S}(\mathbf{q}, \omega)$, the Fourier transform of $\mathscr{G}(\mathbf{r}, t)$. This link was first elucidated by Van Hove (1954a, b, c) for the case of magnetic correlations, for which $\mathscr{I}(\mathbf{q}, \omega)$ is related to the neutron scattering cross-section and $\mathscr{G}(\mathbf{r}, t)$ is the two-spin correlation function. Van Hove's basic theory has been discussed for the specific case of light scattering by many workers, among whom are Rytov (1957), Komarov and Fisher (1962), Pecora (1964), Greytak (1967), Lastovka (1967), Benedek (1968), Swinney (1968), and Dubin (1970).

We conclude this subsection by stating for the sake of completeness, the result for $\mathscr{I}(\mathbf{q}, \omega)$ as obtained from consideration of the vector character of the electric field and the detailed scattering mechanism (van Kampen 1969),

$$\mathscr{I}(\mathbf{q}, \omega) = I_0 \left(\frac{\omega_0}{c}\right)^4 \left[\frac{\sin \varphi}{4\pi R}\left(\frac{\partial \epsilon}{\partial \rho}\right)_T\right]^2 \mathscr{S}(\mathbf{q}, \omega - \omega_0), \qquad (13.28)$$

where I_0 is the incident intensity, φ is the angle between the polarization of the incident light and the wave vector $\mathbf{k}_\mathrm{s}$ of the scattered light, and in this equation ϵ denotes the dielectric constant of the fluid.

13.4. Predictions of hydrodynamics for the spectrum of the scattered radiation

In the previous section we saw that experimental measurements of the spectrum of scattered radiation are directly related to the dynamic structure factor or Fourier transform of the time-dependent correlation function. In this section we therefore consider the *calculation* of the time-dependent correlation function. We shall see, however, that the correlation function can be calculated only when the wave vector of the fluctuation $\mathbf{q}$ is much smaller than the inverse correlation length κ—or, equivalently, $\lambda = 2\pi/q$, is appreciably larger than the correlation length ξ. Since the calculation of the structure factor is relatively complex even for this *hydrodynamic limit* ($q \ll \kappa$ or $\lambda \gg \xi$), we have presented the details in Appendix D. The desired expression for the dynamic structure factor is

$$\mathscr{S}_{nn}(\mathbf{q}, \omega)/S_{nn}(\mathbf{q}) = \left(1 - \frac{C_V}{C_P}\right)\frac{2D_T q^2}{\omega^2 + (D_T q^2)^2}$$

$$+ \frac{C_V}{C_P}\left\{\frac{\tfrac{1}{2}D_s q^2}{(\omega - v_s q)^2 + (\tfrac{1}{2}D_s q^2)^2} + \frac{\tfrac{1}{2}D_s q^2}{(\omega + v_s q)^2 + (\tfrac{1}{2}D_s q^2)^2}\right\}, \qquad \blacktriangleright (13.29)$$

where we have neglected higher-order terms involving the quantities $D_T q/v_s$ and $D_s q/v_s$. The quantities appearing in eqn (13.29) are the thermal diffusion constant or 'thermal diffusivity',

$$D_T \equiv \frac{\Lambda}{mnC_P} = \frac{\Lambda}{\rho C_P},$$ ► (13.30)

the sound-wave damping constant,

$$D_s \equiv D_T \left(\frac{C_P}{C_V} - 1\right) + D_\ell,$$ ► (13.31)

where $D_\ell \equiv (mn)^{-1}(\frac{4}{3}\eta + \zeta)$, and the velocity of sound for $\omega = 0$, given by

$$v_s^2 \equiv \frac{1}{m} \left(\frac{\partial P}{\partial n}\right)_S = \left(\frac{\partial P}{\partial \rho}\right)_S.$$ (13.32)

The result $\mathscr{S}(\mathbf{q}, \omega)/S(\mathbf{q})$ given by eqn (13.29) is plotted in Fig. 13.2(a) as a function of ω for a fixed value of the temperature that is *not* near the critical temperature T_c. From eqn (13.27) we see that the spectrum of the scattered light is predicted to be of the form of Fig. 13.2(b)— which is simply the curve of Fig. 13.2(a) shifted to being centred about the incident frequency ω_0.

The three terms in eqn (13.29) have the general form

$$f(\omega) = \frac{2\Gamma}{\Gamma^2 + (\omega - \omega')^2},$$ (13.33)

which is a normalized Lorentzian centred at the frequency ω' with a half-width at half maximum given by Γ. Thus the dynamic structure factor is predicted by our hydrodynamic model to consist of the sum of three Lorentzian lineshapes. One, called the Rayleigh component, is centred about $\omega = 0$, with a half-width

$$\Gamma_R = D_T q^2,$$ ► (13.34)

while the other two, called the Brillouin doublet, are located symmetrically about $\omega = 0$ at the frequencies

$$\omega_B^\pm = \pm v_s q,$$ (13.35)

each with a half-width given by

$$\Gamma_B = \tfrac{1}{2} D_s q^2.$$ ► (13.36)

Referring to the definitions in eqns (13.30) and (13.31) of the thermal diffusivity and the sound wave damping constant, respectively, we see that a knowledge of the linewidths of the Rayleigh and Brillouin components plus one additional transport coefficient suffices to determine all three transport coefficients.

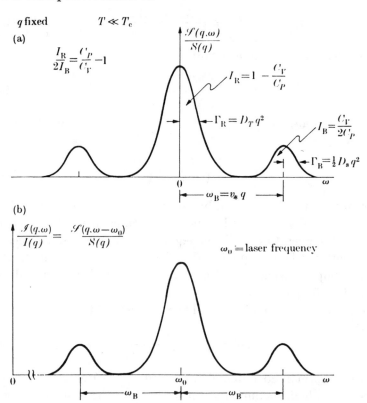

FIG. 13.2. (a) Dependence on ω of the structure factor $\mathscr{S}(\mathbf{q}, \omega)$, as predicted by the results of hydrodynamics, for fixed temperature T and wave vector $\mathbf{q}$. The ordinate has been divided by $S(\mathbf{q})$ in order that the integral over all ω will be unity regardless of T and q: $S(\mathbf{q})$ increases as $q \to 0$ and as $T \to T_c$. The temperature T is far from T_c, as can be judged by the fact that the integrated intensities of the Rayleigh and Brouillon peaks I_R and I_B, are of comparable magnitude ($C_P \simeq C_V$). (b) Dependence on ω of the normalized intensity of light $\mathscr{I}(\mathbf{q}, \omega)/I(\mathbf{q})$ observed in a typical scattering experiment. As shown in § 13.3 $\mathscr{I}(\mathbf{q}, \omega)/I(\mathbf{q})$ is simply $\mathscr{S}(\mathbf{q}, \omega)/S(\mathbf{q})$ translated from being centred about $\omega = 0$ to being centred about the laser frequency ω_0. Actually, in an experiment using optical mixing spectroscopy (either homodyne or heterodyne), this curve is translated back to $\omega = 0$ in order to obtain the resolving power necessary to measure the narrow linewidth of the central Rayleigh peak.

In addition to obtaining information about the transport coefficients from the spectrum we can obtain information about static properties. For example, it is possible to relate the ratio of the constant pressure and constant volume specific heats to the integrated intensities of the separate peaks of the spectrum. (It is meaningful to talk of the integrated intensities since in the derivation of eqn (13.29), we have assumed that $D_T q^2 \ll v_s q$ and $D_s q^2 \ll v_s q$, that is, the peaks are well-separated

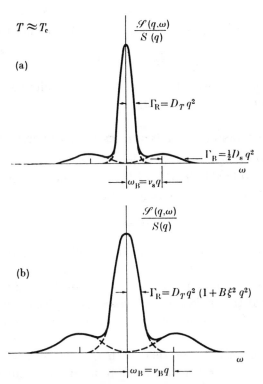

Fig. 13.3. Comparison between (a) the hydrodynamic prediction of eqn (13.29) and (b) the results of typical experiments for a temperature T that is so close to T_c that the approximation $q \ll \kappa$ is no longer valid. Here (13.29) predicts that the Brillouin doublet should slide in under the central Rayleigh peak (i.e. $v_B \to 0$ and $D_s \to \infty$), whereas experimentally we find that the Brillouin doublet remains well defined. Also (13.29) predicts for the intensity ratio $I_R/2I_B \sim \epsilon^{-(\gamma - a)}$, whereas we observe a somewhat different exponent. Finally, the Rayleigh linewidth is found to be slightly broader than predicted by (13.29) due to the correlation length correction, eqn (14.5).

from one another.) From (13.29) and (13.33), we see that the total integrated intensity of the central Rayleigh component is

$$I_R = 1 - \frac{C_V}{C_P}, \tag{13.37}$$

while the intensity of both of the Brillouin doublets is

$$2I_B = \frac{C_V}{C_P}. \tag{13.38}$$

Hence the intensity ratio or, as it has come to be called, the Landau–Placzek ratio, is

$$\frac{I_R}{2I_B} = \frac{C_P - C_V}{C_V} = \frac{C_P}{C_V} - 1. \qquad \blacktriangleright \quad (13.39)$$

It is worth remarking that (13.37) and (13.38) predict that

$$I_R + 2I_B = 1, \qquad (13.40)$$

which is consistent with the normalization requirement of eqn (13.5) for the integral over ω of (13.29).

13.5. Predictions of hydrodynamics near the critical point

Strictly speaking, it is meaningless to study the predictions of hydrodynamics in the limit $T \to T_c$, for as the critical temperature is approached the inverse correlation length

$$\kappa = \kappa_0 \epsilon^\nu \qquad (13.41)$$

approaches zero so that for T arbitrarily close to T_c the fundamental assumption of hydrodynamics $q \ll \kappa$, is no longer valid. This is not to say, however, that it is meaningless to talk of the predictions of hydrodynamics near the critical point, because for a given q there is certainly some range of temperature over which hydrodynamics should be valid.

Let us begin this section, then, with an order-of-magnitude estimate of this temperature range. We recall from Fig. 7.3 and eqn (7.24) that

$$q \equiv \frac{4\pi}{\lambda_0} \sin \tfrac{1}{2}\theta \simeq \frac{4\pi}{\lambda_s} \sin \tfrac{1}{2}\theta, \qquad (13.42)$$

where θ is the scattering angle, and where we have assumed that the wavelength of the incident radiation λ_0 is approximately equal to the wavelength of the scattered radiation λ_s. For example, if we use the $\lambda = 6328$ Å line of a helium–neon laser, and observe the scattered light at $\theta = 60°$, then

$$q \simeq 10^{-3} \text{ Å}^{-1} = 10^5 \text{ cm}^{-1}. \qquad (13.43)$$

We also see that if our highest energy excitations are the phonons that give rise to the Brillouin doublet at $\omega_B = \pm v_s q$, then for $v_s \simeq 10^5$ cm s^{-1} we have $\omega_B \simeq 10^{10}$ Hz, while

$$\omega_0 = c/\lambda_0 \simeq 10^{15} \text{ Hz}. \qquad (13.44)$$

Thus the quasi-elastic approximation $\lambda_s \simeq \lambda_0$ is seen to be excellently satisfied in (13.42).

In a typical one-component fluid the coefficient κ_0 in (13.41) has the value $\kappa_0 \simeq 1$ Å^{-1}, and the exponent ν is roughly $\frac{2}{3}$. Hence

$$\kappa \simeq \epsilon^{2/3} \text{ Å}^{-1} = 10^8 \ \epsilon^{2/3} \text{ cm}^{-1}. \tag{13.45}$$

Therefore, on comparing (13.43) and (13.45), we see that $q \ll \kappa$ providing

$$\epsilon \gg 10^{-5}. \tag{13.46}$$

Hence we might expect hydrodynamics to apply for values of the reduced temperature ϵ which are larger than about 10^{-3} or 10^{-4}, and we are motivated to consider the detailed predictions of eqn (13.29).

Incidentally, eqn (13.42) predicts that the range of applicability of hydrodynamics can be somewhat extended by measuring the spectrum of the scattered radiation at somewhat smaller scattering angles θ. It is also worth pointing out that since the wavelength of thermalized neutrons is only about 1 Å, it is rather more difficult to come close to the critical temperature and yet to remain in the hydrodynamic region in a neutron-scattering experiment. These considerations are summarized in the κ–q diagram of Halperin and Hohenberg (1967, 1969) reproduced in Fig. 12.2.

The ω dependence of the dynamic structure factor (13.29) is displayed schematically in Fig. 13.3(a) for $T \simeq T_c$. There are four features worthy of particular note. These are as follows:

(i) *Width of the Rayleigh component.* We see from eqns (13.34) and (13.30) that the half-width of the Rayleigh line is proportional to the ratio $\Lambda/\rho C_p$. If the thermal conductivity Λ diverges less strongly than the specific heat at constant pressure, then it follows that the Rayleigh linewidth will appear to approach zero.

(ii) *Width of the Brillouin component.* From eqns (13.36), (13.30), and (13.31) it follows that the half-width Γ_B of the Brillouin peaks remains finite providing all three transport coefficients Λ, η, and ζ remain finite at the critical-point. However, if either η or ζ diverge, or if Λ diverges faster than C_V, then D_s and hence the linewidth Γ_B will appear to diverge.

(iii) *Position of the Brillouin doublet.* The Brillouin doublet is located symmetrically about the central peak at frequencies $\omega_B^{\pm}$ given by (13.35), where v_s is given by eqn (13.32). Since the adiabatic compressibility $K_S \equiv \rho^{-1}(\partial \rho/\partial P)_S$ is predicted to diverge as $T \to T_c$, we see that ω_B in

(13.35) should appear to approach zero, i.e. the Brillouin doublet should begin to move in toward the central peak.

(iv) *The Landau–Placzek intensity ratio.* From eqn (13.39) we see that the intensity ratio $I_R/2I_B$ varies as $C_P/C_V = K_T/K_S$, which approaches infinity as $T \to T_c$. Hence the Rayleigh component becomes much more prominent than the Brillouin component as ϵ decreases.

Suggested further reading
Van Hove (1954a, b, c).
Landau and Lifshitz (1960).
Kadanoff and Martin (1963).
Mountain (1966).
Benedek (1968).
van Kampen (1969).

14

MEASUREMENTS OF THE DYNAMIC STRUCTURE FACTOR FOR FLUID SYSTEMS

In Chapter 13 we defined the dynamic structure factor $\mathscr{S}(\mathbf{q}, \omega)$ and demonstrated that it can be directly measured in scattering experiments. In this chapter, we discuss the results of such experiments, and we compare these results with the predictions of the hydrodynamic theory presented in § 13.4 and Appendix D.

14.1. Time-dependent density fluctuations

We begin with an extremely brief description of the early work on light scattering from fluid systems, referring the reader who desires further details to the treatments of Cummins and Swinney (1966), Cummins and Gammon (1966), Mountain (1966), Lastovka (1967), Benedek (1968), and Dubin (1970).

In 1871 Lord Rayleigh (Rayleigh 1871, 1899) solved the problem of light scattered by a gas of particles of sufficiently low density such that the interparticle spacing was larger than the wavelength of the light. He did not include the interactions between the particles of the gas in his calculations. With his famous result that the time average intensity of the scattered light is inversely proportional to the fourth power of the wavelength of the incident radiation, he was able to explain a variety of basic phenomena, including the blue colour of the sky.

Rayleigh's theory failed to account for the observed scattering from media whose interparticle separation is on the order of the wavelength of light. In particular, it could not account for critical opalescence, the dramatic increase in light scattering near the critical point (cf. Fig. 1.6). This behaviour had been observed by Andrews (1869) in carbon dioxide and given a qualitative explanation in his 1869 paper. Andrews' basic argument is that the density inhomogeneities that cause light scattering increase tremendously as the critical point is approached. Four decades later Einstein (1910) gave a more quantitative treatment of the critical opalescence phenomenon. In particular, he applied the principle that the wave vector of the fluctuation giving rise to the scattering is equal to the difference in wave vectors of the incident and scattered light waves. Since the Einstein treatment calculated only the mean square amplitude of the density inhomogeneities it could not predict the spectral distribution of the light scattered by the medium. Debye (1912) argued that the thermal content of a fluid could be considered to consist of adiabatically propagating pressure fluctuations (or sound waves). This analogy has proved to be quite useful. Brillouin (1914, 1922) and, independently, Mandel'shtam (1926), realized that the frequency of light scattered from these fluctuations should differ from that of the incident light by an amount equal to the frequency of a sound wave whose wave vector is equal to the difference in wave vectors of the incident and scattered light waves. These theoretical predictions of the Mandel'shtam–Brillouin doublet were experimentally verified by Gross (1930a, b, c; 1932). Gross' measurements also revealed the presence of a third component of the scattered light which was un-shifted in frequency. Instead of observing only the theoretically-pre-dicted doublet, he found a triplet of lines. The central unshifted line was ascribed to be an experimental artifact, until Landau and Placzek (1934) provided an interpretation justifying the existence of Gross' 'artifact'. The mathematics of the Landau–Placzek result was discussed in § 13.4, but the physical idea is worth emphasizing. The density fluctuations producing the scattering can be described in terms of pressure fluctuations at constant entropy and entropy fluctuations at constant pressure. The former are sound waves, which Brillouin and Mandel'stam found were responsible for the doublet, while the latter give rise to the central unshifted portion or Rayleigh component.

14.2. Optical mixing spectroscopy

Conventional spectrometers are not capable of measuring the ex-tremely narrow linewidths of the Rayleigh and Brillouin peaks. To get an idea of the degree of resolution that is needed even far from the critical point, we shall perform an order-of-magnitude calculation. From the data of Lastovka and Benedek (1966) for a typical liquid (in this case toluene), we have $\Lambda \simeq 10^{-3} J\,s^{-1}\,cm^{-1}\,deg^{-1}$, $\rho \simeq 1\,g/cm^3$, $C_P \simeq 2J/g\,deg$, and $q_{max} \simeq 10^5\,cm^{-1}$. Hence the maximum half-width (in cycles) of the Rayleigh line, Γ_R^{max}, is given by

$$(2\pi)^{-1}\Gamma_R^{max} = (2\pi)^{-1}D_T q^2 = (2\pi)^{-1}(\Lambda/\rho C_P)q^2 \simeq 10^6\,Hz = 1\,MHz.$$
$$(14.1)$$

Thus the measurement of the Rayleigh linewidth with a 6328Å He–Ne laser would require a resolving power of at least

$$(\omega/\Delta\omega) \simeq (3 \times 10^{15})/(2\pi \times 10^{6}) \simeq 10^{9}. \qquad (14.2)$$

Only since the development of the optical heterodyne and self-beating spectrometer has such precise resolution become possible. These new techniques of optical mixing spectroscopy were first proposed by Forrester, Parkins and Gerjuoy (1947) and, independently, by Gorelik (1947). Particularly recommended are the recent review articles of Benedek (1968, 1969) and French, Angus, and Walton (1969).

The basic idea of this new type of spectroscopy is that a suitable non-linear detector can be used to observe the beat note between light waves of slightly different frequencies. The beat notes are produced by beating a standard frequency source (such as a laser) against the scattered light. The result is that the central peak of the spectrum of the scattered light is shifted from being centred about $\omega_0 \simeq 10^{15}$Hz to being centred about $\omega = 0$. By this means, the desired spectral information is translated to lower frequencies where it can be more easily detected. This is the same heterodyne principle that has long been used to obtain high resolution in radio broadcasting.

A second type of high-resolution, optical-beating spectroscopy employs the homodyne or self-beating spectrometer. In this case, only the scattered light is allowed to fall on the photodetector, which uses its non-linear characteristics to mix the spectral components of the signal itself. This technique again leads to a translation of the spectral information from a spectrum centred about the laser frequency to a spectrum centred about $\omega = 0$.

The principle of optical-mixing spectroscopy was first demonstrated in 1955 (Forrester, Gudmundsen, and Johnson 1955; Forrester 1956), but it was not until the advent of the laser in the early 1960s that the technique became of practical significance (Forrester 1961, Townes, 1961, Benedek, 1962). Cummins, Knable, and Yeh (1964), Ford and Benedek (1965), Lastovka and Benedek (1966), and Dubin, Lunacek, and Benedek (1967) developed the techniques still further to the extent that nowadays a resolving power on the order of 10^{14} is not at all uncommon.

14.3. Measurements of the Rayleigh linewidth

We will first discuss measurements of the Rayleigh linewidth far from the critical point, using the experimental techniques described in the previous section. Fig. 14.1 shows a block diagram of the optical heterodyne spectrometer used by Lastovka and Benedek (1966) to study the spectrum of light scattered from liquid toluene. Also, we display a recorder tracing of the spectrum observed at a very small

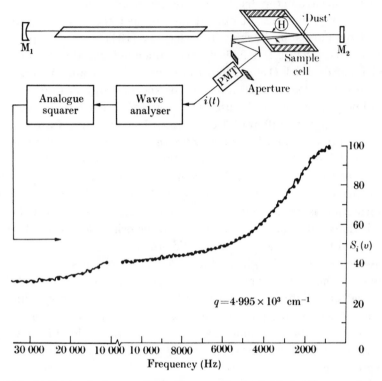

FIG. 14.1. At the top is shown a block diagram of a typical heterodyne spectrometer (Lastovka and Benedek 1966). The local oscillator of frequency equal to the laser frequency ω_0 is provided by the dust on the sample cell windows, for this dust scatters the laser light elastically. At the bottom is shown the measured photocurrent $S_i(\nu)$ as a function of frequency $\nu = \omega/2\pi$, for scattering from toluene far from T_c. For the particular small scattering angle shown, q is only 4995 cm^{-1} and hence the width of the Rayleigh peak, $\Gamma_R = D_T q^2$ is exceedingly narrow. In fact, the heavy dots represent a fit to a Lorentzian line shape whose half-width is only 3374 Hz, so that the effective resolving power for these measurements is approximately 10^{12}; this is several orders of magnitude higher than resolving powers obtained using conventional spectroscopic techniques.

scattering angle $q \simeq 5 \times 10^3$ cm^{-1}. Since q in Fig. 14.1 is reduced by a factor of 20 from the value of the order-of-magnitude calculation presented above, the Rayleigh linewidth of eqn (14.1) should be reduced by a factor of 400. Consequently, our estimate of the linewidth would be $\Gamma_R \simeq 2 \cdot 5 \times 10^3$ Hz instead of 10^6Hz. We see that the data of

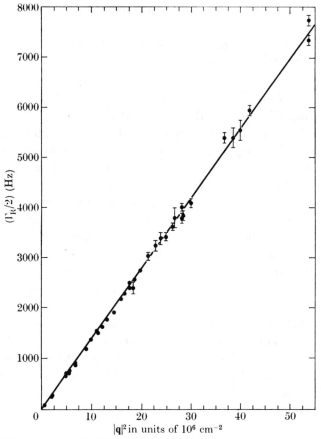

FIG. 14.2. Dependence of the half-width of the Rayleigh component on q^2, the square of the wave vector, for scattering from toluene far from the critical point. These measurements, obtained using the heterodyne spectrometer of Fig. 14.1 extend over two orders of magnitude in linewidth, from widths as narrow as 75 Hz to widths as large as 7500 Hz. The fact that almost all the data lie on the straight line shown indicates that Γ_R is, to within experimental error, linear in q^2. Thus the correlation length correction term of eqn (14.5) is evidently negligible for these measurements on toluene far from the critical point. After Lastovka and Benedek (1966).

Fig. 14.1 have the shape of a Lorentzian with $\Gamma_R = 3374$ Hz; this Lorentzian curve is also just the type that is characteristic of the predictions of hydrodynamic theory. Moreover, as we vary the scattering angle, and thus the momentum transfer vector $q = (4\pi/\lambda_0) \sin \frac{1}{2}\theta$, the hydrodynamic prediction, eqn (13.34), is that the Rayleigh linewidth varies as q^2. In Fig. 14.2 we see that this behaviour is supported by these experiments on toluene at temperatures far from the critical temperature.

Accurate measurements of the Rayleigh linewidth near the critical point have been carried out for only a few fluids such as carbon dioxide, xenon, and sulphur hexafluoride. For simplicity, we discuss principally the work using carbon dioxide.

Experimental set-up

We first describe in some detail the method of measuring the Rayleigh linewidth (Swinney and Cummins 1968). We first fill a sample cell with a known quantity of CO_2 fluid, chosen such that the density ρ exceeds the critical density ρ_c by only a very small amount (about 0·3 per cent). The cell is now sealed off. It is important to stress that even for $T > T_c$. the density of the fluid as a function of position in the cell will vary considerably more than 0·3 per cent from the critical density due to the effect of the earth's gravitational field (i.e. the fluid near the top of the sample cell will be considerably less dense than ρ_c, while the fluid at the bottom will be more dense). The earth's field has such a pronounced effect near the critical point because the isothermal compressibility, $K_T = \rho^{-1}(\partial\rho/\partial P)_T$, takes on values in the critical region as much as a million times larger than in the normal domain. We then focus a low-power 6328Å He–Ne laser to a diameter of 0·2 mm inside the sample and observe the scattered light at the desired angle. As the cell is scanned in the vertical direction the beam samples a range of densities.

Results

Using a self-beating spectrometer, Swinney and Cummins (1968) measured Γ_R/q^2 as a function of beam height for a variety of temperatures. Their results are shown in Fig. 14.3 where the abscissa, 'height', is a measure of the beam position above or below the meniscus. The value of Γ_R/q^2 along the coexistence curve is determined by extrapolation to zero height of the data in each region. We see from Fig. 14.3(a) that the Rayleigh linewidth appears to be smaller on the liquid side of the coexistence curve than on the gas side. In Fig. 14.3(b), we show the analogous data for temperatures above the critical temperature, and we observe that the linewidth dips sharply at the critical density. The values of Γ_R/q^2 for this critical isochore are plotted as a function of $T - T_c$ in Fig. 14.4(a).

Hydrodynamics predicts that

$$\Gamma_R/q^2 = D_T = (\Lambda/\rho C_P) \sim \epsilon^{\gamma-a}. \qquad (14.3)$$

Here we have assumed that C_V diverges with an exponent less than that of C_P in order to infer from an identity analogous to (2.54),

$$C_P - C_V = TVK_T\left(\frac{\partial P}{\partial T}\right)_V^2, \tag{14.4}$$

that C_P diverges with the same exponent as K_T, the isothermal compressibility. Equation (14.3), together with the definition in eqn (3.2) of a critical-point exponent, predicts that a plot of the logarithm of $(\Lambda/\rho C_P)$ against the logarithm of ϵ should, for sufficiently small values of

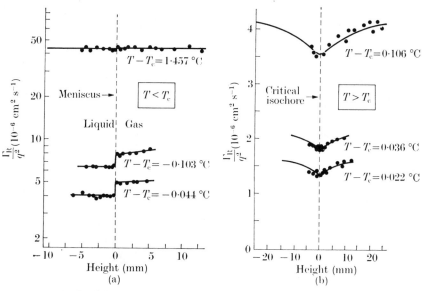

FIG. 14.3. The Rayleigh linewidth Γ_R divided by the square of the momentum transfer vector q^2 as a function of height, where the zero of the height scale is chosen to be the meniscus for the subcritical data (a) and the minimum in Γ_R/q^2 for the supercritical data (b). The curves taken from Swinney and Cummins (1968), work on CO_2, are extended symmetrically below the zero of height because Γ_R/q^2 is symmetrical about the critical density ρ_c. Notice from (a) that as we approach the coexistence curve from the liquid side the limiting value of Γ_R/q^2 is lower than when we approach from the gas side. This difference is due to different behaviour on the liquid and gas sides of the coexistence curve (i.e. $\gamma'_L - a'_L > \gamma'_G - a'_G$).

ϵ, approach a straight line with slope given by $(\gamma - a)$. This is indeed the case for the small q, or small scattering angle ($\theta = 22°$), measurements shown in Fig. 14.4(a), but we see that for larger scattering angles (cf. the $\theta = 90°$ curve), the power law behaviour predicted by hydrodynamics and eqn (14.3) begins to break down for temperatures closer than about $0.1°$ from T_c. Since the value of T_c for CO_2 is about $31°C$ or

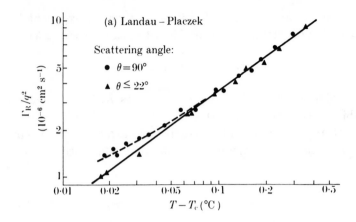

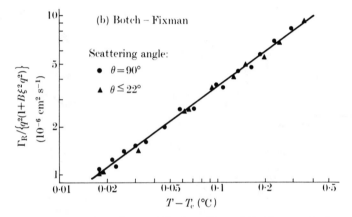

Fig. 14.4. (a) shows the dependence of the logarithm of Γ_R/q^2 upon the logarithm of $(T - T_c)$ for CO_2 (from Swinney and Cummins (1968)). Linearized hydrodynamics predicts that Γ_R/q^2 should be equal to $D_T = \Lambda/\rho C_P \sim \epsilon^{\gamma - a}$ and hence that this plot should be a straight line with slope given by $\gamma - a$. We see that the assumptions of hydrodynamics apparently fail for the large angle measurements, since for large θ ($q \propto \sin \theta/2$) the approximation $q \ll \kappa$ is no longer valid for $T - T_c \lesssim 0.1°$. (b) shows the temperature dependence of the quantity $\Gamma_R/\{q^2(1 + B\xi^2 q^2)\}$, which acccording to the correlation length correction, eqn (14.5), should be given by D_T. The linearity of this plot supports the general validity of eqn (14.5), but does not help in choosing the value of the constant B since the correlation length $\xi = 0.53 \, \epsilon^{-2/3}$ Å was so chosen to obtain a linear fit with $B = 1$. Kawasaki (1970) finds that $B = \frac{3}{5}$.

304 K, this apparent breakdown of the hydrodynamic predictions is occurring for $\epsilon < 10^{-1}/(3 \times 10^2) \simeq 3 \times 10^{-4}$. Thus we see that our order-of-magnitude estimate of eqn (13.46) concerning just how close we could get to T_c before hydrodynamics would break down is roughly confirmed.

14.4. Corrections to the hydrodynamic theory of the Rayleigh linewidth

14.4.1. *Theoretical predictions: the correlation length correction*

For $\epsilon < 3 \times 10^{-4}$, the predictions of the hydrodynamic theory for critical phenomena as developed in § 13.4 and Appendix D evidently fail, and we must seek a better theoretical understanding of the phenomena in the critical region (region II of Fig. 12.2). Since the basic assumptions underlying the hydrodynamic theory (that $q \ll \kappa$ or $q\xi \ll 1$) are no longer valid, there is no particular reason to expect that a simple modification of the hydrodynamic theory will be adequate.

There are, however, other theoretical approaches which have been successful in predicting the Rayleigh linewidth data of Swinney and Cummins. In the first of these, Botch and Fixman (1965)—and subsequently, Felderhof (1966), Mountain (1966), and Cummins and Swinney (1966)—proposed that the q^2 factor in the hydrodynamic result (13.34) for the Rayleigh linewidth, $\Gamma_{\mathrm{R}} = D_T q^2$ is actually only the first term in a series in ascending powers of q^2, and that for large q we should apply instead the modified formula

$$\Gamma_{\mathrm{R}} = D_T q^2 (1 + B\xi^2 q^2), \qquad \blacktriangleright \ (14.5)$$

with $B = 1$, and where ξ is, as before, the correlation length. Equation (14.5) is frequently called the *Fixman correction* to the hydrodynamic result. Perhaps the most pedagogical derivation is that of Cummins and Swinney (1966). These authors reason that since Debye (1959) showed that the Ornstein–Zernike theory was able to provide an improvement over the simpler Einstein theory of static critical phenomena by taking into account the marked increase in the correlation among particles near the critical point, a similar modification of the hydrodynamic theory should be applicable in the critical region for dynamic phenomena. Hence they obtain eqn (14.5) as a modification of the hydrodynamic result for Γ_{R}, eqn (13.34).

The second treatment of the Rayleigh linewidth is the dynamic scaling law argument of Halperin and Hohenberg (1967), which will be discussed in detail in Chapter 15. The dynamic scaling approach is extremely general and appealing. However, in so far as the Rayleigh linewidth is concerned, we shall find that it is capable of predicting the form of eqn (14.5) but it cannot predict the value of the constant B.

Yet a third theoretical discussion of the Rayleigh linewidth is due to Kawasaki (1970), in which the mode–mode coupling approximation is applied. This approach—also discussed in Chapter 15—leads to the prediction that $B = \frac{3}{5}$ and agrees (better than the Botch–Fixman prediction $B = 1$) with the most recent experiments. Although the Fixman correction to the Rayleigh linewidth is a relatively small term in the case of CO_2 and other one-component fluids, it has a much more significant effect in the case of binary mixtures.

14.4.2. Experimental tests of the correlation length correction for CO_2 and Xe

Provided the detailed temperature dependence of ξ is known, the correlation length correction of eqn (14.5) can be put to a rather severe test. Unfortunately this is not the case for carbon dioxide. Hence Swinney and Cummins make the plausible assumption that ξ diverges with a power law form,

$$\xi = \xi_0 \, \epsilon^{-\nu}, \tag{14.6}$$

with two adjustable parameters, ξ_0 and ν. Since the three-dimensional Ising model is a crude lattice-gas model for a one-component fluid (cf. § 1.1 and Appendix A) and since numerical calculations (Jasnow and Wortis, 1968) predict that the exponent ν has a value of about $\frac{2}{3}$ independent of particular details of Ising interaction (such as lattice structure), Swinney and Cummins (1968) choose $\nu = \frac{2}{3}$ in eqn (14.6). The same sort of numerical calculations, when used to obtain a value for the coefficient ξ_0, predict that ξ_0 does depend upon the details of the interaction, and hence might be expected to vary from fluid to fluid. Hence Swinney and Cummins have chosen to regard ξ_0 as an adjustable parameter. They have found that the choice

$$B^{-\frac{1}{2}}\xi_0 = 0 \cdot 53 \pm 0 \cdot 11 \text{Å} \tag{14.7}$$

permits their data to be fitted to eqn (14.5) (they used the Botch–Fixman choice $B = 1$); i.e. when eqns (14.6) and (14.7) are substituted into eqn (14.5), together with the choice $\nu = \frac{2}{3}$, $D_T = \Gamma_R / \{q^2(1 + \xi^2 q^2)\}$ is found to have a simple power law dependence as shown in Fig. 14.4(b).

It is important to remark that although Fig. 14.4(b) provides evidence supporting the general form of eqn (14.5), it says nothing whatsoever regarding the value of the constant B. Certainly much more convincing evidence would be provided if we were able to obtain the quantities ξ_0 and ν in eqn (14.6) from independent experiments on CO_2. Although ξ_0 and ν have not yet been directly measured for CO_2, they have recently been determined (Giglio and Benedek 1969) for xenon, with the results

$\xi_0 = 1\cdot8 \pm 0\cdot2\text{Å}$ and $\nu' = 0\cdot57 \pm 0\cdot05$. For xenon ξ_0 is almost four times larger than the ξ_0 obtained from the CO_2 analysis of (14.7) and, more important, the measured value of ν' in Xe is 15 per cent smaller than the assumed value ($\frac{2}{3}$) of ν for CO_2. This suggests that it would be extremely desirable either (a) to measure $\xi = \xi_0 \, \epsilon^{-\nu}$ independently for CO_2 and hence permit a more searching analysis of the Swinney–Cummins linewidth data, or (b) to perform Rayleigh linewidth measurements on xenon analogous to those of Swinney and Cummins (1968) on CO_2. Alternative (b) has been undertaken by Henry, Cummins and Swinney (1969, 1970), and the data thus far available serve to suggest that the general form of eqn (14.5) is indeed valid, but that the constant B should have the value $\frac{3}{5}$ predicted by the mode–mode coupling theory, rather than the value unity predicted by the Botch–Fixman result.

14.4.3. Critical-point exponents for CO_2, Xe, and SF_6

The final topic that we discuss is the values of the critical-point exponents. The exponent $\gamma - a$, which describes the approach to zero of $D_T = (\Lambda/\rho C_P)$, is simply obtainable as the slope of the straight line in Fig. 14.4(b). In calculating the slope from a log–log plot, such as that shown in Fig. 14.4(b), we must be careful to calculate the differences of the logarithms rather than of the number themselves. Thus, for example, we can obtain for the slope of the straight line

$$\gamma - a \simeq \frac{\ln 10 - \ln 1}{\ln 0\cdot40 - \ln 0\cdot017} = \frac{\ln 10}{\ln 23\cdot5} = 0\cdot73, \qquad (14.8)$$

which agrees with the value $\gamma - a = 0\cdot73 \pm 0\cdot02$ calculated from a least squares fit by Swinney and Cummins (cf. Table 14.1). This value is also consistent with earlier and somewhat less extensive measurement of the Rayleigh linewidth by Siegel and Wilcox (1967) and with some early thermodynamic measurements of D_T by Sengers (1966).

TABLE 14.1

Critical-point parameters of representative fluid systems discussed in this chapter

Fluid	T_c(K)	P_c(atm)	ρ_c(g cm^{-3})	$T > T_c$ $\gamma - a$	$T < T_c$ $\gamma'_L - a'_L$	$\gamma'_G - a'_G$
CO_2	304·23	72·85	0·468	0·73 $\pm$ 0·02	0·72 $\pm$ 0·05	0·66 $\pm$ 0·05
Xe	289·75	57·64	1·105	0·751 $\pm$ 0·004	—	—
SF_6	318·71	37·11	0·730	1·26 $\pm$ 0·02	0·635 $\pm$ 0·003	0·632 $\pm$ 0·002

We must determine independently the isothermal compressibility exponent γ (in order to obtain a numerical value for the thermal conductivity exponent a). A recent analysis of experimental data on CO_2 (Sengers 1971) suggests that γ is very close to the value $\frac{5}{4}$ which is predicted theoretically for the three-dimensional Ising model. Hence we calculate for a the value

$$a = (1 \cdot 26 \pm 0 \cdot 05) - (0 \cdot 73 \pm 0 \cdot 02) \simeq 0 \cdot 53 \qquad (14.9)$$

On the liquid and gas sides of the coexistence curve the measured exponents are equal within the experimental error, but the data strongly suggest that the approach to zero of D_T is somewhat sharper on the gas side than on the liquid side, i.e.

$$\gamma'_G - a'_G < \gamma'_L - a'_L. \qquad (14.10)$$

No intuitive or formal explanation for this asymmetry has been advanced as yet.

For Xe, Henry *et al.* (1969, 1970) find behaviour qualitatively similar to that in CO_2, as we indicate in Table 14.1. However, for SF_6 it appears that the linewidth approaches zero much more rapidly for $T > T_c$ than for $T < T_c$, with $\gamma - a = 1 \cdot 26 \pm 0 \cdot 02$ while $\gamma' - a' \simeq 0 \cdot 63$ (Benedek 1969). The source of this marked asymmetry between the high-temperature indices and the low-temperature indices has been the object of considerable discussion since its discovery (cf. the very recent work of Braun, Hammer, Tscharnuter, and Weinzierl 1970).

14.5. Measurements of the Brillouin peak: velocity and attenuation of hypersonic sound waves

In this section we discuss measurements of the Brillouin doublet. We recall from § 13.5 that hydrodynamics makes predictions concerning (i) the position of the doublet (related to the sound velocity), (ii) the linewidth of the doublet (related to a combination of all three transport coefficients through eqns (13.30) and (13.31)), and (iii) the ratio of the integrated intensity of the central Rayleigh peak to the intensity of Brillouin doublet. We shall find that all three of these predictions of hydrodynamics must be corrected in order to explain recent experimental measurements.

(i) *Position of Brillouin doublet*

According to the discussion in § 13.4, the Brillouin doublet is located symmetrically on either side of the central Rayleigh peak at a frequency

$$\omega_B^{\pm} \equiv \pm v_B q. \qquad (14.11)$$

This expression differs from eqn (13.35) in that v_B is not necessarily the same as the velocity of sound waves in the zero-frequency limit, given by the expression

$$v_s^2 = (\rho K_S)^{-1} = C_P/(\rho C_V K_T) = C_P/\{C_V(\partial\rho/\partial P)_T\} \quad (14.12)$$

where

$$K_S \equiv -\frac{1}{V}\left(\frac{\partial V}{\partial P}\right)_S = +\frac{1}{\rho}\left(\frac{\partial \rho}{\partial P}\right)_S \quad (14.13)$$

is the adiabatic compressibility. The second equality in (14.12) follows from the thermodynamic identity of eqn (2.29), $C_P/C_V = K_T/K_S$.

If v_B were equal to its zero-frequency limit v_s, then we would expect that the magnitude of the Brillouin shift to approach zero as

$$v_s \sim \epsilon^{\alpha/2} \quad (14.14)$$

because the divergences in $C_P(\sim \epsilon^{-\gamma})$ and $K_T(\sim \epsilon^{-\gamma})$ (cf. eqn (14.4)) exactly cancel each other in eqn (14.12). The exponent α is found from specific heat experiments to lie in the range 0·0–0·2. Gammon, Swinney, and Cummins (1967) point out that for conventional ultrasonic measurements at frequencies on the order of $\omega_B \simeq 1$ MHz the sound velocity can in fact be fitted to a logarithmic singularity ($\alpha = 0$),

$$v_s \sim (\ln \epsilon)^{-1/2}, \quad (14.15)$$

over the temperature range $0\cdot4°C \leq T - T_c \leq 10°C$, so that it indeed looks as if the sound velocity is approaching zero. The Brillouin scattering measurements, on the other hand, are concerned with frequencies three orders of magnitude larger (about 500–1000 MHz). Thus, eqn (14.15) is not necessarily applicable. In fact, it is found experimentally that v_B is almost constant for temperatures less than about 1°C from the critical temperature (Gammon *et al.* 1967, Ford, Langley, and Puglielli 1968; Benedek and Cannell 1968). Hydrodynamics is found to be valid for the Rayleigh linewidth for $T - T_c \gtrsim 0\cdot1°C$; for the Brillouin frequency ω_B, the zero-frequency hydrodynamic prediction (14.14) fails ten times farther from the critical temperature! The κ–q diagram of Fig. 12.2 is not sufficient to explain this effect since this breakdown occurs much farther from T_c than the temperature at which we might expect hydrodynamics to fail.

(ii) *Width of the Brillouin doublet*

The experimental problems associated with measuring the width of the Brillouin doublet are complicated by the fact that its intensity is about three orders of magnitude weaker than the Rayleigh peak in the vicinity of the critical point. Nevertheless, extremely accurate measurements have recently been made of the linewidth for temperatures as close as $0\cdot015°C$ from the critical point of CO_2 (Ford *et al*. 1968, Benedek and Cannell 1968). We see from Fig. 14.5(b) that the linewidth ceases to

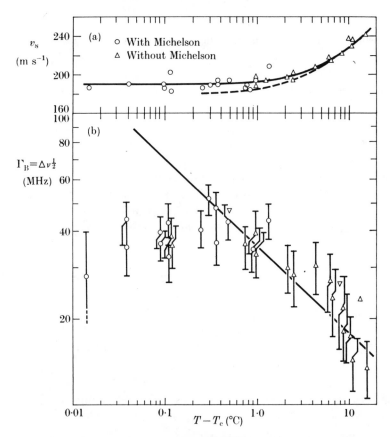

Fig. 14.5. Dependence on $T - T_c$ of (a) the sound velocity v_B as calculated from the frequency of the Brillouin doublet, and (b) the half-width Γ_B of the Brillouin doublet. The fact that the data become essentially constant for $T - T_c \lesssim 1\cdot0°$ is explained by the mode–mode coupling theory. All the data are for CO_2: the inverted triangles in (b) are from the work of Benedek and Cannell (1968) and the remainder of the data are those of Ford *et. al*. (1968).

increase at the same temperature at which the sound velocity ceases to *decrease*. The mode–mode coupling theory is capable of explaining this behaviour as we shall see in Chapter 15.

(iii) *Behaviour of the Landau–Placzek ratio*

To the extent that it is possible to measure the Brillouin linewidth, we can calculate the integrated intensity I_B and hence study the dependence of the Landau–Placzek ratio $I_R/2I_B$ upon temperature. Zero-frequency hydrodynamics predicts that the Landau–Placzek ratio is given by eqn (13.39), so that as $T \to T_c$,

$$\frac{I_R}{2I_B} = \frac{C_P}{C_V} - 1 \sim \epsilon^{-(\gamma - \alpha)}. \qquad (14.16)$$

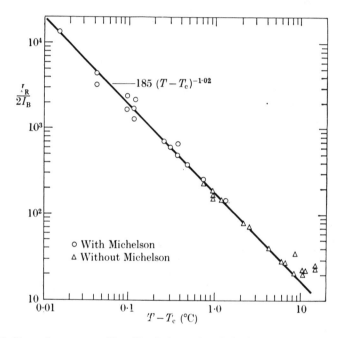

FIG. 14.6. Dependence upon $T - T_c$ of the ratio of the integrated intensity of the Rayleigh line to that of the Brillouin doublet as $T \to T_c^+$ along the critical isochore. The fact that the slope of this log–log plot is a constant ($\simeq 1 \cdot 02$) over three decades of temperature is strong evidence favouring the simple power-law divergence shown. After Ford *et al.* (1968).

There have been two recent experimental measurements of the temperature dependence of $I_R/2I_B$ near the critical-point. In the first of these, Gammon *et al.* (1967) find that the Landau–Placzek ratio diverges with an exponent $0 \cdot 95 \pm 0 \cdot 15$ over the temperature interval $0 \cdot 1 °C < T - T_c < 10 °C$. More recently, Ford *et al.* (1968) have found that an exponent of $1 \cdot 02 \pm 0 \cdot 03$ is sufficient to fit their data (cf. Fig. 14.6) over three decades of temperature ($0.01°C < T - T_c < 10°C$). Using the value for γ suggested by Sengers (1971) and the value of α derived from specific heat measurements, we find that the Landau–Placzek ratio should diverge with an exponent of $(\gamma - \alpha) \simeq \{1.26 \pm 0.05 - (0.2)\} = 1.06 \pm 0.05$. It has not yet been satisfactorily explained why hydrodynamics breaks down for the Brillouin position measurement but not for the Landau–Placzek ratio divergence.

Suggested further reading

Sengers (1966).
Benedek (1968).
Swinney and Cummins (1968).
Benedek (1969).
Chu (1970).
Cummins and Swinney (1970).
Garland (1970).
Sengers (1971).

15

DYNAMIC SCALING LAWS AND THE
MODE–MODE COUPLING APPROXIMATION

In Chapter 13 we presented the classical hydrodynamic theory of dynamic critical phenomena, realizing full well that this theory is expected to fail as soon as we get sufficiently close to T_c that we are no longer in one of the hydrodynamic regions of the κ–q diagram of Fig. 12.2. Not surprisingly, then, when we examined recent experimental measurements—in Chapter 14—we found that the hydrodynamic theory begins to fail approximately when the assumption $q \ll \kappa$ begins to fail. We also noted certain discrepancies between the data and the hydrodynamic theory in the hydrodynamic region itself. It is our purpose in this chapter to describe two recently-proposed theoretical approaches that have been successful in interpreting many—though not all—of the results of these recent experiments. The first of these approaches represents an attempt to generalize the static scaling approach to the case of dynamic phenomena, and is capable of making predictions in all regions of the κ–q diagram. However, like the static scaling laws, these predictions depend on as yet unverified assumptions concerning certain functions. The second of these approaches, called the mode–mode coupling theory by its originators, represents a method of utilizing the principles of the static scaling laws in order to calculate, in an approximate fashion, the divergent part of the transport co-efficients that arises from the interaction among different modes of excitation of the system (such as sound waves, heat modes, and viscous modes).

15.1. Dynamic scaling hypothesis

The dynamic scaling hypothesis was first formulated in connection with the superfluid transition in helium by Ferrell, Menyhárd, Schmidt, Schwabl, and Szépfalusy (1967a, b; 1968). Here we shall present a reformulation of their approach which was developed by Halperin and Hohenberg (1967, 1969). This approach is similar in some respects to the Halperin–Hohenberg formulation of static scaling that was presented in § 12.3. In particular, whereas the static scaling hypothesis involves a characteristic inverse length κ (or length ξ), we shall see that the dynamic scaling hypothesis involves both a characteristic inverse length and a characteristic frequency ω_c (or inverse time). This characteristic frequency will be a function of T, and since a given value of the inverse correlation length κ corresponds in general to two values of the temperature (one value below T_c and the other value above T_c), we can write

$$\omega_c \rightarrow \omega_c(T, \mathbf{q}) \rightarrow \omega_c^{\pm}(\kappa, \mathbf{q}), \tag{15.1}$$

where the superscript $\pm$ denotes the sign of $T - T_c$ and is frequently omitted when no confusion should arise. Thus we are led to consider the characteristic frequency $\omega_c^{\pm}(\kappa, \mathbf{q})$ to be a function defined over the κ–q plane of Fig. 12.2.

A convenient definition of the characteristic frequency is that ω_c is that frequency such that precisely half of the total integrated area under a plot of the dynamic structure factor as a function of frequency arises from frequencies in the interval $-\omega_c \leq \omega \leq \omega_c$. That is, ω_c is defined through the relation

$$\frac{\displaystyle\int_{-\omega_c}^{\omega_c} \frac{d\omega}{2\pi} \mathscr{S}^{\pm}(\kappa, \mathbf{q}, \omega)}{\displaystyle\int_{-\infty}^{\infty} \frac{d\omega}{2\pi} \mathscr{S}^{\pm}(\kappa, \mathbf{q}, \omega)} = \int_{-\omega_c}^{\omega_c} \frac{d\omega}{2\pi} \{\mathscr{S}^{\pm}(\kappa, \mathbf{q}, \omega)/S^{\pm}(\kappa, \mathbf{q})\} = \tfrac{1}{2}. \tag{15.2}$$

In eqn (15.2) we have indicated through the inverse correlation length the temperature dependence of the dynamic structure factor, that is, in analogy with (12.32) we have replaced the dependence on T, $\mathbf{q}$, and ω with a dependence on κ, $\mathbf{q}$, and ω

$$\mathscr{S}(\mathbf{q}, \omega) \rightarrow \mathscr{S}(T, \mathbf{q}, \omega) \rightarrow \mathscr{S}^{\pm}(\kappa, \mathbf{q}, \omega). \tag{15.3}$$

It is worth emphasizing that eqn (15.2) serves to define a characteristic frequency not only for the hydrodynamic regions but for all values of $(\mathbf{q}, \kappa)$.

We further define a dimensionless function $\mathscr{F}^{\pm}(\kappa, \mathbf{q}, \omega/\omega_c)$, called the *shape function*, through the relation

$$\mathscr{S}^{\pm}(\kappa, \mathbf{q}, \omega) = \frac{2\pi}{\omega_c(\kappa, \mathbf{q})} S^{\pm}(\kappa, \mathbf{q}) \mathscr{F}^{\pm}\left(\kappa, \mathbf{q}, \frac{\omega}{\omega_c}\right). \tag{15.4}$$

From eqns (13.5) and (15.4) we see that shape function $\mathscr{F}^{\pm}(\kappa, \mathbf{q}, x)$ when considered as a function of $x \equiv \omega/\omega_c$ is normalized to unit area, and that when $x = 1$, ω is equal to the characteristic frequency of the system. Examples of possible forms of the shape function are shown in Fig. 15.1.

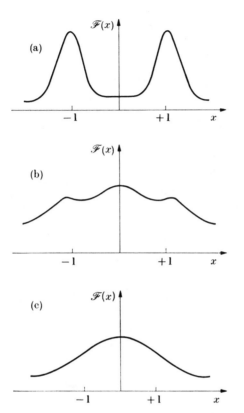

FIG. 15.1. Possible forms for the shape function $\mathscr{F}(\kappa, \mathbf{q}, x)$ defined in eqn (15.4). Since $x \equiv \omega/\omega_c$, the definition of the characteristic frequency ω_c as being that frequency such that half the area of the frequency spectrum lies between $\omega = -\omega_c$ and $\omega = +\omega_c$ is made apparent in this figure. After Halperin and Hohenberg (1969).

We are now in a position to state and appreciate the dynamic scaling hypothesis, which consists of the following two assumptions:

(i) $\mathscr{F}^{\pm}(\kappa, \mathbf{q}, x)$ is a function only of q/κ,

$$\mathscr{F}^{\pm}(\kappa, \mathbf{q}, x) = F^{\pm}(q/\kappa, x), \qquad \blacktriangleright \quad (15.5)$$

(ii) $\omega_c(\kappa, \mathbf{q})$ is a homogeneous function of the variables κ and q of degree z,

$$\omega_c^{\pm}(\kappa, \mathbf{q}) = q^z G^{\pm}(q/\kappa). \qquad \blacktriangleright \quad (15.6)$$

Here $x \equiv \omega/\omega_c$, and the superscripts $\pm$ allow for a different form for the arbitrary functions $F(q/\kappa, x)$ and $G(q/\kappa)$ above and below the critical temperature.

So far we have considered the density–density correlation function $\mathscr{S}^{\pm}_{nn}(\kappa, \mathbf{q}, \omega)$, since the density is the order parameter for the fluid system. In general, the application of the dynamic scaling assumptions (15.5) and (15.6) to the order parameter correlation function is called *restricted* dynamic scaling, while the application of these assumptions to the correlation functions of other microscopic variables (such as the momentum density or the energy density) is known as *extended* dynamic scaling.

15.2. Predictions of the restricted dynamic scaling hypothesis for fluid systems

In general, in order to obtain predictions from the dynamic scaling assumptions (15.5) and (15.6) we must first calculate the shape function $\mathscr{F}^{\pm}(\kappa, \mathbf{q}, x)$ and the characteristic frequency $\omega^{\pm}_{c}(\kappa, \mathbf{q})$ for some region of the κ–q diagram. In practice this region is, by default, the hydrodynamic region ($q/\kappa \ll 1$) because only in this region can we calculate the dynamic structure factor $\mathscr{S}^{nn}_{\pm}(\kappa, \mathbf{q}, \omega)$ and obtain both ω_c and the shape function.

If we follow the prescription given in the defining equations (15.2) and (15.4) we find that in region III, sufficiently near to the critical temperature,

$$\omega^{+}_{c}(\kappa, \mathbf{q}) = \Gamma_{\mathrm{R}} = D_T q^2 = \frac{\Lambda}{\rho C_P}\, q^2 \tag{15.7}$$

and

$$\mathscr{F}^{+}(\kappa, \mathbf{q}, x) = \frac{1}{\pi}\frac{1}{x^2 + 1}. \tag{15.8}$$

The validity of eqns (15.7) and (15.8) is apparent almost by inspection from eqn (13.29), since as $T \to T_c$, the coefficient C_V/C_P of the two terms representing the Brillouin doublet approaches zero. Hence the structure factor is dominated by the Rayleigh peak (eqn (15.8)) whose half-width is, by definition, the Rayleigh linewidth (eqn (15.7)).

Now from eqns (15.7) and (14.3) it follows that

$$\omega^{+}_{c}(\kappa, \mathbf{q}) \sim \epsilon^{\gamma - a} q^2, \tag{15.9}$$

and from eqn (13.41) that $\epsilon \sim \kappa^{1/\nu}$. Therefore eqn (15.9) may be written as

$$\omega^{+}_{c}(\kappa, \mathbf{q}) \sim \kappa^{(\gamma - a)/\nu}\, q^2 = q^{2 + (\gamma - a)/\nu}\, (q/\kappa)^{-(\gamma - a)/\nu}. \tag{15.10}$$

Comparison of (15.10) with the dynamic scaling hypothesis eqn (15.6) leads to the prediction that the homogeneity parameter z appearing in (15.6) has the value

$$z = 2 + (\gamma - a)/\nu. \qquad (15.11)$$

The above argument may be repeated in region I ($T < T_c$), with the result that

$$\omega_c^- (\kappa, \mathbf{q}) \sim q^{2 + (\gamma' - a')/\nu'} (q/\kappa)^{-(\gamma' - a')/\nu'}. \qquad (15.12)$$

Hence, on comparing eqn (15.12) with eqn (15.6), we obtain

$$z = 2 + (\gamma' - a')/\nu.' \qquad (15.13)$$

If we further assume that the static scaling predictions $\gamma = \gamma'$ and $\nu = \nu'$ are valid, then eqns (15.11) and (15.13) predict that

$$a = a'. \qquad (15.14)$$

Although the dynamic scaling hypothesis cannot predict the actual values of the thermal conductivity exponents, it does predict that they are equal above and below the critical temperature. As we have already remarked, this symmetry above and below T_c of the thermal conductivity exponent is not supported by all data (e.g. that on SF_6 (Benedek 1968, 1969)).

The dynamic scaling hypothesis permits predictions in addition to that of eqn (15.14) regarding critical-point exponents. In particular, eqn (15.6) requires that if there is to be a correction to the small-q hydrodynamic result that $\omega_c^{\pm}(\kappa, \mathbf{q}) = D_T q^2$, then it must be of the form

$$\omega_c^{\pm}(\kappa, \mathbf{q}) = D_T q^2 (1 + Bq^2/\kappa^2 + \cdots). \qquad (15.15)$$

This form is in agreement with eqn (14.5), where B is an arbitrary constant. The experimental corroboration of the correction (15.15) to the Rayleigh linewidth was discussed in § 14.4 where we saw that experimental evidence favours the choice $B = \frac{3}{5}$, the value predicted by the mode–mode coupling theory (Kawasaki 1970).

The dynamic scaling hypothesis is not restricted in its range of applicability to the hydrodynamic regions of the κ–q diagram. In region II ($q/\kappa \gg 1$), eqn (15.6) predicts that the characteristic frequency is of the form

$$\omega_c^{\pm}(\kappa, \mathbf{q}) \sim Aq^z \{1 + A'(\kappa/q)^2 + \cdots\}, \qquad (15.16)$$

so that at $T = T_c(\kappa = 0)$,

$$\omega_c^+(\kappa = 0, q) = Aq^z. \qquad (15.17)$$

Thus if we can determine the exponents ν, a, and ν (or, γ', a', and ν') from independent measurements, we can predict, using eqn (15.11), the value of z. A rough estimate may be obtained by choosing $\gamma = 1\cdot25$, $a = 0\cdot58$, and $\nu \cong 0\cdot67$ (Swinney and Cummins 1968), so that

$$z \cong 2 + \frac{\gamma - a}{\nu} \cong 3. \tag{15.18}$$

For one-component fluid systems, there has not yet been any experimental verification that the linewidth is proportional to q^3 in the critical region. However for a binary mixture, for which the dynamic scaling prediction (15.18) is essentially unchanged, it is possible to make measurements in region II because the correlation length parameter ξ_0 is larger than for one-component fluids. Very recently Bergé, Calmettes, Laj, Tournarie, and Volochine (1970) have confirmed the prediction (15.18) from measurements on the binary mixture 53 per cent cyclohexane–47 per cent aniline. Their data are shown in Fig. 15.2.

In summary, then, restricted dynamic scaling (i.e. the dynamic scaling hypotheses (15.5) and (15.6) when applied to the dynamic structure factor for the order parameter $\mathscr{S}_{nn}^{\pm}(\kappa, \mathbf{q}, \omega)$), makes three predictions:

(1) the thermal conductivity critical-point exponent a is identical above and below $T_c (a = a')$, providing $\gamma = \gamma'$ and $\nu = \nu'$ (as static scaling predicts);

(2) the form of the Botch–Fixman correction to the Rayleigh linewidth, Γ_R, in the critical region is correct; and

(3) the characteristic frequency in the critical region, region II, of the κ–q diagram is proportional to the wave vector raised to a power $z = 2 + (\gamma - a)/\nu = 2 + (\gamma' - a')/\nu'$.

Dynamic scaling cannot predict either (1) the magnitude of the thermal conductivity exponent a, (2) the magnitude of the Botch–Fixman coefficient B in eqn (15.15), or (3) the value of the critical region linewidth exponent z (except in terms of static exponents).

15.3. Predictions of extended dynamic scaling for fluid systems

Restricted dynamic scaling is not capable, in principle, of making any predictions whatsoever concerning the position, width, or intensity of the Brillouin doublet. This is because near the critical point the entire weight of the spectrum is concentrated in the central Rayleigh component (recall that $I_R/2I_B \sim C_P/C_V \to \infty$) and hence the characteristic

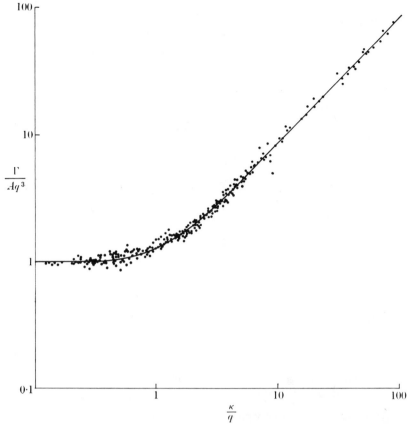

FIG. 15.2. Dependence upon $(\kappa/q) = 1/(q\xi)$ of the linewidth divided by Aq^3. Small values of $1/(q\xi)$ correspond to region II Fig. 12.2. The initial constancy of the plot thus supports (15.17), with $z = 3$ (cf. (15.18)). The linearity of this log–log plot for larger values of the abscissa is consistent with the correction term $A'(\kappa/q)^2$ shown in eqn (15.16). After Bergé et al. (1970). The solid curve is a plot of the 'Kawaski function', $\Gamma = (kT\kappa^3/8\pi\eta)$ $\times \{1 + q^2/\kappa^2 + (q^3/\kappa^3 - \kappa/q) \arctan (q/\kappa)\}$ (Kawasaki 1970).

frequency defined in eqn (15.2) is essentially the Rayleigh linewidth, Γ_R.

We can, however, obtain predictions concerning the Brillouin doublet by applying the dynamic scaling hypothesis to the momentum density–momentum density correlation function. This calculation represents an application of *extended* dynamic scaling and results in the prediction that the sound wave damping constant D_s varies as

$$D_s \sim v_s \, \xi = v_s \, \xi_0 \epsilon^{-\nu}; \qquad (15.19)$$

that is, it predicts that the Brillouin linewidth $\Gamma_B = \frac{1}{2}D_s q^2$ should diverge with about the same critical-point exponent as the correlation length (assuming $\alpha \cong 0$). This prediction is not borne out by the semi-microscopic calculations of Kadanoff and Swift (1968b) using the mode–mode coupling approximation, or by recent measurements on CO_2 and Xe that are as yet unpublished.

It is perhaps worth noting that the result (15.19) of extended dynamic scaling can also be obtained by means of a simple heuristic argument (Kadanoff 1968). In the hydrodynamic region, the dispersion relation for sound is

$$\omega_B = v_s q + \tfrac{1}{2} i D_s q^2. \tag{15.20}$$

If one assumes that q enters into the frequency only through the combination q/κ, and, moreover, that there is only one characteristic frequency, $v_s q$, then we can write (15.20) in the form

$$\omega_B = v_s q f(q/\kappa). \tag{15.21}$$

On comparing eqns (15.20) and (15.21) we have

$$f(q/\kappa) = 1 + i(\text{const.}) (q/\kappa), \tag{15.22}$$

so that, from (15.22) and (15.20) $D_s \propto v_s/\kappa = v_s \xi$, which agrees with (15.19).

15.4. Evidence supporting the dynamic scaling hypothesis provided by magnetic systems

The dynamic scaling hypothesis has also received support from recent measurements on magnetic systems. In this section we shall describe the very recent work of Schulhof, Heller, Nathans, and Linz (1970a, b) on the anisotropic antiferromagnetic material manganese fluoride (MnF_2); other measurements that provide evidence for the dynamic scaling hypothesis are those of Lau, Corliss, Delapalme, Hastings, Nathans, and Tucciarone (1969) on the isotropic antiferromagnet rubidium manganese fluoride ($RbMnF_3$), those of Collins, Minkiewicz, Nathans, Passell, and Shirane (1969) on the ferromagnetic metal iron, and those of Minkiewicz, Collins, Nathans, and Shirane (1969) on the ferromagnet nickel.

The measurements on all four magnetic materials (MnF_2, $RbMnF_3$, Fe, and Ni) were carried out using the inelastic scattering of neutrons. Since the relation between the scattering cross-section for neutrons and the dynamic structure factor (and hence the time-dependent two-spin correlation function defined in eqn (13.3b)) is quite similar to that discussed for light scattering in § 13.3, we shall refer the interested reader to Marshall and Lovesey (1971) for the details of the derivation. Suffice it to say that experimentally one must determine not only the wave vector of the scattered neutron beam, but also the energy distribution of the scattered neutrons. Such an experiment is conventionally referred to as a three-crystal experiment (cf. Fig. 15.3): Bragg reflection from

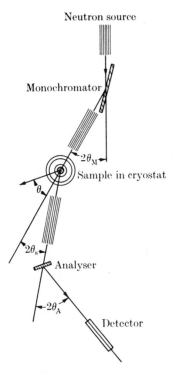

Fig. 15.3. The triple-axis neutron spectrometer used by Schulhof, Heller, Nathans, and Linz (1970a, b) in their inelastic neutron scattering experiments in the critical region of the anisotropic antiferromagnet MnF_2. Higher intensity at the detector is obtained by removing the analyser crystal, so that measurements of time-independent or equilibrium properties of the system (such as the static susceptibility) can frequently be made with greater accuracy than can measurements of time-dependent properties for which energy analysis of the scattered radiation is necessary. After Schulhof (1970).

the first crystal provides a monochromatic beam of neutrons, the second crystal is the sample, and Bragg reflection from the third crystal serves to energy analyse the scattered radiation. The scattering vector $\mathbf{q}$ is varied by changing the angle θ_s (cf. eqn (7.24)) while the energy loss (or gain) is determined by varying the angle θ_A. Not surprisingly, only a neutron source of rather large intensity is sufficient to produce intensities at the detector of sufficient magnitude as to represent statistically reliable information. In fact, all four experiments referred to in the previous paragraph were carried out at the High-Flux reactor of the Brookhaven National Laboratory.

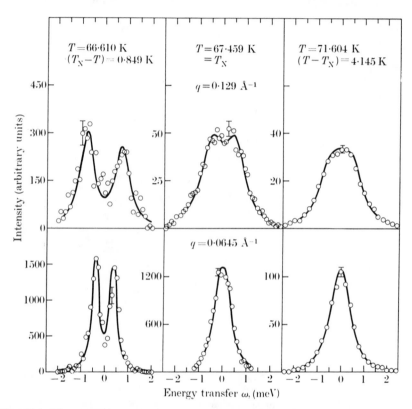

FIG. 15.4. The energy (or frequency) spectrum of the scattered radiation in the inelastic neutron scattering experiments of Schulhof, Heller, Nathans, and Linz (1970b). In this figure the scattering angle θ_s of Fig. 15.3 is chosen so that the scattering vector $\mathbf{q}$ is parallel to the (001) direction. Hence the spectrum is a measure of solely the transverse spin fluctuations. Data are shown for two different values of $\mathbf{q}$ and for temperatures slightly less than, equal to, and greater than the critical temperature ($T_c = 67\cdot459$ K). After Schulhof (1970).

Manganese fluoride is an *anisotropic* antiferromagnetic material; the net magnetization present in the ordered phase, $T < T_c$, is not free to choose an arbitrary orientation to the crystalline axes. Dipolar interactions among the constituent magnetic moments give rise to the effective anisotropy field (roughly 10^4 Oe), which favours alignment of the moments parallel to what we shall call the c-axis. Hence fluctuations in the c-direction (longitudinal fluctuations) might be expected to be qualitatively different from fluctuations in the a- or b-directions (transverse fluctuations). If we position the neutron detector so as to measure the scattering corresponding to a momentum transfer vector $\mathbf{q}$ which lies along the crystal (001) direction, we will observe only the effects of these transverse fluctuations (cf. Fig. 15.4), while if $\mathbf{q}$ lies along

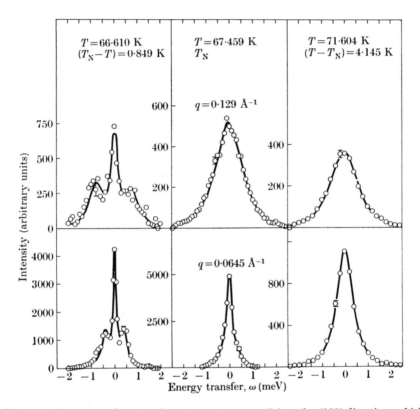

FIG. 15.5. Spectrum of scattered neutrons for $\mathbf{q}$ parallel to the (100) direction, which arises from both transverse and longitudinal spin fluctuations. As in Fig. 15.4, data are shown for two different values of q and three different values of temperature. The characteristic frequency is given by the width of the central component. After Schulhof (1970).

the (100) direction we will observe both the transverse and the longitudinal fluctuations (cf. Fig. 15.5).

The reader will observe from Fig. 15.5 that the spectrum observed with $\mathbf{q}$ along (100) is not altogether unlike the triplet structure predicted for the spectrum of inelastic light scattering from a fluid system (cf.

Fig. 13.2). In particular, the central component of the spectrum in Fig. 15.5 corresponds to longitudinal fluctuations, and as the critical point is approached, the integrated intensity of this component increases.

It should therefore come as little surprise that the characteristic frequency ω_c defined in eqn (15.2) corresponds to the linewidth $\Gamma_L(T, \mathbf{q})$

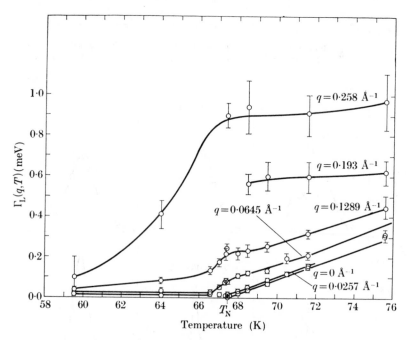

FIG. 15.6. Dependence upon temperature of the characteristic frequency, which is the width of the central component of the scattering for $\mathbf{q}$ parallel to the (100) direction. After Schulhof *et al.* (1970b).

of this central component (we call $\Gamma_L(T, \mathbf{q}) \equiv \Gamma_L^\pm(\kappa, \mathbf{q})$ the *relaxation rate for longitudinal spin fluctuations*). Hence the second assumption, eqn (15.6), of the dynamic scaling hypothesis implies that $\Gamma_L^\pm(\kappa, \mathbf{q})$ is a homogeneous function of the variables κ and $\mathbf{q}$,

$$\Gamma_L^\pm(\lambda\kappa, \lambda\mathbf{q}) = \lambda^z \Gamma_L^\pm(\kappa, \mathbf{q}) \qquad (15.23)$$

for all values of λ. Now if we choose $\lambda \equiv 1/\kappa$, then (15.23) becomes

$$\Gamma_L^\pm(\kappa, \mathbf{q}) = \kappa^z\, G^\pm(q/\kappa), \qquad (15.24)$$

where the function $G^\pm(q/\kappa)$ is an as yet unspecified scaling function.

Note that in this case we used the homogeneity argument to extract the κ dependence of $\Gamma_L^{\pm}$, while in eqn (15.6) we extracted the q dependence.

Fig. 15.6 shows the temperature dependence of $\Gamma_L^{\pm}(T, \mathbf{q})$, the relaxation rate for the longitudinal spin fluctuations, as measured by the linewidth of the central component of Fig. 15.5. There is, of course, a different curve for each scattering vector $\mathbf{q}$. In Fig. 15.7 we show

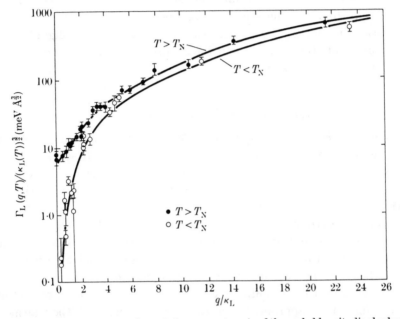

Fig. 15.7 Dependence upon the scaled wave vector q/κ of the scaled longitudinal relaxation rate or characteristic frequency divided by κ^z, where $z = \frac{3}{2}$. The data shown are those of Fig. 15.6, and the two distinct curves on which the data appear to lie are presumably the scaling functions $G^+(q/\kappa)$ and $G^-(q/\kappa)$ defined in eqn (15.24). Thus these data would seem to provide rather striking evidence supporting the dynamic scaling hypothesis. After Schulhof *et al.* (1970*b*).

the same data plotted in such a fashion as to provide a test of the dynamic scaling prediction (15.24). To this end we have divided $\Gamma_L^{\pm}$ $(\kappa, \mathbf{q})$ by κ^z and plotted this function against q/κ. The value $z = \frac{3}{2}$ was chosen on the basis of arguments presented by Halperin and Hohenberg (1969), and we have allowed for different values of the inverse correlation length for longitudinal and transverse spin fluctuations (κ_L and κ_T respectively). According to the dynamic scaling prediction (15.24), the quantity $\Gamma_L(\kappa, \mathbf{q})/\kappa^z$ should be the scaling function $G^{\pm}(q/\kappa)$, where the $\pm$ sign indicates that there are actually two functions,

$G^+(q/\kappa)$ for $T > T_c$ and $G^-(q/\kappa)$ for $T < T_c$. The data shown in Fig. 15.7 do indeed appear to fall on two separate curves, according as to whether the sign of $T - T_c$ is positive or negative. Thus the data provide experimental support for the validity of the dynamic scaling prediction (15.24), much as the static data of Fig. 11.4 provide support for the validity of the static scaling hypothesis. Of course, neither set of data constitutes a proof of the validity of the respective scaling hypotheses, but they certainly do serve to lend them credibility. Attempts to place the scaling hypotheses on an alternative mathematical foundation are under way (Polyakov 1968, Migdal 1968, Matsuno and Stanley 1970), but the results as yet are only preliminary and will not be discussed here.

15.5. Spirit of the mode–mode coupling approach

We have seen that the application of the dynamic scaling hypothesis to the case of time-dependent critical phenomena leads to several specific predictions. We emphasized, however, that these predictions do not include all the information that we might wish to know. In particular, while the dynamic scaling hypothesis does predict that the exponents a and a' describing the behaviour of the thermal conductivity are identical above and below T_c (i.e. $a = a'$), it cannot predict the value of a; in fact it cannot even tell us if Λ diverges or remains finite at T_c. The theory to be outlined in this section succeeds in showing which of the transport coefficients are expected to diverge, and in predicting the values of the appropriate critical-point exponents for the transport coefficients by relating these exponents to the exponents for the static functions. For example, the thermal diffusivity $D_T = \Lambda/\rho C_P \sim \epsilon^{\gamma - a}$ is predicted to approach zero in the same fashion as the inverse correlation length, $\kappa \sim \epsilon^\nu$, so that the exponent $a = \gamma - \nu$ will be obtained.

Perhaps a more fundamental shortcoming of the scaling theory is that its underlying principles—the scaling hypotheses—have not been justified, except in so far as the predictions are borne out by many, though not all, experiments. Clearly it would be desirable to corroborate these predictions by means of microscopic calculations. The theoretical developments to be described in the remainder of this chapter, though by no means a complete treatment, do represent a first step in this direction.

Actually, there have been several treatments (Fixman 1962a, b, 1964, 1967; Kawasaki 1966d, 1968a, b, c, d, 1969, 1970; Kawasaki and Tanaka

1967; Deutch and Zwanzig, 1967; Mountain and Zwanzig, 1968; Villain, 1968a, b, c; Ferrell, 1970; Kadanoff and Swift 1968b; Kadanoff 1969) which embody essentially the same physical ideas and are now commonly called *mode-mode coupling approximations*. These physical ideas were perhaps first enunciated by Fixman (1962a) who considered the critical behaviour of the viscosity in a binary mixture. Fixman noted that because near the critical point the long wavelength part of the spectrum of composition fluctuations (the analogue of density fluctuations in a single-component fluid) is very intense, a velocity gradient created by exerting viscous shear forces at the boundary of the fluid can easily induce inhomogeneities in concentration. The return to uniform composition (i.e. the dying away through diffusion of the inhomogeneities) dissipates energy. From a macroscopic point of view, however, we interpret this energy dissipation caused by coupling between the viscous and diffusive modes as being the result of an anomalously large viscosity.

We can view this physical process in the following, slightly more mathematical, fashion. Consider first the linearized equations of motion that are discussed in Appendix D. Suppose we choose Λ_0, η_0, and ζ_0 to be the values calculated for Λ, η, and ζ far from the critical-point, which are non-divergent. If we now argue that near the critical point we must add additional non-linear terms to these equations, the normal modes of the system are certainly no longer those given by the original equations. We cannot even be certain that the system possesses normal modes at all, but we might calculate the best normal mode-type solution to these non-linear equations. If we restrict ourselves to this type of solution and find that, for example, a heat wave propagates as $\exp(-D_T q^2 t)$, we will identify the macroscopically observable thermal conductivity Λ through the equation $\Lambda = \rho C_P D_T$; note that Λ will, in general, be a function of other parameters in the original equations (e.g. Λ_0, η_0, ζ_0, and ξ). We could, of course, now work backwards and determine the *linear* equations that our normal mode solutions satisfy. These equations would be the equations of hydrodynamics that we give in Appendix D. However we would have explicit expressions for the quantities Λ, η and ζ that were introduced phenomenologically in that section.

Now the fundamental problem is not only solving the non-linear equations but also determining what the non-linear corrections are. Usually these non-linear terms are not displayed explicitly and their effect is taken into account in other ways more amenable to calculations.

For example, Kawasaki (1966) and Deutch and Zwanzig (1967) make use of the Kubo formulae to extract the transport coefficients.

Kadanoff and Swift (1968*b*) have recently developed a formalism to describe the transport modes, which can be given a schematic interpretation (cf. Fig. 15.8(a)). The transport modes are assumed to be coupled to each other non-linearly, and disturbances to the fluid can be transmitted back and forth between the various modes. For instance, sound waves can be transformed into thermal and viscous modes, as

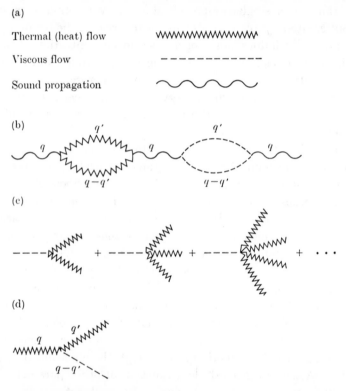

FIG. 15.8. (a) Schematic representation of the transport modes heat flow, viscous flow, and sound propagation. (b) Schematic representation of processes (such as the emission and reabsorption of heat modes or viscous modes) that might be expected to contribute to an increase in the damping of a sound wave. (c) Schematic representation of the decay of a viscous mode into two, three, and four thermal modes. (d) The contribution $\Lambda_{T\eta}$ to the thermal conductivity Λ obtained from the decay of a heat flow mode into a viscous mode plus a thermal mode.

Fig. 15.8(b) illustrates. Kadanoff and Swift performed semi-microscopic calculations in which they evaluated the net relaxation rates for these decay processes by using approximation procedures based upon the ideas of the static scaling laws.

A detailed presentation of the Kadanoff–Swift calculation would be beyond the scope of this short book. Therefore we shall restrict our remarks in this chapter to a brief discussion of the theoretical results obtained and a comparison of these results with recent experimental evidence.

15.6. Predictions of mode–mode coupling approximation

The principal results of Kadanoff and Swift (1968b) are contained in Table 15.1, which concerns predicted temperature behaviour for each of the three transport coefficients in each of three regions of the frequency–temperature plane. In order to understand the origin of these regions, it is necessary to consider somewhat the nature of the calculation.

15.6.1. Damping rates for heat flow, viscous flow, and sound modes

We will see in Appendix D that, for $q \ll \kappa$, heat flows through a diffusion process with a relaxation rate or decay rate $\Omega_T(\epsilon, \mathbf{q})$ given by the simple expression

$$\Omega_T(\epsilon, \mathbf{q}) = (\Lambda/\rho C_P)q^2, \tag{15.25}$$

as follows from eqns (13.34) and (13.30). In Appendix D we consider only the specific case of a local relationship between the energy current and the temperature gradient. More generally, we must allow for a non-local relationship in space and time and consider a wave number and frequency-dependent thermal conductivity $\Lambda(\mathbf{q}, \Omega)$ where $\Omega \equiv i\omega$. In this more general situation we identify the decay rate for heat flow $\Omega_T(\epsilon, \mathbf{q})$ as the solution to the equation

$$\Omega_T = \frac{\Lambda(\mathbf{q}, \Omega_T)}{\rho C_P} q^2. \tag{15.26}$$

From simple hydrodynamic arguments we can show that a transverse velocity flow relaxes exponentially with a decay rate proportional to the shear viscosity η. We thus identify the decay rate for this mode, $\Omega_\eta(\epsilon, \mathbf{q})$, as the solution of

$$\Omega_\eta = \frac{\eta(\mathbf{q}, \Omega_\eta)q^2}{\rho}. \tag{15.27}$$

Finally, we identify the sound wave decay rate, $\Omega_s(\epsilon, \mathbf{q})$, which is complex because the mode is a propagating one, as the solution of

$$\Omega_s = \pm iv_s q + \tfrac{1}{2}D_s(\mathbf{q}, \Omega_s)q^2 \tag{15.28}$$

TABLE 15.1

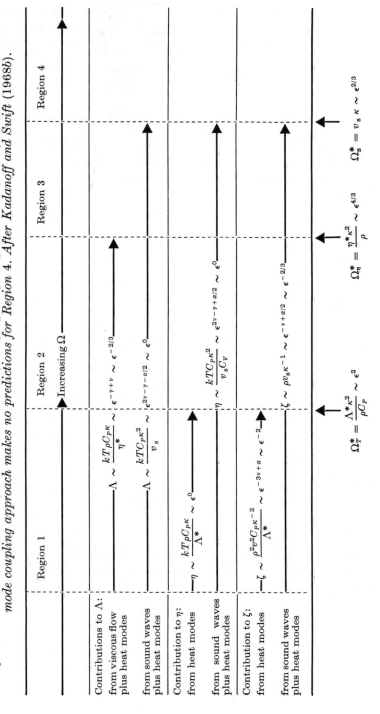

Contributions to transport coefficients. Here the numerical values for the exponents are calculated by assuming $\gamma = \frac{4}{3}$, $\nu = \frac{2}{3}$, and $\alpha = 0$. The regions referred to are regions of the frequency-temperature diagram of Fig. 15.9. The mode-mode coupling approach makes no predictions for Region 4. After Kadanoff and Swift (1968b).

Region 1 Region 2 Region 3 Region 4

Increasing Ω

Contributions to Λ:

from viscous flow plus heat modes
$$\Lambda \sim \frac{kT\rho C_P \kappa}{\eta^*} \sim \epsilon^{-\gamma+\nu} \sim \epsilon^{-2/3}$$

from sound waves plus heat modes
$$\Lambda \sim \frac{kTC_P\kappa^2}{v_s} \sim \epsilon^{2\nu-\gamma-\alpha/2} \sim \epsilon^0$$

Contribution to η:

from heat modes
$$\eta \sim \frac{kT\rho C_P \kappa}{\Lambda^*} \sim \epsilon^0$$

from sound waves plus heat modes
$$\eta \sim \frac{kTC_P\kappa^2}{v_s C_V} \sim \epsilon^{2\nu-\gamma+\alpha/2} \sim \epsilon^0$$

Contribution to ζ:

from heat modes
$$\zeta \sim \frac{\rho^2 v_s^2 C_P \kappa^{-2}}{\Lambda^*} \sim \epsilon^{-3\nu+\alpha} \sim \epsilon^{-2}$$

from sound waves plus heat modes
$$\zeta \sim \rho v_s \kappa^{-1} \sim \epsilon^{-\nu+\alpha/2} \sim \epsilon^{-2/3}$$

$$\Omega_T^* = \frac{\Lambda^* \kappa^2}{\rho C_P} \sim \epsilon^2 \qquad \Omega_\eta^* = \frac{\eta^* \kappa^2}{\rho} \sim \epsilon^{4/3} \qquad \Omega_s^* = v_s \kappa \sim \epsilon^{2/3}$$

where

$$D_s(\mathbf{q}, \Omega) \equiv \frac{\frac{4}{3}\eta(\mathbf{q}, \Omega) + \zeta(\mathbf{q}, \Omega)}{\rho} + \frac{\Lambda(\mathbf{q}, \Omega)}{\rho}\left(\frac{1}{C_V} - \frac{1}{C_P}\right). \quad (15.29)$$

Note that the decay rates $\Omega_T(\epsilon, \mathbf{q})$, $\Omega_\eta(\epsilon, \mathbf{q})$ and $\Omega_s(\epsilon, \mathbf{q})$ defined by eqns (15.26)–(15.28) are functions of both ϵ and $\mathbf{q}$. We shall later have occasion to consider these decay rates when $q = \kappa$, so we are led to define the auxiliary functions

$$\Omega_T^*(\epsilon) \equiv \Omega_T(\epsilon, q = \kappa) = \frac{\Lambda^*}{\rho C_P}\kappa^2, \quad (15.30)$$

$$\Omega_\eta^*(\epsilon) \equiv \Omega_\eta(\epsilon, q = \kappa) = \frac{\eta^*}{\rho}\kappa^2, \quad (15.31)$$

and

$$\Omega_s^*(\epsilon) \equiv \Omega_s(\epsilon, q = \kappa). \quad (15.32)$$

Here the second equality in (15.30) serves to define Λ^* as being $\Lambda(q = \kappa, \Omega = \Omega_T^*)$; similarly, in (15.31), $\eta^* \equiv \eta(q = \kappa, \Omega = \Omega_\eta^*)$. The temperature dependences of $\Omega_T^*(\epsilon)$, $\Omega_\eta^*(\epsilon)$, and $\Omega_s^*(\epsilon)$ are shown in Fig. 15.9 for a typical one-component fluid.

The basis of the Kadanoff–Swift approach is to calculate these decay rates in the neighbourhood of the critical point by considering the various ways in which one mode can decay into other modes. The transport coefficients can then be obtained from the decay rates using eqns (15.26)–(15.29). Actually, this approach does not give us the exact values of the transport coefficients, but rather it predicts their temperature dependences within certain regions of the frequency-temperature plane (cf. Fig. 15.9). In particular, it does not predict the temperature dependence arbitrarily close to the critical point, i.e. in region 4 of Fig. 15.9.

We shall catalogue by means of the schematic representation shown in Fig. 15.8(a), the various processes that might be expected to contribute to the decay rates of the transport modes.

For example, that contribution to the viscosity from decay processes in which a viscous flow breaks up into heat modes can be written, schematically, as

$$\eta = \eta_{TT} + \eta_{TTT} + \eta_{TTTT} + \cdots \quad (15.33)$$

and represented by the series of diagrams shown in Fig. 15.8(c). Now Kadanoff and Swift (1968b) have argued on the basis of scaling laws that the contributions η_{TT}, η_{TTT}, η_{TTTT}, ... are all of the same order of magnitude and hence it is sufficient to calculate one of them—say η_{TT}. Of course, one must calculate, in addition to η_{TT}, the contribution from processes in which a viscous mode breaks up into modes other than thermal modes.

15.6.2. Calculation for the decay of a thermal mode into a thermal mode plus a viscous mode

We shall now illustrate the principal features of the Kadanoff–Swift approach by outlining their treatment of the contribution to Λ of the process $\Lambda_{T\eta}$ represented schematically in Fig. 15.8(d), in which a thermal mode of wave vector $\mathbf{q}$ breaks up into a thermal mode of wave vector $\mathbf{q}'$ plus a viscous mode of wave vector $\mathbf{q} - \mathbf{q}'$. A particularly physical result which we shall obtain is that the contribution to the thermal diffusion coefficient, $D_T = \Lambda/\rho C_P$, corresponding to this particular process, has the form

$$D_T \sim \frac{kT}{\eta\xi},\tag{15.34}$$

which is what one would obtain if one were to use Stokes law for the diffusion coefficient of a sphere of radius $R \equiv \xi$.

Kadanoff and Swift argue that the contribution from this process to the decay rate $\Omega_T(\epsilon, \mathbf{q})$ of eqn (15.26) is of the form

$$\frac{\Lambda_{T\eta}(\mathbf{q}, \Omega)}{\rho C_P} q^2 = \int \frac{d\mathbf{q}'}{(2\pi)^3} \frac{|M(\mathbf{q}; \mathbf{q}', \mathbf{q} - \mathbf{q}')|^2}{\Omega_T(\epsilon, \mathbf{q}') - \Omega_\eta(\epsilon, \mathbf{q} - \mathbf{q}') - \Omega},\tag{15.35}$$

where the quantity $M(\mathbf{q}; \mathbf{q}', \mathbf{q} - \mathbf{q}')$ in the numerator can be viewed as matrix element for this process in which, speaking loosely, a thermal mode of wave vector $\mathbf{q}$ is annihilated and thermal and viscous modes of wave vectors $\mathbf{q}$ and $\mathbf{q} - \mathbf{q}'$ are created (cf. Fig. 15.8(d)).

Using arguments based on the static scaling hypothesis, Kadanoff and Swift make some estimates of the behaviour of the integral (15.35) near the critical point. They find that

$$\Lambda_{T\eta}(\mathbf{q}, \Omega) \cong \begin{cases} \dfrac{kT\,\rho C_P\kappa}{\eta^*} & q \lesssim \kappa' \text{ and } \Omega \lesssim \Omega_\eta^* \\ \text{non-divergent} & \text{otherwise,} \end{cases}\tag{15.36}$$

where η^* is defined by eqn (15.31). Performing a separate though similar calculation for η, they find that η^* is at most weakly divergent, i.e. its exponent is approximately zero.

Equation (15.36) is essentially the final result for the contribution to the thermal conductivity from the process described schematically in Fig. 15.8(d); it tells us that for frequencies less than a characteristic viscous frequency Ω_η^*, the thermal conductivity should have a contribution $\Lambda_{T\eta}(\mathbf{q}, \Omega)$ which diverges as

$$\Lambda_{T\eta} \sim \epsilon^{\nu - \gamma}, \tag{15.37}$$

where $\nu - \gamma \cong -\frac{2}{3}$ typically. This behaviour is shown schematically in the first line of Table 15.1, where we have indicated the frequency range over which the prediction (15.37) is valid by means of the arrow extending from $\Omega = 0$ to $\Omega = \Omega_\eta^*$ (ϵ). It is important to stress that the frequencies $\Omega_T^*(\epsilon)$, $\Omega_\eta^*(\epsilon)$, and Ω_s^* (ϵ) which form the boundaries between Regions 1, 2, and 3 are themselves functions of temperature.

15.6.3. *Calculations for other decay processes*

Kadanoff and Swift (1968b) have carried out calculations similar to that outlined above for other processes, and the results of these calculations for the principal processes are also shown in Table 15.1. For example, we see in the second line that there is a second contribution to the thermal conductivity that arises for all frequencies less than $\Omega_s^*(\epsilon)$. However, the fact that this contribution is of order $\epsilon^{2\nu - \gamma}$ and typically $2\nu - \gamma \cong 0$, while the contribution $\Lambda_{T\eta}$ calculated above diverges as $\epsilon^{\nu - \gamma}$ with $\nu - \gamma \cong -\frac{2}{3}$, means that results of measurements of the thermal conductivity in regions 1 and 2 will be dominated by the contribution $\Lambda_{T\eta}$ and we should expect to see a divergent thermal conductivity with an exponent roughly equal to $\frac{2}{3}$. This is in fact what we found in § 14.4 from the data of Swinney and Cummins (1968) on CO_2, but it is not what the data of Benedek (1968, 1969) show for SF_6. In this respect the predictions of the mode–mode coupling theory agree with some but not all of the experimental evidence. In support of this theory, however, is the fact that directly from eqn (15.36) we can verify eqn (15.34) which, as we remarked above, is a particularly appealing expression for the thermal diffusivity.

In fact, quite recently Kawasaki (1970) has verified eqn (15.34) using a slightly different mode–mode coupling approach, and he has found that the proportionality constant has the same value as in the Stokes law expression,

$$D_T = \frac{kT}{6\pi\eta^*\xi}. \qquad \blacktriangleright \tag{15.38}$$

Ferrell (1970) has obtained, using a simple argument, the analogue of eqn (15.38) for the case of a binary liquid mixture.

15.7. Application of the mode–mode coupling approach to the interpretation of experimental results on fluids

We shall attempt to illustrate the application of the mode–mode coupling theory to the interpretation of experimental results by considering one prototype experiment—the Brillouin doublet measurements on carbon dioxide of Ford *et al.* (1968) already briefly discussed in § 14.5.

We begin by referring to Fig. 15.9, which shows the dependence on temperature of the characteristic frequencies $\Omega_T^*(\epsilon)$, $\Omega_\eta^*(\epsilon)$, and $\Omega_s^*(\epsilon)$. The typical frequency of the Brillouin doublet measurements is about $\omega_B \equiv 2\pi \nu_B \cong 2\pi \times 700$ MHz, which we show by a horizontal dashed

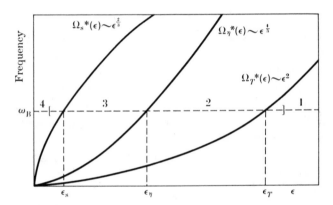

FIG. 15.9. Dependence upon reduced temperature $\epsilon \equiv (T - T_c)/T_c$ of the critical frequencies $\Omega_T^*(\epsilon)$, $\Omega_\eta^*(\epsilon)$ and $\Omega_s^*(\epsilon)$ as defined in eqns (15.30), (15.31), and (15.32) respectively. The actual form of the plots was chosen for CO_2, for which the exponents are approximately 2, $\frac{4}{3}$, and $\frac{2}{3}$ respectively, and the coefficients are all of order 10^{12} in c.g.s. units. These three curves serve to define three regions of the frequency–temperature plane which Kadanoff and Swift (1968b) call regions 1, 2, and 3. In the experiments of Ford *et al.* (1968), the Brillouin frequency was $\nu_B = 700$ MHz, and we obtain the values of the cross-over temperatures ϵ_s, ϵ_η, and ϵ_T shown. Note that the measurements at $\nu_B = 700$ MHz span all four frequency regions (cf. brackets).

line in Fig. 15.9. If the boundaries between the frequency regions were not temperature dependent, then measurements at one frequency would correspond to one and only one region. However, since the boundaries are temperature dependent, measurements at the Brillouin frequency range can, in principle, yield information about all four frequency regions. As we approach the critical-point, at a temperature ϵ_T we will cross from region 1 to region 2; at a temperature ϵ_η we will cross from region 2 to region 3, and for $\epsilon < \epsilon_s$ we will be inside region 4.

Let us, then, obtain numerical estimates for the temperatures ϵ_T, $\epsilon\eta$, and ϵ_s in order to see if all three regions are experimentally accessible. By definition, ϵ_T is that temperature such that

$$\Omega_T^*(\epsilon_T) = 2\pi \, \nu_B = \omega_B, \tag{15.39}$$

where $\Omega_T^*(\epsilon)$ is defined in eqn (15.30) so that (15.39) becomes

$$\frac{\Lambda^*}{\rho C_P} \kappa^2 \simeq 2\pi \times 700 \times 10^6 \, s^{-1}. \tag{15.40}$$

Now Swinney and Cummins (1968) have found that for CO_2

$$\Lambda/\rho C_P \simeq (2 \times 10^{-5}) \, (T - T_c)^{0.73} \simeq (2 \times 10^{-5}) \, (304)^{2/3} \epsilon^{2/3} \tag{15.41}$$

in c.g.s. units, and we have, from eqn (13.45), that $\kappa = \kappa_0 \epsilon^\nu$ with $\kappa_0 \simeq 1 \text{Å}^{-1}$ and $\nu \simeq \frac{2}{3}$. If we assume $\Lambda^* \simeq \Lambda$, then

$$\Omega_T^*(\epsilon) \simeq 9 \times 10^{12} \, \epsilon^2 \, s^{-1}. \tag{15.42}$$

Equating (15.40) and (15.42), we obtain

$$\epsilon_T \simeq 2.2 \times 10^{-2}, \tag{15.43}$$

which, since $T_c = 304$ K, corresponds to $T - T_c \simeq 7°$.

The second cross-over temperature ϵ_η may be calculated in a similar fashion. Using $\eta^* \simeq \eta = 4 \times 10^{-4} \text{g cm}^{-1} \text{s}^{-1}$ and $\rho = 0.5 \text{ g cm}^{-3}$ (Sengers 1966), we have, from eqn (15.31), that

$$\Omega_\eta^*(\epsilon) \simeq 8 \times 10^{12} \, \epsilon^{4/3} \, s^{-1}, \tag{15.44}$$

so that $\Omega_\eta^*(\epsilon = \epsilon_\eta) = \omega_B$ for

$$\epsilon_\eta \simeq 3.6 \times 10^{-3}, \tag{15.45}$$

which corresponds to $T - T_c \simeq 1°$.

Finally, we obtain from eqn (15.32) that for CO_2

$$\Omega_s^*(\epsilon) \simeq 2 \times 10^{12} \, \epsilon^{2/3}, \tag{15.46}$$

where we have chosen $v_s \simeq 2 \times 10^4 \text{ cm s}^{-1}$ from Fig. 14.5(a). Setting $\Omega_s(\epsilon = \epsilon_s) = \omega_B$, we obtain

$$\epsilon_s \simeq 10^{-4}, \tag{15.47}$$

which corresponds to $T - T_c = 0.03°$.

Since the data of Ford et al. (cf. Fig. 14.5) extend from $0.015° \leq (T - T_c) \leq 12°$, or

$$5 \times 10^{-5} \leq \epsilon \leq 4 \times 10^{-2}, \tag{15.48}$$

TABLE 15.2

Comparison of mode–mode coupling predictions with experimental results. After Kadanoff (1969).

Phase transition	Physical quantity	Theoretical Value of index for $(T - T_c)$ dependence	Theoretical Reference	Experimental Value of index	Material	Reference	Agreement with theory?
Liquid-gas	$\dfrac{\Lambda}{\rho C_P}$ on coexistence curve	$\sim \tfrac{2}{3}$	1	0.66 ± 0.05 (gas) 0.72 ± 0.05 (liquid)	CO_2	8	yes yes
				~ 0.64	SF_6	9	yes
	$\dfrac{\Lambda}{\rho C_P}$ on critical isochore	$\sim \tfrac{2}{3}$	1	0.73 ± 0.02 ~ 1.27	CO_2 SF_6	8 9	yes no
Fluid mixture	particle diffusivity $T > T_c$	$\sim \tfrac{2}{3}$	2	0.68 ± 0.04	isobutyric acid in H_2O	10	yes
				0.65 ± 0.05	nitrobenzene and n-hexane	11	yes
				0.55 ± 0.05	analine-cyclohexane	12	yes
Heisenberg anti-ferromagnet	spin wave frequency $T = T_c$	(frequency $\sim q^{3/2}$) q = wave number	3 4		$RbMnF_3$	13	yes
	spin wave frequency $T > T_c$	-1	3 4	no simple fit	$RbMnF_3$	13	no
	ultrasonic attenuation $T > T_c$	-1	5 4	$\sim -\tfrac{1}{3}$ -1.0 ± 0.1	$RbMnF_3$ Ho	14 15	no yes

ferromagnet	attenuation constant $T > T_c$						
Single Axis Anti-ferromagnet		$\approx -\tfrac{4}{3}$	4	$-1\cdot24 \pm 0\cdot1$ $-1\cdot37 \pm 0\cdot1$	Tb Dy	15 15	yes yes
Superfluid	$\dfrac{\Lambda}{\rho C_P}$ bulk ($T > T_c$)	$\tfrac{1}{3}$	6 7	$\sim \tfrac{1}{3}$ $0\cdot34 \pm 0\cdot02$	He^4 He^4	17 18	yes yes
	ultrasonic attenuation constant $T < T_c$	-1	6 7	-1	He^4	19	yes
	$\dfrac{\Lambda}{\rho C_P}$ (small cell $\sim$ 3×10^{-3} cm)	-1	6 7	$0\cdot68 \pm 0\cdot02$	He^4	20	no
	ultrasonic attenuation constant $T > T_c$	-1	6 7	$\sim -\tfrac{1}{2}$	He^4	19	no

1. Kadanoff and Swift (1968b).
2. Swift (1968).
3. Halperin and Hohenberg (1967, 1969).
4. Laramore (1968).
5. Swift and Kadanoff (1968).
6. Kawasaki (1968e, f).
7. Ferrell et al. (1967, 1968).
8. Swinney and Cummins (1968).
9. Benedek (1969).
10. Chu, Schoenes, and Kao (1968).

11. Chen and Polonsky (1968).
12. Bergé and Volochine (1968).
13. Nathans, Menzinger, and Pickart (1968).
14. Golding (1968).
15. Luthi and Pollina (1969).
16. Luthi and Pollina (1968).
17. Kerresk and Keller (1967).
18. Ahlers (1969).
19. Barmatz and Rudnick (1968).
20. Archibald, Mochel, and Weaver (1969).

it appears from eqns (15.43), (15.45), and (15.47) that all four regions of the frequency–temperature diagram of Fig. 15.9 will be covered.

The Brillouin linewidth is $\frac{1}{2}D_s q^2$, where from eqn (13.31) we have

$$D_s = D_T \left(\frac{C_P}{C_V} - 1\right) + \frac{\frac{4}{3}\eta + \zeta}{\rho} \simeq \frac{1}{\rho}\left(\frac{\Lambda}{C_V} + \zeta\right). \qquad (15.49)$$

We can write the second equality because near the critical point, $C_P/C_V \gg 1$ and $\zeta \gg \eta$, and the transport coefficients are understood to depend on q and Ω (cf. eqn (15.29)). From (15.49) we expect that the behaviour of D_s and hence the Brillouin linewidth will be controlled by either Λ/C_V or by ζ, whichever is dominant. Perhaps the most striking feature of Fig. 14.5 is that the data for $T - T_c \lesssim 1°$ have a slope approximately equal to zero. This is surprising because from eqn (15.45) this temperature range corresponds to $\epsilon < \epsilon_\eta$ or region 3 of Fig. 15.9. Hence we are led from the bottom line of Table 15.1 to expect that the bulk viscosity diverges as $\epsilon^{-(\nu - \alpha/2)}$ in this region and hence to expect that the slope would be $-(\nu - \alpha/2) \simeq \frac{2}{3}$ rather than zero. Thus the data would appear to suggest that the behaviour in region 3 is dominated more by the thermal conductivity term which, from the second line of Table 15.1, has a much weaker singularity than ζ in region 3. Thus either the mode–mode coupling approach would seem to fail, or else we must assume that the coefficient in front of the $\epsilon^{-2/3}$ divergence in ζ is sufficiently smaller than the coefficient in front of the Λ/C_V singularity that the latter dominates. Of course, we could test this assumption if it were possible to make measurements arbitrarily close to T_c and if the $\epsilon^{-2/3}$ prediction were to still apply (which it might not since region 3 does not apply for $\epsilon \lesssim \epsilon_s$).

Now consider the data in region 2, i.e. for $\epsilon_\eta \lesssim \epsilon \lesssim \epsilon_T$ corresponding to $1° \lesssim T - T_c \lesssim 5°$. Here Table 15.1 tells us that the slope should be either $-(\gamma - \nu - \alpha)$ if Λ/C_V dominates or $-(\nu - \alpha/2)$ if ζ dominates. The analysis of Ford et al. for the Landau–Placzek ratio suggests that if $\alpha \gtrsim 0$, then $\gamma \gtrsim 1·02$ and $\nu \simeq 0·73$, so that $-(\gamma - \nu - \alpha) \simeq -0·29$, whereas $-(\nu - \alpha/2) \simeq -0·73$. The straight line shown in Fig. 14.5(b) corresponds to a slope of $-0·29$, suggesting that the thermal conductivity dominates in region 2 also.

There appears to be no marked change in the data as we pass into region 1 $(T - T_c \gtrsim 5°)$, although Table 15.1 predicts that ζ should diverge in region 1 with an exponent $3\nu - \alpha \simeq 2$, which is much larger than $0·29$. For this data to be consistent with the theory, we must again assume that the coefficient of the Λ/C_V divergence in region 1 is much larger than the coefficient of the ζ divergence.

15.8. Applications of the mode–mode coupling approach to other systems

So far we have discussed the mode–mode coupling theory only within the context of fluid systems. It has also been applied to magnetic systems (Laramore 1969; Kawasaki 1968e, f), to binary fluid mixtures (Swift 1968), and to the superfluid phase transition in ^{4}He (Swift and Kadanoff 1969). In most cases the agreement between the theoretical predictions and the experimental results (cf. Table 15.2) has been encouraging.

Suggested further reading
Kadanoff (1968).
Halperin and Hohenberg (1969).
Kadanoff (1969).
Craig and Goldburg (1969).
Kawasaki (1970).

APPENDIX A

THE LATTICE–GAS MODEL OF A FLUID SYSTEM

IN Chapter 1 we introduced the lattice–gas model, and we saw that it provided an appealing analogy between an Ising ferromagnet and a simple fluid. Since we have made use throughout the book of the analogies between fluid and magnetic systems, it might seem appropriate to consider the lattice–gas model in greater detail.

We recall from the discussion in Chapter 1 (and, especially, from Figs. 1.4 and 1.5) that the lattice–gas model for a fluid confined to a volume V involves partitioning V into cells of volume v, where v is roughly the volume of one of the constituent molecules of the fluid. We say that a cell is in the occupied state if it is occupied by the centre of a molecule. It is worth emphasizing that we need not assume that the molecules themselves are restricted to cells, because some authors (erroneously) state that the atoms of the lattice–gas are restricted to the discrete sites of a lattice. Since the volume v is roughly the size of the molecules, we do not allow a cell to be occupied by more than one molecule. Such multiple occupancy is forbidden by assuming that there is a hard-core potential, i.e. we assume that the interaction between cells i and j is

$$
u(i,j) \equiv
\begin{cases}
\infty & \text{if } i = j \text{ and both } i \text{ and } j \text{ are occupied} \\
-J & \text{if } i \text{ and } j \text{ are neighbouring cells and both are} \\
& \text{occupied,} \\
0 & \text{otherwise.}
\end{cases}
\tag{A.1}
$$

The potential of (A.1) is in a sense a very crude approximation to the Lennard–Jones potential of Fig. 7.1.

Now since each cell is capable of existing in only one of two states (occupied by one molecule or empty) we introduce a cell variable e_j through the relation

$$
e_j \equiv
\begin{cases}
1 & \text{if cell } j \text{ is singly occupied,} \\
0 & \text{if cell } j \text{ is empty.}
\end{cases}
\tag{A.2}
$$

The reader will recall that the dichotomic variable s_j in the Ising model is defined through a similar relation,

$$
s_j \equiv
\begin{cases}
1 & \text{if the spin on lattice site } j \text{ is 'up',} \\
-1 & \text{if the spin on lattice site } j \text{ is 'down'.}
\end{cases}
\tag{A.3}
$$

We can identify an occupied cell of the lattice gas with an Ising spin in the 'up' orientation provided we set

$$e_j = \tfrac{1}{2}(1 + s_j).$$ (A.4)

Thus one might expect that there exists a formal correspondence between the Ising model and the lattice–gas model. This is in fact the case, and one can show that the canonical partition function of the former is proportional to the grand partition function of the latter. The Ising model partition function is extensively discussed in Chapter 8, and for our purposes here we write the expression

$$Z_{\text{Ising}} = \sum_{\{s\}} \exp\left(\mathscr{J} \sum_{\langle ij \rangle} s_i s_j + h \sum_{i=1}^{N} s_i \right),$$ (A.5)

where $\mathscr{J} \equiv J/kT$ and $h \equiv \mu H/kT$ as before. For the lattice gas the partition function is quite similar in form

$$Z = \sum_{\{e\}}^{*} \exp\left(\mathscr{J} \sum_{\langle ij \rangle} e_i e_j \right);$$ (A.6)

where the asterisk denotes that the summation is restricted to those configurations satisfying

$$\sum_{j=1}^{n} e_j = N$$ (A.7)

where n is the number of cells in the system. If we replace (A.6) by the grand canonical ensemble, we obtain an expression of the form

$$\mathscr{Z} = \sum_{N=0}^{\infty} z^N \exp\left(\mathscr{J} \sum_{\langle ij \rangle} e_i e_j \right)$$ (A.8)

where $z \equiv \exp(\beta\mu)$ is the fugacity and μ is the chemical potential. Hence if we use eqn (A.7), we can write (A.8) in the form

$$\mathscr{Z} = \sum_{\{e\}} \exp\left(\ln z \sum_{i=1}^{\mathcal{N}} e_i + \mathscr{J} \sum_{\langle ij \rangle} e_i e_j \right)$$ (A.9)

where the summation in (A.9), unlike that in (A.6), is unrestricted. Equation (A.9) can be reduced to an expression quite similar to the Ising partition function (A.5) by substituting

$$e_i e_j = \tfrac{1}{4} s_i s_j + \tfrac{1}{4}(s_i + s_j) + \tfrac{1}{4}.$$ (A.10)

It is worth remarking that many relations which can be proved for fluid systems have their analogues for the case of magnetic systems. For example, we showed in Chapter 7 that the response function K_T was simply related to the pair correlation function $G(r)$. Similarly, we can relate the magnetic susceptibility χ_T to the two-spin correlation function $\Gamma(r)$. We begin with a Hamiltonian of the form

$$\mathscr{H} = -\bar{\mu}H \sum_{i=1}^{N} S_i^z + \mathscr{H}_0, \qquad (A.11)$$

where, as before, $\bar{\mu} \equiv g\mu_B$, H is the magnetic field, S_i^z is the z component of a spin vector, and $\mathscr{H}_0$ is the Hamiltonian for $H = 0$. The spontaneous magnetization $M(T, H)$ is given by

$$M(T, H) = \frac{1}{Z(T, H)} \operatorname{tr} (\bar{\mu} S_{TOT}^z e^{-\beta\mathscr{H}}), \qquad (A.12)$$

where we have defined

$$S_{TOT}^z \equiv \sum_{i=1}^{N} S_i^z. \qquad (A.13)$$

Hence we have for the isothermal susceptibility $\chi_T \equiv (\partial M/\partial H)_T$ the expression

$$\chi_T = \bar{\mu} \frac{\partial}{\partial H} \left\{ \frac{1}{Z(T, H)} \operatorname{tr} (S_{TOT}^z e^{-\beta\mathscr{H}}) \right\}_T. \qquad (A.14)$$

Now both $\mathscr{H}$ and Z depend upon H. Hence

$$\chi_T = \bar{\mu} \left\{ -\frac{\operatorname{tr} (S_{TOT}^z e^{-\beta\mathscr{H}})}{Z^2} \frac{\partial}{\partial H} Z + \frac{1}{Z} \operatorname{tr} \left(S_{TOT}^z \frac{\partial}{\partial H} e^{-\beta\mathscr{H}} \right) \right\}_T. \qquad (A.15)$$

From (A.11) it follows that

$$\frac{\partial}{\partial H} e^{-\beta\mathscr{H}} = \beta\bar{\mu} S_{TOT}^z e^{-\beta\mathscr{H}}. \qquad (A.16)$$

Hence we find

$$\chi_T = \beta\bar{\mu}^2 \{ \langle (S_{TOT}^z)^2 \rangle - \langle S_{TOT}^z \rangle^2 \}. \qquad (A.17)$$

Since (A.17) can be written in the form

$$\chi_T = \beta\bar{\mu}^2 \langle (S_{TOT}^z - \langle S_{TOT}^z \rangle)^2 \rangle, \qquad (A.18)$$

it follows that

$$\chi_T \geq 0 \qquad (A.19)$$

as we had argued in Chapter 2. Equation (A.18) can be written in the somewhat more familiar form

$$kT \chi_T = \bar{\mu}^2 \sum_{i=1}^{N} \sum_{j=1}^{N} \langle S_i^z S_j^z \rangle - M^2. \qquad (A.20)$$

Equation (A.20), or, equivalently, eqn (A.18), is frequently called a *fluctuation–dissipation result*, as it relates the isothermal susceptibility to the fluctuations of the magnetization.

An analogous expression can be obtained for the specific heat at constant magnetic field, which from eqn (2.48) is given by

$$C_H = \left(\frac{\partial E}{\partial T}\right)_H = -\frac{1}{kT^2}\left(\frac{\partial E}{\partial \beta}\right)_H. \tag{A.21}$$

For the 'constant T and constant H ensemble' (Wannier 1966, Kubo 1965), we recall that the quantity $\mathscr{H}$ in the partition function $Z(T, H) = \mathrm{tr}\,[\exp(-\beta\mathscr{H})]$ is the enthalpy, not the energy. Hence we have

$$C_H = -\frac{1}{kT^2}\frac{\partial}{\partial \beta}\left[\frac{1}{Z(T, H)}\,\mathrm{tr}\,(\mathscr{H}e^{-\beta\mathscr{H}})\right], \tag{A.22}$$

or, in complete analogy with eqn (A.18),

$$\begin{aligned}
kT^2\,C_H &= \langle\mathscr{H}^2\rangle - \langle\mathscr{H}\rangle^2 \\
&= \langle(\mathscr{H} - \langle\mathscr{H}\rangle)^2\rangle.
\end{aligned} \tag{A.23}$$

Equations (A.20) and (A.23) are used frequently in calculations of χ_T and C_H. In Chapter 8 we illustrated the use of eqn (A.20) in calculating χ_T for the $d = 1$ Ising model, so we shall now use (A.23) to calculate the specific heat C_H. For the sake of simplicity, we treat here the uniform case $J_i = J$, so that eqn. (8.5) becomes simply

$$\mathscr{H} = -J\sum_{i=1}^{N-1} s_i s_{i+1}. \tag{A.24}$$

Hence from (A.23) it follows that

$$kT^2\,C_H = J^2\sum_{i=1}^{N-1}\sum_{j=1}^{N-1}\left(\langle s_i s_{i+1} s_j s_{j+1}\rangle - \langle s_i s_{i+1}\rangle\langle s_j s_{j+1}\rangle\right). \tag{A.25}$$

Now for a uniform interaction eqn (8.15) reduces to

$$\langle s_i s_{i+1}\rangle = \tanh\mathscr{J}, \tag{A.26}$$

where, as before, $\mathscr{J} \equiv \beta J \equiv J/kT$. The calculation of the four-spin correlation function $\langle s_i s_{i+1} s_j s_{j+1}\rangle$ proceeds exactly as for the two-spin correlation function, and we have

$$Z_N\langle s_i s_{i+1} s_j s_{j+1}\rangle = \frac{\partial^2}{\partial\mathscr{J}^2}\,Z_N, \tag{A.27}$$

where

$$\langle s_i s_{i+1} s_j s_{j+1}\rangle = \begin{cases} 1 & i = j \\ \tanh^2\mathscr{J} & i \neq j \end{cases}. \tag{A.28}$$

Thus we obtain, on substituting (A.26) and (A.28) into (A.25),

$$\begin{aligned}
kT^2 C_H &= J^2\sum_{i=1}^{N-1}\sum_{j=1}^{N-1}\{\delta_{ij} + (1 - \delta_{ij})\tanh^2\mathscr{J} - \tanh^2\mathscr{J}\} \\
&= J^2\sum_{i=1}^{N-1}\sum_{j=1}^{N-1}\delta_{ij}(1 - \tanh^2\mathscr{J}).
\end{aligned} \tag{A.29}$$

The summation in (A.29) may be carried out and, on using the trigonometric identity $\operatorname{sech}^2 \mathscr{J} \equiv 1 - \tanh^2 \mathscr{J}$, we obtain

$$C_H = k \mathscr{J}^2 (N - 1) \operatorname{sech}^2 \mathscr{J} \tag{A.30}$$

in agreement with eqn (8.27b).

Suggested further reading

Fisher (1965).
Thompson (1971).

APPENDIX B

EXACT SOLUTION OF THE ZERO-FIELD ISING MODEL FOR A TWO-DIMENSIONAL LATTICE

IN § 9.3 we obtained a solution for the partition function of the Ising model for a linear chain lattice by associating with each term in the partition function a graph on the lattice. The graphs that correspond to terms that make a non-vanishing contribution to the partition function were found to be small in number and hence the problem was rather elementary to solve. We emphasized that the method presented was not restricted to one-dimensional lattices, and that in fact one can with this technique derive Onsager's solution for two-dimensional lattices with considerably less labour than is required using Onsager's original method (Onsager 1944). Since the Onsager solution of the two-dimensional Ising model (Onsager 1944) is a landmark in the history of phase transitions and critical phenomena, we shall devote this appendix to a presentation of the derivation of his result.

We begin by recalling eqn (9.21),

$$Z_N(T, 0) = (\cosh \mathscr{J})^{\mathscr{P}} \, 2^N \sum_{\ell=0}^{\mathscr{P}} g(\ell) v^\ell, \qquad (\text{B.1})$$

where $\mathscr{J} \equiv J/kT$, $v \equiv \tanh \mathscr{J}$, and $\mathscr{P} \equiv Nq/2$ is the number of pairs of nearest-neighbour sites on the lattice, and $g(\ell)$ is the number of graphs that one can draw on the lattice using ℓ lines such that each vertex of the lattice is even; here

$$g(0) \equiv 1.$$

We have seen in Chapter 9 that the function $g(\ell)$ increases rather rapidly with order ℓ, so that in practice one can calculate only the first 15–20 coefficients $g(\ell)$ for, say, a square lattice. In particular, no one has succeeded in calculating the general coefficient $g(\ell)$ of the series by *direct enumeration* of closed graphs, all of whose vertices are even. However, it has recently become possible (Vdovichenko 1964, Glasser 1970) to solve essentially the same problem. The principal idea is to replace the problem of calculating $g(\ell)$ by a similar problem of calculating closed paths in the lattice. These closed paths, which we shall call loops, stand in a many-to-one correspondence with the coefficients $g(\ell)$, in the sense that there can be more than a single loop corresponding to a given graph. To illustrate this fact we consider the graph shown in Fig. B.1(a), which enters in order $\ell = 8$. (The only graphs entering in lower orders are the four-sided and six-sided polygons, and these enter in orders 4 and 6, respectively.) The graph in Fig. B.1(a) has a self-intersection, and we can, in general, describe a self-intersection by three topologically different loops. In order to insure that the contributions from all three loops are identical, it is convenient to associate with each vertex a factor $\exp\left(\frac{1}{2}i\theta\right)$, where $\theta\left[= -\frac{1}{2}\pi, 0, \frac{1}{2}\pi\right]$ is the angle of rotation of the path as it passes through

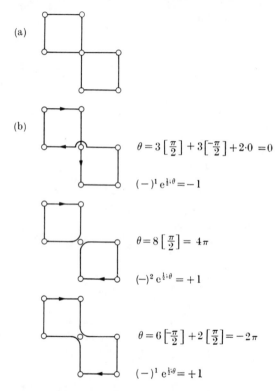

(a)

(b)

$$\theta = 3\left[\frac{\pi}{2}\right] + 3\left[\frac{-\pi}{2}\right] + 2\cdot 0 = 0$$

$$(-)^1 e^{\frac{1}{2}i\theta} = -1$$

$$\theta = 8\left[\frac{\pi}{2}\right] = 4\pi$$

$$(-)^2 e^{\frac{1}{2}i\theta} = +1$$

$$\theta = 6\left[\frac{-\pi}{2}\right] + 2\left[\frac{\pi}{2}\right] = -2\pi$$

$$(-)^1 e^{\frac{1}{2}i\theta} = +1$$

FIG. B.1. (a) Shows a graph with a self-intersection which enters in order $\ell = 8$ in the calculation of the series (B.1) of the partition function for the Ising model on a square lattice. (b) Represents closed paths or loops which correspond to the graph of (a). The phase factor for each loop is calculated in accordance with the rules described in the text, and we see that the first loop enters with an overall minus sign, while the second and third loops enter with plus signs. Thus the net contribution of all three loops is $+1$.

the vertex. If we associate with each loop an additional factor of $(-1)^n$, where n is the order of the loop (the number of independent closed paths), then we will obtain a contribution of $+1$ from the sum of the three loops which correspond to a self-intersection. Thus we see in Fig. B.1(b) that two of the three loops enter with plus signs and one with a minus sign, so that the net contribution is $+1$.

The summation over all loops will also contain configurations corresponding to multiple bonds (two or more lines between a pair of vertices). However,
using the same argument, we can show that such contributions cancel one
another. Hence we can write $g(\ell)$ as a summation over all loops, containing ℓ
lines.

Because of the presence of the combinatorial factors [such as the factor
$(-1)^n$ referred to above] it is convenient to collect together the loops of the
same order n, and to write

$$g(\ell)v^\ell = \sum_n (-1)^n L(n), \tag{B.2}$$

where $L(n)$ is the contribution from all the n-fold loops,

$$L(n) = \frac{1}{n!} \frac{1}{2^n} \sum_{\ell_1,\ldots,\ell_n}^* D(\ell_1) D(\ell_2) \ldots D(\ell_n) v^{\ell_1 + \cdots + \ell_n}, \tag{B.3}$$

where $D(\ell_1)$ is the contribution of a simple loop of ℓ_1 lines and is simply
given by the product of the phase factors $\exp(\tfrac{1}{2}i\theta)$ for each vertex in the loop.
The factor $1/n!$ in eqn (B.3) occurs because the same n–fold loop is obtained
if we permute the indices $\ell_1, \ell_2, \ldots, \ell_n$. Similarly, the factor 2^{-n} arises from
the fact that the same graph is obtained when we change the direction of
the paths constituting the loop. The asterisk on the summation symbol in
(B.3) indicates that the summation is over all values of the indices $\ell_1, \ell_2, \ldots, \ell_n$
subject to the constraint that $\ell_1 + \ell_2 + \ldots + \ell_n = \ell$.

From eqns (B.2) and (B.3) it follows that

$$\sum_{\ell=1}^\infty g(\ell)v^\ell = \sum_{n=1}^\infty \frac{1}{n!}\left(-\frac{1}{2}\right)^n \sum_{\ell_1=1}^\infty \cdots \sum_{\ell_n=1}^\infty D(\ell_1) \ldots D(\ell_n) v^{\ell_1 + \cdots + \ell_n}$$
$$= \sum_{n=1}^\infty \frac{1}{n!}\left(-\frac{1}{2}\right)^n \left\{\sum_{\ell=1}^\infty D(\ell) v^\ell\right\}^n . \tag{B.4}$$

The right-hand side of eqn (B.4) is recognized as being the series for the exponential function, providing we add the term unity to both sides of the
equation. Since $g(\ell = 0) = 1$, we have

$$\sum_{\ell=0}^\infty g(\ell)v^\ell = \exp\left\{-\tfrac{1}{2}\sum_{\ell=1}^\infty D(\ell)v^\ell\right\}. \tag{B.5}$$

Finally, we can replace the upper limit of the summation in eqn (B.1) by
$\ell = \infty$ since for all $\ell > \mathscr{P} \equiv Nq/2$, $g(\ell) = 0$ (since it is impossible to construct a graph without multiple bonds which has more lines than the total
number of bonds in the lattice). Hence the zero-field partition function can
be written

$$Z_N(T, 0) = 2^N(\cosh \mathscr{J})^{Nq/2} \exp\left\{-\tfrac{1}{2}\sum_{\ell=1}^\infty D(\ell)\, v^\ell\right\}. \tag{B.6}$$

Equation (B.6) expresses the partition function of the two-dimensional
Ising model in terms of $D(\ell)$, the sum of the products of the vertex functions,

where each vertex on the loop has been weighted by the phase factor $\exp\left(\frac{1}{2}i\theta\right)$.

Although we considered the square lattice in the preceding discussion, an analogous (though not identical) argument can be presented for any two-dimensional lattice. The second stage of our calculation is restricted to the case of a square lattice, although similar arguments can be developed for other two-dimensional lattices.

The problem now is to calculate the function $D(\ell)$. To this end it is convenient to introduce a matrix $M_l(\mathbf{p}, \mathbf{p}')$ defined as the sum over all paths of ℓ bonds from point $\mathbf{p}$ to point $\mathbf{p}'$, and where each path is weighted by the appropriate phase factors. Here $\mathbf{p}$ stands for an ordered pair, (i, α), where i indexes the site in the lattice and α indexes the direction of the path leaving

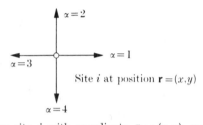

FIG. B.2. Single lattice site i with coordinate $\mathbf{r} = (x, y)$, and the four directions ($\alpha = 1,2,3,4$) from which directed paths might emanate from the lattice site.

site i (cf. Fig. B.2). Since there are N sites ($i = 1, 2, \ldots, N$) and four directions in the square lattice ($\alpha = 1, 2, 3, 4$), it follows that the matrix $M_\ell(\mathbf{p}, \mathbf{p}')$ has $4N \times 4N$ elements. For small values of ℓ, most of these elements will be zero as there will be no paths linking points $\mathbf{p}$ and $\mathbf{p}'$. For example, $M_0(\mathbf{p}, \mathbf{p}')$ $= \delta_{\mathbf{p},\mathbf{p}'}$. For a closed loop, we clearly have $\mathbf{p} = \mathbf{p}'$, whence

$$D(\ell) = \frac{1}{\ell}\sum_{p} M_\ell(\mathbf{p}, \mathbf{p}). \tag{B.7}$$

The factor $1/\ell$ appears in eqn (B.7) since each ℓ-step loop can be started at any of ℓ vertices.

Now, in general, $M_\ell(\mathbf{p}, \mathbf{p}')$ can be regarded as a sequence of single-step walks, and we have

$$M_\ell(\mathbf{p}, \mathbf{p}') = \sum_{\mathbf{p}_1,\ldots\mathbf{p}_{\ell-1}} M_1(\mathbf{p}, \mathbf{p}_1) M_1(\mathbf{p}_1, \mathbf{p}_2)\ldots M_1(\mathbf{p}_{l-1}, \mathbf{p}'). \tag{B.8}$$

Equation (B.8) leads to the simple expression

$$M_\ell(\mathbf{p}, \mathbf{p}') = M_1^\ell(\mathbf{p}, \mathbf{p}'). \tag{B.9}$$

On substituting (B.9) into (B.7) we obtain

$$D(\ell) = \frac{1}{\ell}\,\text{trace }M_1^\ell$$

$$= \frac{1}{\ell}\sum_{j=1}^{4N} (\lambda_j)^\ell \tag{B.10}$$

where λ_j denotes the eigenvalues of M_1, and where we have used the result that the eigenvalues of M_1' are simply λ_j'. Combination of eqns (B.10) and (B.6) results in the expression

$$Z_N(T, 0) = 2^N(\cosh \mathcal{J})^{2N} \exp\left(-\tfrac{1}{2} \sum_{j=1}^{4N} \sum_{\ell=1}^{\infty} \frac{1}{\ell} \lambda_j'^{\ell} v^{\ell}\right).$$ (B.11)

Now the summation over the index ℓ is recognized as being the series for the logarithm,

$$-\tfrac{1}{2} \sum_{j=1}^{4N} \sum_{\ell=1}^{\infty} \frac{1}{\ell} (v\lambda_j)^{\ell} = \tfrac{1}{2} \sum_{j=1}^{4N} \ln(1 - v\lambda_j)$$

$$= \sum_{j=1}^{4N} \ln(1 - v\lambda_j)^{1/2}$$

$$= \ln \prod_{j=1}^{4N} (1 - v\lambda_j)^{1/2}.$$ (B.12)

Hence we obtain the remarkably simple expression for the partition function in terms of the eigenvalues λ_j of the matrix $M_1(\mathbf{p}, \mathbf{p})$,

$$Z_N(T, 0) = 2^N(\cosh \mathcal{J})^{2N} \left\{\prod_{j=1}^{4N} (1 - v\lambda_j)\right\}^{\frac{1}{2}}.$$ (B.13)

Thus all that is needed to complete the calculation is to find the eigenvalues of the $4N \times 4N$ matrix $M_1(\mathbf{p}, \mathbf{p})$. Fortunately the matrix $M_1(\mathbf{p}, \mathbf{p}')$ has a simple form; the reader can verify almost by inspection that

$$M_1(x, y, 1; x', y', \alpha') = \delta_{x,x'+1} \delta_{y,y'} \delta_{1,\alpha'} + i^{-\frac{1}{2}} \delta_{x,x'} \delta_{y,y'+1} \delta_{2,\alpha'}$$
$$+ i^{\frac{1}{2}} \delta_{x,x'} \delta_{y,y'-1} \delta_{4,\alpha'},$$ (B.14)

$$M_1(x, y, 2; x', y', \alpha') = i^{\frac{1}{2}} \delta_{x,x'+1} \delta_{y,y'} \delta_{1,\alpha'} + \delta_{x,x'} \delta_{y,y'+1} \delta_{2,\alpha'}$$
$$+ i^{-\frac{1}{2}} \delta_{x,x'-1} \delta_{y,y'} \delta_{3,\alpha'},$$ (B.15)

$$M_1(x, y, 3; x', y', \alpha') = i^{\frac{1}{2}} \delta_{x,x'} \delta_{y,y'+1} \delta_{2,\alpha'} + \delta_{x,x'-1} \delta_{y,y'} \delta_{3,\alpha'}$$
$$+ i^{-\frac{1}{2}} \delta_{x,x'} \delta_{y,y'-1} \delta_{4,\alpha'},$$ (B.16)

$$M_1(x, y, 4; x', y', \alpha') = i^{-\frac{1}{2}} \delta_{x,x'+1} \delta_{y,y'} \delta_{1,\alpha'} + i^{\frac{1}{2}} \delta_{x,x'-1} \delta_{y,y'} \delta_{3,\alpha'}$$
$$+ \delta_{x,x'} \delta_{y,y'-1} \delta_{4,\alpha'},$$ (B.17)

where (x, y) are the Cartesian coordinates of a vector $\mathbf{r}$ to the lattice site i.

The matrix M_1 can be diagonalized by taking advantage of the lattice periodicity and transforming M_1 to its Fourier representation. Thus we have

$$\hat{M}^{\alpha\alpha'}(\mathbf{q}, \mathbf{q}') = \sum_{r,r'} \exp(-i\mathbf{q} \cdot \mathbf{r}) M_1(\mathbf{p}, \mathbf{p}') \exp(i\mathbf{q}' \cdot \mathbf{r}').$$ (B.18)

From (B.14)–(B.17), we see that $\hat{M}^{\alpha\alpha'}(\mathbf{q}, \mathbf{q}')$ has a block diagonal form, $\hat{M}^{\alpha\alpha'}(\mathbf{q}, \mathbf{q}') = \hat{M}^{\alpha\alpha'}(\mathbf{q})\, \delta_{\mathbf{q},\mathbf{q}'}$, with

$$\hat{M}(\mathbf{q}) = \begin{cases} Q_1 & i^{-\frac{1}{2}}Q_2 & 0 & i^{\frac{1}{2}}Q_2^* \\ i^{\frac{1}{2}}Q_1 & Q_2 & i^{-\frac{1}{2}}Q_1^* & 0 \\ 0 & i^{\frac{1}{2}}Q_2 & Q_1^* & i^{-\frac{1}{2}}Q_2^* \\ i^{-\frac{1}{2}}Q_1 & 0 & i^{\frac{1}{2}}Q_1^* & Q_2^* \end{cases}, \tag{B.19}$$

where

$$Q_i \equiv \exp\,(-iq_i),\, Q_i^* \equiv \exp\,(iq_i), \tag{B.20}$$

and q_1, q_2 are the Cartesian coordinates of $\mathbf{q}$.

From (B.19) we see that for each 4×4 block, we have

$$\prod_{j=1}^{4} (1 - v\lambda_j) = \det |I - v\hat{M}(\mathbf{q})|$$
$$= (1 + v^2)^2 - 2v(1 - v^2)\, Re(Q_1 + Q_2). \tag{B.21}$$

From (B.20) we have $Re\, Q_j = \cos q_j$, whence from (B.13),

$$Z_N(T, 0) = 2^N(1 - v^2)^{-N} \left[\prod_{\mathbf{q}} \{(1 + v^2)^2 - 2v(1 - v^2)(\cos q_1 + \cos q_2)\} \right]^{\frac{1}{2}} \tag{B.22}$$

where we have used the identity $\cosh^2 \mathcal{J} = (1 - v^2)^{-1}$.

Hence we have

$$\ln Z_N(T, 0) = N \ln 2 - N \ln (1 - v^2) + \tfrac{1}{2} \sum_q \ln \{(1 + v^2)^2 - 2v(1 - v^2)\,(\cos q_1 + \cos q_2)\}. \tag{B.23}$$

In the limit $N \to \infty$, we can change the summation to an integral, with the result

$$\ln Z_N(T, 0) \sim \ln 2 - \ln (1 - v^2)$$
$$+ \tfrac{1}{2} \left(\frac{1}{2\pi}\right)^2 \iint dq_1\, dq_2 \ln \{(1 + v^2)^2 - 2v(1 - v^2)\,(\cos q_1 + \cos q_2)\}. \tag{B.24}$$

We frequently eliminate the variable $v \equiv \tanh \mathcal{J}$ in favour of $\mathcal{J}$,

$$\ln Z_N(T, 0) \sim \ln 2 + \tfrac{1}{2} \frac{1}{(2\pi)^2} \iint dq_1\, dq_2 \ln \{\cosh^2 2\mathcal{J} - 2 \sinh 2\mathcal{J}\,(\cos q_1 + \cos q_2)\}. \tag{B.25}$$

Equations (B.24) and (B.25) are the forms in which the solution to the Ising model for a square lattice is usually written.

It is actually not possible to carry out the integration in (B.25) in closed form. However we *can* express the enthalpy per spin, $\bar{E} = -J(\partial/\partial\mathscr{J})(N^{-1}\ln Z)$, in terms of an elliptic integral. We thereby find, on differentiating eqn (B.25), that

$$\bar{E} = J \coth 2\mathscr{J} \left\{ 1 \pm \frac{2}{\pi}(1 - z^2)^{\frac{1}{2}} K(z) \right\} \tag{B.26}$$

where

$$K(z) \equiv \int_0^{2\pi} (1 - z^2 \sin^2 q)^{-\frac{1}{2}} \, dq \tag{B.27}$$

and

$$z \equiv \frac{2 \sinh 2\mathscr{J}}{\cosh^2 2\mathscr{J}} = \frac{2 \tanh 2\mathscr{J}}{\cosh 2\mathscr{J}}. \tag{B.28}$$

The $\pm$ sign in eqn (B.26) refers to whether $z > 1$ or $z < 1$. Note that for $z = 1$, the integrand in eqn (B.27) diverges. As z approaches unity from either side the specific heat $C_H = (\partial E/\partial T)_H$ becomes logarithmically infinite. Thus the solution of eqn (B.28) when $z = 1$ corresponds to $\mathscr{J}_c \equiv J/kT_c$. We can easily show that

$$v_c \equiv \tanh \mathscr{J}_c = 2^{\frac{1}{2}} - 1. \tag{B.29}$$

or, since $\tanh^{-1} x = \frac{1}{2} \ln \{(1 + x)/(1 - x)\}$,

$$\mathscr{J}_c = -\frac{1}{2} \ln (2^{\frac{1}{2}} - 1). \tag{B.30}$$

Comparison of (B.30) with the result of the mean field theory for the critical temperature of the Ising model, $kT_M/J = q = 4$, results in

$$\frac{T_c}{T_M} = 0{\cdot}5673 \ldots \tag{B.31}$$

(cf. also Fig. 9.5). Thus we see that the molecular field approximation *overestimates* the value of the critical temperature by almost 50 per cent for the square lattice.

Suggested further reading

Onsager (1944).
Kac and Ward (1952).
Burgoyne (1963).
Vdovichenko (1964, 1965).
Schultz, Mattis, and Lieb (1964).
Landau and Lifschitz (1969).
Glasser (1970).

APPENDIX C

GEOMETRIC INTERPRETATION OF THE STATIC SCALING HYPOTHESIS FOR THERMODYNAMIC POTENTIALS

In Chapter 11 we presented the static scaling hypothesis, eqn (11.30), without attempting to provide justification of any sort. Here we shall present a particularly appealing geometric argument due to Griffiths.

The Griffiths argument supports the assumption that singular part of the Helmholtz potential $A_s(\epsilon, M)$ is a generalized homogeneous function. Begin by considering the ϵ–M plane and choose any two values of reduced temperature ϵ_1 and ϵ_2 such that $\epsilon_2 < \epsilon_1 < 0$ (cf. Fig. C.1). Form the rectangles,

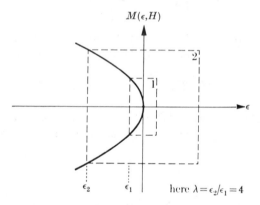

FIG. C.1. A portion of the M-ϵ plane near the critical-point, illustrating the construction of the rectangles labelled 1 and 2. The curve shown is $M(\epsilon, H = 0) = \epsilon^\beta$ with $\beta = \frac{1}{2}$ as predicted by the molecular field theory. The rectangles are chosen such that $\lambda \equiv \epsilon_2/\epsilon_1 = 4$; note that in accordance with eqn (C.2), $M(\epsilon_2)/M(\epsilon_1) = \lambda^\beta = 4^{1/2} = 2$.

labelled 1 and 2 in Fig. C.1, and imagine that the $A_s(\epsilon, M)$ surface is a sort of topographic map that can be stretched (as, e.g., a rubber sheet). We now consider two portions of this topographic map: portion 1 which is over rectangle 1 and portion 2 which is over rectangle 2. Now further suppose that this rubber sheet can be stretched more easily in the ϵ-direction than in the M-direction. In particular, suppose that a stretch of λ units in the

ϵ-direction corresponds to only λ^β units in the M-direction ($\beta < 1$). Then the scaling assumption says that if we stretch the smaller map (map No. 1, the portion over rectangle 1), by an amount such that it covers rectangle 2 (i.e. λ units in the ϵ-direction and λ^β units in the M-direction), then map No. 1 will be identical to map No. 2. This assumption is, then, that the Helmholtz potential over rectangle 1 is identical to the Helmholtz potential over rectangle 2 except for a change of scale—a change that will, in general, be different in a direction parallel to the ϵ-axis than in a direction parallel to the M-axis. This assumption leads, in particular, to the expression

$$\frac{A_s\{\epsilon_2, M(\epsilon_2)\}}{A_s\{\epsilon_1, M(\epsilon_1)\}} = g(\epsilon_2/\epsilon_1) \equiv g(\lambda), \tag{C.1}$$

where λ denotes the ratio ϵ_2/ϵ_1. Since near T_c,

$$\frac{M(\epsilon_2)}{M(\epsilon_1)} = \frac{(-\epsilon_2)^\beta}{(-\epsilon_1)^\beta} = \lambda^\beta \tag{C.2}$$

we may replace the arguments $\{\epsilon_2, M(\epsilon_2)\}$ in eqn (C.1) by $\{\lambda\epsilon_1, \lambda^\beta M(\epsilon_1)\}$ and hence obtain

$$A_s(\lambda\epsilon, \lambda^\beta M) = g(\lambda)\, A_s\, (\epsilon, M). \tag{C.3}$$

Equation (C.3) is recognized as being equivalent to a statement that $A_s(\epsilon, M)$ is homogeneous. Of course, we have not derived the scaling hypothesis by this method, but rather we have shown a way by which this assumption can be arrived at.

We conclude by demonstrating that if a function of two variables $f(x, y)$ is a generalized homogeneous function,

$$f(\lambda^a x, \lambda^b y) = \lambda f(x, y), \tag{C.4}$$

then the Legendre transforms of $f(x, y)$ are also generalized homogeneous functions (Hankey and Stanley, 1971). Let

$$g\{u(x, y), y\} = f(x, y) - x\, u(x, y) \tag{C.5}$$

be the Legendre transform of $f(x, y)$ with respect to the variable x, where $u(x, y) \equiv (\partial f/\partial x)_y$ as in eqn (2.6a). Therefore

$$u(\lambda^a x, \lambda^b y) = \lambda^{1-a} u(x, y). \tag{C.6}$$

Now from eqn (C.5) we can write

$$f(\lambda^a x, \lambda^b y) = g\{u(\lambda^a x, \lambda^b y), \lambda^b y\} + \lambda^a x\, u(\lambda^a x, \lambda^b y). \tag{C.7}$$

On using eqns (C.4) and (C.6), eqn (C.7) simplifies to

$$g\{\lambda^{1-a}\, u(x, y), \lambda^b y\} = \lambda f(x, y) - \lambda^a x\, \lambda^{1-a}\, u(x, y). \tag{C.8}$$

Now the right hand side of (C.8) is simply $\lambda g\{u(x, y), y\}$ whence we have that

$g(u, y)$ is a generalized homogeneous function with scaling parameters $(1 - a)$ and b.

This result has the immediate consequence that if one of the thermodynamic potentials is a generalized homogeneous function, then all four thermodynamic potentials are. For example, if we assume that for the Gibbs potential,

$$G(\lambda^{a_\varepsilon} \epsilon, \lambda^{a_H} H) = \lambda G(\epsilon, H), \tag{C.9}$$

then it follows from (C.8) that, for the Helmholtz potential,

$$A(\lambda^{a_\varepsilon} \epsilon, \lambda^{a_M} M) = \lambda A(\epsilon, M), \tag{C.10}$$

for the internal energy,

$$U(\lambda^{a_S} S, \lambda^{a_M} M) = \lambda U(S, M), \tag{C.11}$$

and for the enthalpy,

$$E(\lambda^{a_S} S, \lambda^{a_H} H) = \lambda E(S, H), \tag{C.12}$$

where the scaling parameters in eqns (C.9)–(C.12) are related through

$$a_M = 1 - a_H \tag{C.13}$$

and

$$a_S = 1 - a_\epsilon. \tag{C.14}$$

Suggested further reading

Stell (1968*b*)

APPENDIX D

THE DYNAMIC STRUCTURE FACTOR IN
THE HYDRODYNAMIC LIMIT

IN this appendix we outline the derivation of eqn (13.29) for the dynamic structure factor $\mathscr{S}(\mathbf{q}, \omega)$ in the hydrodynamic limit. We must keep in mind the fact that the arguments used in this section are expected to be valid only in a limited range of temperature (for example, in the range $\epsilon \gg 10^{-5}$ for the case of 60° quasi-elastic scattering, from a typical fluid, of laser light of wavelength 5000 Å).

The conservation laws of hydrodynamics are discussed in many texts (Huang 1963; Reif 1965; Kittel 1958, 1969); the treatment we follow is most similar to that of Kadanoff and Martin (1963). There is a conservation law for each of the five conserved quantities: number density $n(\mathbf{r}, t)$, momentum density $\mathbf{g}(\mathbf{r}, t)$, and energy density $e(\mathbf{r}, t)$. Here $\mathbf{g}(\mathbf{r}, t)$ is a three-component vector, each of whose components is independently conserved. These conservation laws are

$$\frac{\partial}{\partial t} n(\mathbf{r}, t) + m^{-1} \mathbf{\nabla} \cdot \mathbf{g}(\mathbf{r}, t) = 0 \tag{D.1}$$

for number conservation,

$$\frac{\partial}{\partial t} \mathbf{g}(\mathbf{r}, t) + \mathbf{\nabla} \cdot \mathbf{T}(\mathbf{r}, t) = 0 \tag{D.2}$$

for momentum conservation, and

$$\frac{\partial}{\partial t} e(\mathbf{r}, t) + \mathbf{\nabla} \cdot \mathbf{j}_e(\mathbf{r}, t) = 0 \tag{D.3}$$

for energy conservation.

In eqn (D.2) $\mathbf{T}(\mathbf{r}, t)$ is the momentum current (or stress tensor), and $\mathbf{j}_e(\mathbf{r}, t)$ is called the energy current; these are given by the relations $T^0_{ij}(\mathbf{r}, t) = \delta_{ij} P(\mathbf{r}, t)$ and $\mathbf{j}^0_e(\mathbf{r}, t) = (e + P)\mathbf{v}(\mathbf{r}, t)$ for the limiting case of a non-viscous fluid ($\Lambda = \eta = \zeta = 0$). Here $P(\mathbf{r}, t)$ is the pressure (the stress tensor is often called the pressure tensor), and $\mathbf{v}(\mathbf{r}, t)$ is the average velocity given by the expression

$$< \mathbf{g}(\mathbf{r}, t) > \ = m < n(\mathbf{r}, t) > \mathbf{v}(\mathbf{r}, t). \tag{D.4}$$

For the more general cases there is an additional energy current $-\Lambda\mathbf{\nabla}\mathbf{T}(\mathbf{r}, t)$ and an additional contribution to the momentum current produced by the forces arising from the shear and bulk viscosities. Hence we have, in general, the expressions

$$T_{ij}(\mathbf{r}, t) = \delta_{ij} P(\mathbf{r}, t) - \eta\{\partial v_i(\mathbf{r}, t)/\partial r_j + \partial v_j(\mathbf{r}, t)/\partial r_i\} - \delta_{ij} \boldsymbol{\nabla} \cdot \mathbf{v}(\mathbf{r}, t)\, (\zeta - \tfrac{2}{3}\eta) \tag{D.5}$$

and

$$\mathbf{j}_{\mathrm{e}}(\mathbf{r}, t) = (e + P)\, \mathbf{v}(\mathbf{r}, t) - \Lambda\boldsymbol{\nabla}T(\mathbf{r}, t). \tag{D.6}$$

Equations (D.1)–(D.3), together with eqns (D.4)–(D.6), constitute a complete description of a fluid in the hydrodynamic region.

We will next obtain the linearized equations of motion of a non-equilibrium disturbance in a fluid. To this end, we define the three new functions

$$n_1(\mathbf{r}, t) \equiv n(\mathbf{r}, t) - n_0, \tag{D.7}$$

$$\mathbf{v}_1(\mathbf{r}, t) \equiv \mathbf{v}(\mathbf{r}, t) - v_0, \tag{D.8}$$

and

$$e_1(\mathbf{r}, t) \equiv e(\mathbf{r}, t) - e_0, \tag{D.9}$$

where n_0, $\mathbf{v}_0$, and e_0 are the appropriate equilibrium values $\langle n\rangle$, $\langle \mathbf{v}\rangle$, and $\langle e\rangle$ respectively. (Without loss of generality we can choose $\mathbf{v}_0 = 0$ since we can always choose a reference frame in which the fluid as a whole is at rest.) Comparison of eqns (D.7) and (13.2) would seem to indicate that $\delta n(\mathbf{r}, t)$ and $n_1(\mathbf{r}, t)$ are identical functions. This is not strictly true, because $n(\mathbf{r}, t)$ denotes an operator in (13.2) and is the expectation value of the density in the non-equilibrium state in (D.7). However one can argue that the fluid does not 'know' whether a small deviation in the equilibrium parameters has been caused by an external perturbation or by a spontaneous fluctuation so that the subsequent motion of these two types of deviations, $n_1(\mathbf{r}, t)$ (defined in eqn (D.7)) and $\delta n(\mathbf{r}, t)$ (defined in eqn (13.2)) should be the same. Therefore we assume that the equations of motion we shall obtain for $n_1(\mathbf{r}, t)$ are also valid for the fluctuation operator $\delta n(\mathbf{r}, t)$, when we are considering the equilibrium ensemble. That this assumption, originally due to Onsager (1931a, b), is in fact correct is a consequence of the fluctuation–dissipation theorem employed by Kadanoff and Martin (1963).

We now proceed to find these equations of motion. It will be convenient (Kadanoff and Martin 1963) to replace the energy density $e(\mathbf{r}, t)$ with the heat density $h(\mathbf{r}, t)$, defined by

$$h(\mathbf{r}, t) \equiv e(\mathbf{r}, t) - \left(\frac{e_0 + P_0}{n_0}\right) n(\mathbf{r}, t). \tag{D.10}$$

The identification of the function $h(\mathbf{r}, t)$ defined in eqn (D.10), as a heat energy density follows because at constant particle number N, the first law of thermodynamics yields

$$\frac{T}{V}\,dS = \frac{dE}{V} + \frac{1}{V}\,PdV = \frac{d(eV)}{V} + \frac{1}{V}\,PdV$$

$$= de - \left(\frac{e_0 + P_0}{n_0}\right)dn$$

$$= dh. \tag{D.11}$$

Utilizing eqns (D.7)–(D.10), we can rewrite the hydrodynamic equations (D.1)–(D.3) in the form

$$\frac{\partial}{\partial t}\,n_1(\mathbf{r}, t) + n_0\boldsymbol{\nabla} \cdot \mathbf{v}_1(\mathbf{r}, t) = 0, \tag{D.12}$$

$$\frac{\partial}{\partial t}\,\mathbf{v}_1(\mathbf{r}, t) + \frac{1}{mn}\,\boldsymbol{\nabla}P_1(\mathbf{r}, t) - \frac{\eta}{mn}\,\boldsymbol{\nabla}^2\mathbf{v}_1(\mathbf{r}, t) - \frac{(\zeta + \frac{1}{3}\eta)}{mn}\,\boldsymbol{\nabla}\{\boldsymbol{\nabla} \cdot \mathbf{v}_1(\mathbf{r}, t)\} = 0 \tag{D.13}$$

and

$$\frac{\partial}{\partial t}\,h_1(\mathbf{r}, t) - \Lambda\boldsymbol{\nabla}^2 T_1(\mathbf{r}, t) = 0. \tag{D.14}$$

Because we want to obtain equations in terms of two independent local thermodynamic variables—in this case $n_1(\mathbf{r}, t)$ and $h_1(\mathbf{r}, t)$—we will use the fact that the system is in local equilibrium to write

$$P_1(\mathbf{r}, t) = \left(\frac{\partial P}{\partial n}\right)_S n_1(\mathbf{r}, t) + \left(\frac{\partial P}{\partial S}\right)_n h_1(\mathbf{r}, t)\,\frac{V}{T}, \tag{D.15}$$

$$T_1(\mathbf{r}, t) = \left(\frac{\partial T}{\partial n}\right)_S n_1(\mathbf{r}, t) + \left(\frac{\partial T}{\partial S}\right)_n h_1(\mathbf{r}, t)\,\frac{V}{T}, \tag{D.16}$$

where $P_1(\mathbf{r}, t)$, $T_1(\mathbf{r}, t)$, and $h_1(\mathbf{r}, t)$ are defined in analogy with eqns (D.7) and (D.9).

Next we substitute eqns (D.15) and (D.16) into eqns (D.12)–(D.14), and eliminate the velocity from these equations by taking the divergence of both sides of eqn (D.13), and combining (D.13) with (D.12) we obtain,

$$\left\{-\frac{\partial^2}{\partial t^2} + \left(v_s^2 + D_\ell\frac{\partial}{\partial t}\right)\boldsymbol{\nabla}^2\right\}n_1(\mathbf{r}, t) + \frac{1}{m^2nC_V}\left(\frac{\partial P}{\partial T}\right)_V \boldsymbol{\nabla}^2 h_1(\mathbf{r}, t) = 0 \tag{D.17}$$

and

$$\left(\frac{\partial}{\partial t} - \frac{C_P}{C_V}\,D_T\,\boldsymbol{\nabla}^2\right)h_1(\mathbf{r}, t) - D_T\frac{C_P}{C_V}\frac{T}{n}\left(\frac{\partial P}{\partial T}\right)_V \boldsymbol{\nabla}^2\,n_1(\mathbf{r}, t) = 0. \tag{D.18}$$

In writing eqns (D.17) and (D.18) we have introduced the thermal diffusion constant D_T, the sound wave damping constant D_ℓ, and the velocity of sound for $\omega = 0$ as given by eqns (13.30)–(13.32).

We now make the substitutions $n_1(\mathbf{r}, t) \to \delta n(\mathbf{r}, t)$ and $h_1(\mathbf{r}, t) \to \delta h(\mathbf{r}, t)$ in eqns (D.17) and (D.18), since these equations are assumed in the above discussion to be also valid for equilibrium fluctuations. In order to obtain equations for the density–density correlation function, we multiply both sides of eqns (D.17) and (D.18) by $\delta n(0, 0)$ and take the ensemble average, with the result

$$\left\{ -\frac{\partial^2}{\partial t^2} + \left(v_s^2 + D_\ell \frac{\partial}{\partial t} \right) \mathbf{\nabla}^2 \right\} \mathscr{G}_{nn}(\mathbf{r}, t) + \frac{1}{m^2 n C_V} \left(\frac{\partial P}{\partial T} \right)_V \mathbf{\nabla}^2 \mathscr{G}_{hn}(\mathbf{r}, t) = 0$$

(D.19)

and

$$\left(\frac{\partial}{\partial t} - \frac{C_P}{C_V} D_T \mathbf{\nabla}^2 \right) \mathscr{G}_{hn}(\mathbf{r}, t) - D_T \frac{C_P}{C_V} \frac{T}{n} \left(\frac{\partial P}{\partial T} \right)_V \mathbf{\nabla}^2 \mathscr{G}_{nn}(\mathbf{r}, t) = 0, \quad \text{(D.20)}$$

where $\mathscr{G}_{nn}(\mathbf{r}, t)$ was defined in eqn (13.1) and $\mathscr{G}_{hn}(\mathbf{r}, t)$ is defined similarly,

$$\mathscr{G}_{hn}(\mathbf{r}, t) \equiv \langle \delta h(\mathbf{r}, t)\, \delta n(0, 0) \rangle. \qquad \text{(D.21)}$$

The coupled equations (D.19) and (D.20) can be solved by transforming them into algebraic equations through the introduction of the Fourier–Laplace transforms

$$\mathscr{S}_{nn}^{\mathrm{L}}(\mathbf{q}, \omega) \equiv \int \mathrm{d}\mathbf{r} \int_0^\infty \mathrm{d}t\; \mathscr{G}_{nn}(\mathbf{r}, t)\, \mathrm{e}^{-\mathrm{i}\mathbf{q}\cdot\mathbf{r} + \mathrm{i}\omega t}, \qquad \text{(D.22)}$$

with an analogous equation for $\mathscr{S}_{hn}^{\mathrm{L}}(\mathbf{q}, \omega)$. Notice that although $\mathscr{S}_{nn}(\mathbf{q}, \omega)$ differs from the Fourier–Laplace transform $\mathscr{S}_{nn}^{\mathrm{L}}(\mathbf{q}, \omega)$ defined in (D.22) in that the time integration in eqn (13.4) is carried out from $-\infty$ to $+\infty$ instead of only from 0 to $+\infty$, the two functions are simply related by

$$\mathscr{S}_{nn}(\mathbf{q}, \omega) = 2\, \mathrm{Re}\, \mathscr{S}_{nn}^{\mathrm{L}}(\mathbf{q}, \omega). \qquad \text{(D.23)}$$

In terms of the transformed variables $\mathscr{S}_{nn}^{\mathrm{L}}(\mathbf{q}, \omega)$ and $\mathscr{S}_{hn}^{\mathrm{L}}(\mathbf{q}, \omega)$, the coupled equations (D.19) and (D.20) become

$$\{\mathrm{i}\omega(-\mathrm{i}\omega + D_\ell q^2) - v_s^2 q^2\}\, \mathscr{S}_{nn}^{\mathrm{L}}(\mathbf{q}, \omega) - \frac{1}{m^2 n C_V} \left(\frac{\partial P}{\partial T} \right)_V q^2 \mathscr{S}_{hn}^{\mathrm{L}}(\mathbf{q}, \omega)$$

$$= -(-\mathrm{i}\omega + D_\ell q^2)\, S_{nn}(\mathbf{q}) \qquad \text{(D.24)}$$

and

$$\left(-\mathrm{i}\omega + q^2 D_T \frac{C_P}{C_V} \right) \mathscr{S}_{hn}^{\mathrm{L}}(\mathbf{q}, \omega) + q^2 D_T \frac{C_P}{C_V} \frac{T}{n} \left(\frac{\partial P}{\partial T} \right)_V \mathscr{S}_{nn}^{\mathrm{L}}(\mathbf{q}, \omega) = S_{hn}(\mathbf{q}),$$

(D.25)

where $S_{hn}(\mathbf{q})$ is defined in analogy with the function $S_{nn}(\mathbf{q})$ in eqn (13.5).

The inhomogeneous terms that appear on the right-hand sides of eqns (D.24) and (D.25) arise from the boundary term at $t = 0$ in the partial integration. An additional boundary term, given by the value for $t = 0$ of the quantity $\partial/\partial t\{\exp(-i\mathbf{q}\cdot\mathbf{r})\langle\delta n(\mathbf{r}, t)\,\delta n(\mathbf{0}, 0)\rangle\}$, does not appear because it is identically zero due to time reversal invariance of the ensemble. This can be seen by using the continuity equation and the expression for the correlation function in the canonical ensemble to write

$$\frac{\partial}{\partial t}\langle\delta n(\mathbf{r}, t)\,\delta n(\mathbf{0}, 0)\rangle\,|_{t=0} = -\mathbf{\nabla}\cdot\langle v(\mathbf{r}, 0)\,\delta n(\mathbf{0}, 0)\rangle$$

$$\propto -\mathbf{\nabla}\cdot\left\{\int d\mathbf{r}_1\dots d\mathbf{r}_N\dots d\mathbf{p}_1\dots d\mathbf{p}_N\dots\right.$$

$$\left.\left(\sum_{j=1}^{N}\frac{1}{m}p_j\delta(\mathbf{r}-\mathbf{r}_j)\right)\left(\sum_{k=1}^{N}\delta(\mathbf{r}_k)-\langle n\rangle\right)e^{-\beta\mathscr{H}}\right\}.$$

$$(D.26)$$

The right-hand side of eqn (D.26) is zero since

$$\mathscr{H}(\mathbf{r}_1,\dots\mathbf{r}_N;\mathbf{p}_1,\dots\mathbf{p}_N) = \mathscr{H}(\mathbf{r}_1,\dots\mathbf{r}_N;-\mathbf{p}_1,\dots-\mathbf{p}_N). \quad (D.27)$$

The quantities $S_{nn}(\mathbf{q})$ and $S_{hn}(\mathbf{q})$ appearing on the right-hand side of eqns (D.24) and (D.25) can be evaluated in the long-wavelength $(q\to 0)$ limit using classical fluctuation theory (Landau and Lifschitz 1969), with the results

$$S_{nn}(\mathbf{q}) = \frac{1}{V}\langle\delta N\,\delta N\rangle = nkT\left(\frac{\partial n}{\partial P}\right)_{T,V} \quad (D.28)$$

and

$$S_{hn}(\mathbf{q}) = \frac{T}{V}\langle\delta S\,\delta N\rangle = kT^2\left(\frac{\partial n}{\partial T}\right)_{P,V}. \quad (D.29)$$

Hence eqns (D.24) and (D.25) become a pair of coupled algebraic equations for the quantities $\mathscr{S}_{nn}^{L}(\mathbf{q}, \omega)$ and $\mathscr{S}_{hn}^{L}(\mathbf{q}, \omega)$. Carrying out their solution, we obtain, using eqn (D.23), the desired expression for the structure factor,

$$\mathscr{S}_{nn}(\mathbf{q}, \omega)/S_{nn}(\mathbf{q}) = \left(1-\frac{C_V}{C_P}\right)\frac{2D_{\mathrm{T}}q^2}{\omega^2+(D_{\mathrm{T}}q^2)^2}$$

$$+\frac{C_V}{C_P}\left\{\frac{\frac{1}{2}D_{\mathrm{s}}q^2}{(\omega-v_{\mathrm{s}}q)^2+(\frac{1}{2}D_{\mathrm{s}}q^2)^2}+\frac{\frac{1}{2}D_{\mathrm{s}}q^2}{(\omega+v_{\mathrm{s}}q)^2+(\frac{1}{2}D_{\mathrm{s}}q^2)^2}\right\}, \quad (D.30)$$

where we have neglected higher-order terms involving the quantities $D_{\mathrm{T}}q/v_{\mathrm{s}}$ and $D_{\mathrm{s}}q/v_{\mathrm{s}}$. Equation (D.30) is sometimes credited to Landau and Placzek (1934), although they did not explicitly derive the result. More complete derivations of eqn (D.30) are presented in Kadanoff and Martin (1963) and Mountain (1966).

Suggested further reading

Kadanoff and Martin (1963).
Mountain (1966).

APPENDIX E

MODEL SYSTEMS USEFUL IN THE STUDY OF TIME-DEPENDENT COOPERATIVE PHENOMENA: THE GLAUBER MODEL

As we saw in Chapters 8 and 9, much of our understanding of the equilibrium aspects of critical phenomena arises from the study of model systems. We shall therefore consider in this appendix the use of mathematical models to study time-dependent aspects of cooperative behaviour. We shall see that although considerable progress has occurred in recent years, the current 'state of the art' as regards dynamic model systems is by no means as advanced as it is for static systems. We have already seen that the Ising model provides a useful description for physical systems in which a localized variable can take on either of two discrete values. Here we shall consider the adaptation of the Ising model to the treatment of dynamic phenomena; we shall follow the classic treatment of Glauber (1963).

E.1. The Glauber model for a single-spin system

We begin our discussion of time-dependent processes with an extremely simple system, a system consisting of a single Ising spin in the absence of a magnetic field. This is the case of eqn (8.1) for $D = 1$ and $N = 1$. We allow this system to interact with a heat reservoir which induces spontaneous flips of the spins from one state to the other, with a given transition probability per unit time, w. A complete description of the system consists of the knowledge of the probability $p(s, t)$ that the system is in state $s(s = \pm 1)$ at time t. Of course, it would actually be sufficient to calculate $p(+1, t)$, since $p(-1, t)$ is directly obtainable from the normalization requirement,

$$p(-1, t) + p(+1, t) = 1. \tag{E.1}$$

To calculate $p(s, t)$ we construct a differential equation and solve it. Now the time derivative of $p(s, t)$ is given by two terms. The first term arises from the possibility of a spin flipping out of state s, and hence is given by the product of the probability that the system is in state s, $p(s, t)$, and the transition probability per unit time w. This term enters with a minus sign because it serves to decrease the probability function $p(s, t)$. The second term arises from the possibility of a spin 'flipping in' to the state $p(s, t)$ from the other state, $p(-s, t)$. Hence the appropriate differential equation for our $N = 1$ Ising model is

$$\frac{\mathrm{d}}{\mathrm{d}t} p(s, t) = -w\, p(s, t) + w\, p(-s, t); \quad s = \pm 1. \tag{E.2}$$

Equation (E.2) is generally called the master equation (correctly speaking the master equation is a set of 2^N equations, one for each of the 2^N states of the system). As might be expected, exact solutions of the master equation are impossible except for extremely simple systems. Our single-spin system is such a system, however. To solve (E.2), it is convenient to introduce the variable

$$Q(t) = \sum_{j=1}^{2} s_j p(s_j, t), \tag{E.3}$$

where the summation in eqn (E.3) is over the $2^N = 2$ states of our system— that is, $Q(t)$ is just the expectation value of the spin. Hence in our $N = 1$ case,

$$Q(t) \equiv p(1, t) - p(-1, t). \tag{E.4}$$

Notice that a knowledge of $Q(t)$ is sufficient to obtain both $p(1, t)$ and $p(-1, t)$ through the relation

$$p(\pm 1, t) = \tfrac{1}{2}\{1 \pm Q(t)\}. \tag{E.5}$$

The differential equation obeyed by $Q(t)$ which is analogous to the master equations (E.2) is found by inspection to be

$$\frac{\mathrm{d}}{\mathrm{d}t} Q(t) = -2w\, Q(t), \tag{E.6}$$

whence

$$Q(t) = Q(0)\mathrm{e}^{-2wt}, \tag{E.7}$$

and the master equations are thereby solved. The solution, (E.7), tells us that the mean spin decays exponentially with a relaxation time $1/2w$ from the initial state (i.e. the state at time $t = 0$) to a final state of the system (at $t = \infty$) given by $Q(t = \infty) = 0$, which corresponds to a zero value for the mean spin. For more complicated Ising systems, there is reason to believe that the expectation values analogous to $Q(t)$ do not decay to their equilibrium ($t = \infty$) values with a single relaxation time (Stanley, Paul, Milošević, 1971).

E.2. The Glauber model for a spin system of N spins

The Ising Hamiltonian for a system of N spins is

$$\mathscr{H} \equiv - \sum_{j,\,k=1}^{N} J_{jk} s_j s_k - \bar{\mu} H \sum_{j=1}^{N} s_j, \tag{E.8}$$

where the first summation extends over all pairs of lattice sites j and k

excluding $j = k$, s_j is the dichotomic variable ($s_j = \pm 1$) for site j, J_{jk} is the interaction parameter between sites j and k, $\bar{\mu}$ is the magnetic moment per spin, and H is the external magnetic field.

A physical system described by the Ising Hamiltonian (E.8) can not change its state spontaneously and thus inherently has no dynamic properties. This is because all operators s_i commute with the Hamiltonian (E.8) and hence are constant as functions of time. Therefore in order to obtain a dynamic model system we must introduce external perturbations in addition to the Hamiltonian (E.8). With few exceptions, this has been done *not* by adding an explicit term to (E.8), but rather by a probabilistic argument first stated by Glauber (1963). It is assumed that the Ising system described by (E.8) is in contact with a heat bath that induces spontaneous flips of the spins from one state to the other. In a real system such perturbations are provided by other non-magnetic degrees of freedom—such as lattice vibrations— which are not taken into account in a purely magnetic Hamiltonian such as (E.8).

This heat bath is not treated explicitly—it is, however, assumed that there is a definite probability per unit time $w_j(s_1, s_2, \ldots s_N)$ that the jth spin flips from the value s_j to $-s_j$, and in this sense the Glauber model is stochastic. The dynamics of this model are thus determined once we specify the form of $w_j(s_1, \ldots s_N)$. We suppose that $w_j(s_1, \ldots s_n)$ does not depend on the previous history of the system. Although $w_j(s_1, \ldots s_N)$ depends on the configuration $\{s_1, \ldots s_N\}$ of the spin system, we shall write this transition probability per unit time as $w_j(s_j)$ for simplicity in what follows.

A complete statistical description of this dynamical Ising model would consist of the knowledge of the probability $p(s_1, s_2, \ldots s_N, t)$, that at time t the spin system is in the state $\{s_1, s_2, \ldots s_N\}$. The time dependence of this probability function is assumed to be governed by the master equation

$$\frac{d}{dt} p(s_1, s_2, \ldots s_N, t) = - \sum_{j=1}^{N} w_j(s_j) p(s_1, \ldots s_j, \ldots s_N, t) \qquad \text{(E.9)}$$

$$+ \sum_{j=1}^{N} w_j(-s_j) p(s_1, \ldots -s_j, \ldots s_N, t).$$

The first summation in (E.9) corresponds to the total number of ways that the system can 'flip out' of the state $\{s_1, s_2, \ldots s_N\}$, whereas the second summation corresponds to the total number of ways that the system can 'flip in' to the state $\{s_1, s_2, \ldots s_N\}$.

The detailed predictions of eqn (E.9) depend strongly upon our choice for the transition probability per unit time, $w_j(s_j)$. This choice will be guided by the requirement that the $w_j(s_j)$ function has such a form that it is capable of bringing our stochastic model to the same equilibrium configuration as that of the conventional static Ising model.

In equilibrium, the left-hand side of eqn (E.9) is, by definition, equal to zero. A stronger condition, however, is the principle of detailed balance, which asserts that the terms on the right-hand side of eqn (E.9) are equal to zero in pairs,

$$w_j(s_j)p_0(s_1, \ldots s_j, \ldots s_N) = w_j(-s_j)p_0(s_1, \ldots -s_j, \ldots s_N), \quad \text{(E.10)}$$

where $p_0(s_1, \ldots s_N)$ denotes the probability of finding the spins in the configuration $\{s_1, \ldots s_N\}$ when the system is in equilibrium.

If we observe that $p_0(s_1, \ldots s_N) \propto \exp(-\beta\mathcal{H})$, then eqns (E.8) and (E.10) lead to

$$\frac{w_j(s_j)}{w_j(-s_j)} = \frac{\exp(-\beta E_j s_j)}{\exp(\beta E_j s_j)} \quad \text{(E.11)}$$

where

$$E_j = \bar{\mu}H + \sum_{k=1}^{N} J_{jk}s_k. \quad \text{(E.12)}$$

Using the identity $\exp(\pm As_j) = \cosh A \pm s_j \sinh A = \cosh A(1 \pm s_j \tanh A)$, we can write (E.11) in the form

$$\frac{w_j(s_j)}{w_j(-s_j)} = \frac{1 - s_j \tanh \beta E_j}{1 + s_j \tanh \beta E_j}. \quad \text{(E.13)}$$

Following Suzuki and Kubo (1968), we choose for $w_j(s_j)$ a form that is consistent with (E.13),

$$w_j(s_j) = \frac{1}{2\alpha}(1 - s_j \tanh \beta E_j), \quad \text{(E.14)}$$

where the constant α^{-1} has the dimensions of inverse time and thus effectively determines the time scale of the dynamic processes. In order to extend our model to the case for which the system is in a time-dependent magnetic field, we will assume that the form of the transition probability in eqn (E.14) is unchanged but that it now becomes a function of time through the magnetic field imbedded in E_j (cf. eqn (E.12)).

In practice, we cannot calculate the probability functions $p(s_1, \ldots s_N, t)$ because of the complexity of the system of equations in (E.9). It is, however, possible to calculate certain expectation values such as

$$\langle s_j \rangle = \sum_{\{s\}} s_j p(s_1, \ldots s_N, t) \quad \text{(E.15)}$$

The summation in eqn (E.15) extends over all 2^N possible configurations of the spin system. It is straightforward to see from (E.9) and (E.14) that the expectation value $\langle s_j \rangle$ satisfies the differential equation

$$\alpha \frac{\mathrm{d}}{\mathrm{d}t}\langle s_j \rangle = -\langle s_j \rangle + \langle \tanh \beta E_j \rangle, \quad \text{(E.16)}$$

Thus the calculation problem has been reduced to the problem of solving differential equations of the sort (E.16), subject to certain initial conditions. This has been done exactly only in the case of one-dimensional lattices (Glauber 1963); for lattices of higher dimensionality we must make approximations in order to obtain any explicit predictions.

E.3. The molecular field approximation

The molecular field approximation provides one of the simplest approximation procedures in the study of dynamic cooperative phenomena, just as it does in the case of static cooperative phenomena (cf. Chapter 6). Following Suzuki and Kubo (1968), we assume that in eqn (E.12) we can make the approximation that

$$s_k = \langle s_k \rangle, \; k \neq j \tag{E.17}$$

and, moreover, that

$$\langle s_k \rangle = \langle s \rangle. \tag{E.18}$$

Equation (E.18) corresponds to assuming translational invariance of the spin system; that is, we assume that each spin behaves, on the average, like every other spin. This might correspond to the situation in which the system is initially in equilibrium and then is subjected to, for example, a sudden change of temperature (by placing it in contact with a new heat bath at a different temperature), or a sudden 'turning off' of a homogeneous magnetic field.

On substitution of eqns (E.17), (E.18), and (E.12) into eqn (E.16) we obtain the equation of motion in the molecular field approximation,

$$\alpha \frac{d}{dt} \langle s \rangle = - \langle s \rangle + \tanh \beta \bar{\mu}(H + \lambda \langle s \rangle), \tag{E.19}$$

where the molecular field parameter λ is defined through the relation

$$\lambda = \frac{1}{\bar{\mu}} \sum_{k=1}^{N} J_{jk}. \tag{E.20}$$

We notice that in equilibrium, $d\langle s \rangle/dt = 0$ and the equation of motion (E.19) reduces to the familiar molecular field equation of static cooperative phenomena

$$\langle s \rangle = \tanh \beta \bar{\mu}(H + \lambda \langle s \rangle). \tag{E.21}$$

When $H = 0$, eqn (E.21) possesses non-zero solutions only for temperatures less than a critical temperature

$$T_c = \frac{\bar{\mu}\lambda}{k}. \tag{E.22}$$

For temperatures very near the critical temperature and for small departures from equilibrium, we can approximate the hyperbolic tangent in eqn (E.19) by the first two terms in its Taylor expansion. Hence we find, for $H = 0$, that

$$\alpha \frac{d}{dt} \langle s \rangle = -\tilde{\epsilon} \langle s \rangle - Q \langle s \rangle^3, \tag{E.23}$$

where

$$\tilde{\epsilon} \equiv 1 - \frac{T_c}{T} = \frac{T_c}{T} \epsilon, \tag{E.24}$$

and $Q \equiv \frac{1}{3}(\beta m \lambda)^3$. Equation (E.20) is the Bernoulli differential equation and can be solved by dividing both sides by $\langle s \rangle^3$ and introducing the new quantity $y \equiv \langle s \rangle^{-2}$. Thus

$$\left(-\frac{1}{2}\right) \alpha \frac{d}{dt} y = -\tilde{\epsilon} y - Q. \tag{E.25}$$

The solution for $T \neq T_c$ is

$$\langle s \rangle = \left\{ \left(\frac{Q}{\tilde{\epsilon}} + \frac{1}{s_0^2}\right) e^{2\tilde{\epsilon}t/\alpha} - \frac{Q}{\tilde{\epsilon}} \right\}^{-\frac{1}{2}}, [T \neq T_c] \tag{E.26}$$

which becomes as $T \to T_c$,

$$\langle s \rangle = \left(s_0^{-2} + \frac{2Qt}{\alpha} \right)^{-\frac{1}{2}} [T = T_c], \tag{E.27}$$

where s_0 is the value of $\langle s \rangle$ when $t = 0$.

The qualitative behaviour predicted by eqn (E.26) is shown in Fig. E.1. for temperatures less than, equal to, and greater than the critical temperature; in particular, we see from eqn (E.26) that as we approach T_c the relaxation time approaches infinity inversely with $T - T_c$,

$$\tau = \frac{\alpha}{\tilde{\epsilon}} = \frac{\alpha T}{T - T_c}. \tag{E.28}$$

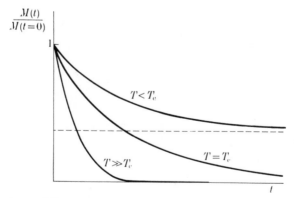

FIG. E.1 The decay of the average magnetization per spin $\langle M(t) \rangle / N$ by the molecular field approximation, eqn (E.26), for $T \lesssim T_c$, $T = T_c$, and $T \gtrsim T_c$.

This general behaviour predicted by the molecular field approximation can be compared with the computer simulation results of Ogita, Ueda, Matsubara, Matsuda, and Yonezawa (1969) (cf. Fig. 1.5). These authors have studied the dynamic critical phenomena of a ferroelectric system which, after certain simplifications, they represent by the time-dependent Ising model. We refer the interested reader to their work for a discussion of the extent to which the mean field theory is adequate.

Suggested further reading

Glauber (1963).
Suzuki and Kubo (1968).
Ogita, Ueda, Matsubara, Matsuda, and Yonezawa (1969).
Suzuki, Ikari, and Kubo (1969).
Yahata and Suzuki (1969).
Bedeaux, Milošević, and Paul (1971).
Njus and Stanley (1971).
Stanley, Paul, and Milošević (1971).

APPENDIX F

TWO-DIMENSIONAL FERROELECTRIC AND ANTIFERROELECTRIC MODELS

We present here a brief discussion of Lieb's exact solution of certain models of ferroelectric and antiferroelectric systems for the case of a two-dimensional lattice. Just as the magnetization of the Ising ferromagnet arises from the fact that the spins are capable of assuming only two distinct orientations, so also the polarization of a hydrogen bonded ferroelectric (such as potassium di-hydrogen phosphate, KH_2PO_4, commonly called KDP) is assumed to arise from the hydrogen atoms being capable of occupying a discrete number of individual sites. The PO_4 groups will be assumed to form a tetrahedral, or diamond, lattice, with coordination number $q = 4$. Now if there are N PO_4 groups, then there are $2N$ hydrogen atoms and also $Nq/2 = 2N$ bonds between nearest-neighbour pairs of PO_4 groups. We associate with each of these $2N$ nearest-neighbour bonds a single hydrogen atom, and we assume that there are two equivalent positions on each bond at which the hydrogen atom can be found. For the sake of specificity we can take these positions to be $\frac{1}{3}$ and $\frac{2}{3}$ of the distance between two nearest-neighbour PO_4 groups. Thus if we regard the PO_4 groups as vertices, then each vertex has $2^4 = 16$ possible configurations corresponding to the arrangement of the 4 hydrogen atoms on its four nearest-neighbour bonds.

Next we assume that two and only two of the four hydrogen atoms can be in the 'close' position; this is called the *ice constraint*. Hence the 16 possible configurations are reduced to the six shown in Fig. F.1, where we have re-presented a hydrogen atom in the 'close' position by an arrow directed toward the vertex. Notice that vertices 1, 2, 3, and 4 have a net polarization (shown by the dashed arrow), but the polarization direction of vertices 1 and 2 is perpendicular to that of vertices 3 and 4.

In the Slater KDP model of a ferroelectric, we regard the polarization axis of vertices 1 and 2 as being favourable, and hence we associate with vertices 3, 4, 5, and 6 a positive energy J (i.e., a Boltzmann factor $\exp(-J/kT)$ that is smaller than unity). The Rys model of an antiferroelectric (named the F-model after Rys' thesis advisor, M. Fierz) favours vertices 5 and 6, the two vertices with no net polarization, by associating with vertices 1, 2, 3, and 4 the positive energy J. The last line of Fig. F.1 shows the additional vertex energies which exist when the vertex is placed in an external vertical electric field.

The Slater KDP model of a ferroelectric and the Rys F model of an antiferroelectric have not been solved in the case of a three-dimensional diamond lattice, but they have been solved exactly—even in the presence of an external electric field—for the case of a two-dimensional square lattice (Lieb 1967a, b, 1969). Lieb's solution utilizes the transfer matrix method which we have illustrated in § 8.5; unfortunately the solution is complex so we shall not display the details or even the answer here.

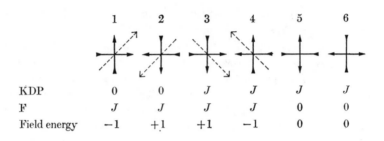

	1	2	3	4	5	6
KDP	0	0	J	J	J	J
F	J	J	J	J	0	0
Field energy	-1	$+1$	$+1$	-1	0	0

FIG. F.1. The six configurations of a vertex which satisfy the ice constraint, and the vertex energy assignments for the Slater KDP model of a ferroelectric and the Rys F model of an antiferroelectric. The net polarization of vertices 1, 2, 3, and 4 is indicated by the dashed arrow. The additional energy with an external electric field in the vertical direction is shown in units of electric field times dipole moment.

There are, however, many unusual and surprising features of the solution, some of which we shall now discuss. When the electric field $\mathscr{E}$ is zero, both models have identical ordering temperatures, given by $kT_c/J = (\ln 2)^{-1}$. Here the similarity with the Ising model ends. The KDP model of a ferroelectric undergoes a first-order phase transition at T_c, with the polarization being 'perfect' (equal to its $T = 0$ value) for all values of $T < T_c$. Therefore the specific heat is zero for $T < T_c$, but it diverges to infinity as $T \to T_c^+$, with a critical point exponent $\alpha = \frac{1}{2}$. Thus the Slater KDP model provides an exactly soluble example of a phase transition for which $\alpha' \neq \alpha$.

On the other hand, the F model of an antiferroelectric undergoes an infinite-order phase transition in that each and every derivative of the internal energy remains finite and continuous at $T = T_c$, yet the critical point is an essential singularity. In particular, then, the specific heat is a smoothly varying function of temperature, displaying a broad maximum at a temperature $T \simeq 0.8T_c$.

The effect of a non-zero external electric field is also surprising. Unlike the Ising ferromagnet, for which the phase transition disappears in the presence of an external magnetic field, the transition persists, but it becomes second order for both the ferroelectric and the antiferroelectric models and T_c changes smoothly with applied electric field. More precisely, T_c increases with the field for the ferroelectric and decreases with field for the antiferro-electric. Also the specific heat of the antiferroelectric diverges at $T_c(\mathscr{E})$ as $T \to T_c^+$ with $\alpha = \frac{1}{2}$, just as the ferroelectric model does for all values of $\mathscr{E}$.

These are but a few of the surprising and unusual features of the phase transitions in the two-dimensional Slater KDP model of a ferroelectric and the Rys F model of an antiferroelectric; the interested reader is referred to the review of Lieb (1969) for further discussion.

BIBLIOGRAPHY

ACZÈL, J. (1966) *Lectures on functional equations and their applications.* Academic Press, New York.

—— (1969) *On applications and theory of functional equations.* Academic Press, New York.

AHLERS, G. (1969) *Proc. eleventh int. Conf. low Temp. Phys.* Vol. 1 (eds. J. F. Allen, D. M. Finlayson, and D. M. McCall). St. Andrews Univ. Press, Scotland, p. 203.

ANDREWS, T. (1869) *Phil. Trans. R. Soc.* **159**, 575.

ARCHIBALD, M., MOCHEL, J. M., and WEAVER, L. (1969). *Proc. eleventh int. Conf. low Temp. Phys.* Vol. 1 (eds. J. F. Allen, D. M. Finlayson, and D. M. McCall). St. Andrews Univ. Press, Scotland, p. 211.

ARFKEN, G. (1970) *Mathematical methods for physicists*, 2nd edn. Academic Press, New York.

ARROTT, A. and NOAKES, J. E. (1967) *Phys. Rev. Lett.* **19**, 786.

BAKER, G. A., Jr. (1961) *Phys. Rev.* **124**, 768.

—— (1963a) *Phys. Rev.* **129**, 99.

—— (1963b) *Phys. Rev.* **130**, 1406.

—— (1965) *Adv. theor. Phys.* **1**, 1.

—— and GAUNT, D. S. (1967) *Phys. Rev.* **155**, 545.

—— GILBERT, H. E., EVE, J., and RUSHBROOKE, G. S. (1967) *Phys. Rev.* **164**, 800.

—— EVE, J., and RUSHBROOKE, G. S. (1970) *Phys. Rev.* **B 2**, 706

BARMATZ, M. and RUDNICK, I. (1968) *Phys. Rev.* **170**, 224.

BEDEAUX, D., MILOŠEVIĆ, S., and PAUL, G. (1971) *J. stat. Phys.* **3** (in press).

BENEDEK, G. B. (1962) In *Research in materials science and engineering, MIT CMSE Annual Report*, p. 55.

—— (1968) In *Statistical physics, phase transitions, and superfluidity*, Vol. 2 (eds. M. Chrétien, E. P. Gross, and S. Deser). Gordon and Breach, New York. p. 1.

—— (1969) In *Polarization, matter and radiation.* Presses Universitaire de France, Paris. p. 49.

—— and CANNELL, D. S. (1968) *Bull. Am. phys. Soc.* **13**, 182.

BERGÉ, P., CALMETTES, P., LAJ, C., TOURNARIE, M., and VOLOCHINE, B. (1970) *Phys. Rev. Lett.* **24**, 1223.

—— and VOLOCHINE, B. (1968) *Phys. Lett.* **26A**, 267.

BERLIN, T. H. and KAC, M. (1952) *Phys. Rev.* **86**, 821.

BETTS, D. D. and DITZIAN, R. V. (1968) *Can. J. Phys.* **46**, 971.

———— ———ELLIOTT, C. J., and LEE, M. H. (1971) *J. Phys., Paris* **32S** (in press).

—— ELLIOTT, C. J., and DITZIAN, R. V. (1971) *Can. J. Phys.* **49** (in press).

BIRGENEAU, R. J., SKALYO, J., and SHIRANE, G. (1970). *J. appl. Phys.* **41**, 1303.

BOTCH, W. and FIXMAN, M. (1962) *J. chem. Phys.* **36**, 3100.

———— —— (1965) *J. chem. Phys.* **42**, 199.

BOWERS, R. G. and WOOLF, M. E. (1969) *Phys. Rev.* **177**, 917.

BRAUN, P., HAMMER, D., TSCHARNUTER, W., and WEINZIERL, P. (1970) *Phys. Lett.* **32A**, 390.

BRILLOUIN, L. (1914) *C.r. habd. Séanc. Acad. Sci., Paris* **158**, 1331.

—— (1922) *Annls Phys.* **17**, 88.

BROUT, R. (1965) *Phase transitions*. Benjamin, New York.

BRUSH, S. G. (1967) *Rev. mod. Phys.* **39**, 883.

BUCKINGHAM, M. J. and FAIRBANK, W. M. (1961) In *Progress in low Temperature Physics*, Vol. 3 (ed. C. J. Gorter). North-Holland, Amsterdam, p. 80.

—— and GUNTON, J. D. (1969) *Phys. Rev.* **178**, 848.

BURGOYNE, P. N. (1963) *J. math. Phys.* **4**, 1320.

BURLEY, D. M. (1960) *Phil. Mag.* **5**, 909.

CALLEN, H. B. (1960) *Thermodynamics*. Wiley, New York.

CHANG, C. H. (1952) *Phys. Rev.* **88**, 1422.

CHEN, S. H. and POLONSKY, N. (1968) *Phys. Rev. Lett.* **20**, 909.

CHU, B., (1970) *Ann Rev. Phys. Chem.* **21**, 145.

—— SCHOENES, F. J., and KAO, W. P. (1968) *J. Am. chem. Soc.* **90**, 3042.

COLLINS, M. F., MINKIEWICZ, V. J., NATHANS, R., PASSELL, L., and SHIRANE, G. (1969) *Phys. Rev.* **179**, 417.

COOPER, M. J. (1968) *Phys. Rev.* **168**, 183.

COOPERSMITH, M. H. (1968*a*) *Phys. Rev.* **167**, 478.

—— (1968*b*) *Phys. Rev. Lett.* **20**, 432.

CRAIG, P. P. and GOLDBURG, W. I. (1969) *J. appl. Phys.* **40**, 964.

CUMMINS, H. Z. and GAMMON, R. W. (1966) *J. chem. Phys.* **44**, 2785.

—— KNABLE, N. and YEH, Y. (1964) *Phys. Rev. Lett.* **12**, 150.

—— and SWINNEY, H. L. (1966) *J. chem. Phys.* **45**, 4438.

———— ———— (1970) In *Progress in optics*, Vol. 8 (ed. E. Wolf). North-Holland, Amsterdam.

DALTON, N. W. and WOOD, D. W. (1968) *Phys. Lett.* **28A**, 417.

—— (1969) *J. math. Phys.* **10**, 1271.

DANIELIAN, A. and STEVENS, K. W. H. (1957) *Proc. phys. Soc.* **70**, 326.

DEBYE, P. J. W. (1912) *Annln Phys.* **39**, 789.

—— (1959) *J. chem. Phys.* **31**, 680.

DEUTCH, J. M. and ZWANZIG, R. (1967) *J. chem. Phys.* **46**, 1612.

DOMB, C. (1960) *Adv. Phys.* **9**, 149.

—— (1968) *Phys. Rev. Lett.* **20**, 1425.

—— (1970) *Adv. Phys.* **19**, 339.

—— and HUNTER, D. L. (1965) *Proc. phys. Soc.* **86**, 1147.

DUBIN, S. B. (1970) Ph.D. Thesis, M.I.T.

—— LUNACEK, J. H., and BENEDEK, G. B. (1967) *Proc. natn. Acad. Sci. U.S.A.* **57**, 1164.

DWIGHT, K., MENYUK, N., and KAPLAN, T. A. (1965) *J. appl. Phys.* **36**, 1090.

EGELSTAFF, P. A. (1967) *An introduction to the liquid state.* Academic Press, London.

EINSTEIN, A. (1910) *Annln Phys.* **33**, 1275.

ESSAM, J. W. and FISHER, M. E. (1963) *J. chem. Phys.* **38**, 802.

—— and HUNTER, D. L. (1968) *J. Phys.* (*C*) **1**, 392.

—— and SYKES, M. F. (1963) *Physica* **29**, 378.

FAY, J. A. (1965) *Molecular thermodynamics.* Addison–Wesley, Reading, Massachusetts.

FELDERHOF, B. U. (1966) *J. chem. Phys.* **44**, 602.

FERER, M., MOORE, M. A., and WORTIS, M. (1969) *Phys. Rev. Lett.* **22**, 1382.

FERRELL, R. A. (1968) In *Fluctuations in superconductors* (eds. W. S. Goree and F. Chilton). Stanford Research Institute.

—— (1970) *Phys. Rev. Lett.* **24**, 1169.

—— MENYHÁRD, N., SCHMIDT, H., SCHWABL, F., and SZÉPFALUSY, P. (1967a) *Phys. Rev. Lett.* **18**, 891.

—— —— —— —— —— (1967b) *Phys. Lett.* **24A**, 493.

—— —— —— —— —— (1968) *Ann. Phys., N.Y.* **47**, 565.

FISHER, M. E. (1962) *Physica* **28**, 172.

—— (1964) *J. math. Phys.* **5**, 944.

—— (1965) In *Lectures in theoretical physics*, Vol. 7C. University of Colorado Press, Boulder, Colorado.

—— (1967) *Rep. Prog. Phys.* **30**, 615.

—— (1969) *Phys. Rev.* **180**, 594.

—— and BURFORD, R. J. (1967) *Phys. Rev.* **156**, 583.

FIXMAN, M. (1962a) *J. chem. Phys.* **36**, 310.

—— (1962b) *J. chem. Phys.* **36**, 1961.

—— (1964) In *Advances in chemical physics*, Vol. I (ed. I. Prigogine) Interscience, New York.

—— (1967) *J. chem. Phys.* **47**, 2808.

FORD, N. C., Jr. and BENEDEK, G. B. (1965) *Phys. Rev. Lett.* **15**, 649.

—— LANGLEY, K. H., and PUGLIELLI, V. G. (1968) *Phys. Rev. Lett.* **21**, 9.

FORRESTER, A. (1956) *Am. J. Phys.* **24**, 192.

—— (1961) *J. opt. Soc. Am.* **51**, 253.

—— GUDMUNDSEN, R. A., and JOHNSON, P. O. (1955) *Phys. Rev.* **99**, 1691.

—— PARKINS, W. E., and GERJUOY, E. (1947) *Phys. Rev.* **72**, 728.

FRENCH, M. J., ANGUS, J. C., and WALTON, A. G. (1969) *Science, N.Y.* **163**, 345.

GAMMEL, J., MARSHALL, W., and MORGAN, L. (1963) *Proc. R. Soc.* **275**, 257.

GAMMON, R. W., SWINNEY, H. L., and CUMMINS, H. Z. (1967) *Phys. Rev. Lett.* **19**, 1467.

GARELICK, H. and ESSAM, J. W. (1968) *J. Phys.* (*C*) **1**, 1588.

GARLAND, C. W. (1970) In *Physical acoustics*, Vol. 7 (ed. W. P. Mason and R. N. Thurston). Academic Press, New York.

GAUNT, D. S. (1967) *Proc. phys. Soc.* **92**, 150.

—— and DOMB, C. (1968) *J. Phys.* (*C*) **1**, 1038.

—— FISHER, M. E., SYKES, M. F., and ESSAM, J. W. (1964) *Phys. Rev. Lett.* **13**, 713.

GIGLIO, M. and BENEDEK, G. B. (1969) *Phys. Rev. Lett.* **23**, 1145.

GLASSER, M. L. (1970) *Am. J. Phys.* **38**, 1033.

GLAUBER, R. J. (1963) *J. math. Phys.* **4**, 294.

GOLDING, B. (1968) *Phys. Rev. Lett.* **20**, 5.

GORELIK, G. (1947) *Dokl. Akad. Nauk SSSR* **58**, 45.

GREYTAK, T. J. (1967) Ph.D. Thesis, M.I.T.

GRIFFITHS, R. B. (1964) *J. math. Phys.* **5**, 1215.

—— (1965*a*) *Phys. Rev. Lett.* **14**, 623.

—— (1965*b*) *J. chem. Phys.* **43**, 1958.

—— (1967*a*) *J. math. Phys.* **8**, 478, 484.

—— (1967*b*) *Phys. Rev.* **158**, 176.

—— (1968) Unpublished lectures delivered at the 1968 Banff Summer Institute on Critical Phenomena.

—— HURST, C. A., and SHERMAN, S. (1970) *J. math. Phys.* **11**, 790.

GROSS, E. (1930a) *Nature, Lond.* **126**, 201.

—— (1930*b*) *Nature, Lond.* **126**, 400.

—— (1930*c*) *Nature, Lond.* **126**, 603.

—— (1932) *Nature, Lond.* **129**, 722.

GUGGENHEIM, E. A. (1945) *J. chem. Phys.* **13**, 253.

—— (1967) *Thermodynamics*, 5th edn. North-Holland, Amsterdam.

GUNTON, J. D. and BUCKINGHAM, M. J. (1968) *Phys. Rev.* **166**, 152.

GUTTMANN, A. J. (1969*a*) Ph.D. Thesis, University of New South Wales.

—— (1969*b*) *J. Phys.* (*C*) **2**, 1900.

—— and THOMPSON, C. J. (1969) *Phys. Lett.* **28A**, 679.

—— —— and NINHAM, B. W. (1970) *J. Phys.* (*C*) **3**, 1641.

HABGOOD, H. W. and SCHNEIDER, W. G. (1954) *Can. J. Chem.* **32**, 98.

HALPERIN, B. I. and HOHENBERG, P. C. (1967) *Phys. Rev. Lett.* **19**, 700.
—— —— (1969) *Phys. Rev.* **177**, 952.
HANDLER, P., MAPOTHER, D. E., and RAYL, M. (1967) *Phys. Rev. Lett.* **19**, 356.
HANKEY, A. and STANLEY, H. E. (1971) *to be published.*
HEISENBERG, W. (1928) *Z. Phys.* **49**, 619.
HELFAND, E. (1969) *Phys. Rev.* **183**, 562.
HELLER, G. and KRAMERS, H. A. (1934) *Proc. Sect. Sci. K. ned. Akad. Wet.* **37**, 378.
HELLER, P. (1967) *Rep. Prog. Phys.* **30**, 731.
—— and BENEDEK, G. B. (1962) *Phys. Rev. Lett.* **8**, 428.
HENRY, D. L., CUMMINS, H. Z., and SWINNEY, H. L. (1969) *Bull. Am. phys. Soc.* **14**, 73.
—— SWINNEY, H. L., and CUMMINS, H. Z. (1970) *Phys. Rev. Lett.* **25**, 1170.
HILEY, B. J. and JOYCE, G. S. (1965) *Proc. phys. Soc.* **85**, 493.
HO, J. T. and LITSTER, J. D. (1969) *Phys. Rev. Lett.* **22**, 603.
HOPKINSON, J. (1890) *Proc. R. Soc.* **48**, 1.
HUANG, K. (1963) *Statistical mechanics.* Wiley, New York.

ISING, E. (1925) *Z. Phys.* **31**, 253.

JANCOVICI, B. (1967) *Phys. Rev. Lett.* **19**, 20.
JASNOW, D. and WORTIS, M. (1968). *Phys. Rev.* **176**, 739.
JOSEPHSON, B. D. (1967) *Proc. phys. Soc.* **92**, 269, 276.
—— (1969) *J. Phys.* (*C*) **2**, 1113.
JOYCE, G. S. (1966) *Phys. Rev.* **146**, 349.
—— and BOWERS, R. G. (1966a) *Proc. phys. Soc.* **88**, 1053.
—— —— (1966b) *Proc. phys. Soc.* **89**, 776.

KAC, M. (1964). *Physics today* **17**, No. 10, p. 40.
—— (1968) In *Statistical physics, phase transitions, and superfluidity*, Vol. 1 (eds. M. Chrétien, E. P. Gross, and S. Deser). Gordon and Breach, New York, p. 241.
—— and THOMPSON, C. J. (1971) *Proc. Norwegian Acad. Science* (in press).
—— and WARD, J. C. (1952) *Phys. Rev.* **88**, 1332.
—— UHLENBECK, G. E., and HEMMER, P. C. (1963) *J. math. Phys.* **4**, 216.
KADANOFF, L. P. (1966) *Physics* **2**, 263.
—— (1968) *Communs Solid St. Phys.* **1**, 5.
—— (1969) *J. phys. Soc. Japan* **26S**, 122.
—— and MARTIN, P. C. (1963) *Ann. Phys., N.Y.* **24**, 419.
—— and SWIFT, J. (1968a) *Phys. Rev.* **165**, 310.
—— —— (1968b) *Phys. Rev.* **166**, 89.

—— GÖTZE, W., HAMBLEN, D., HECHT, R., LEWIS, E. A. S., PALCIAUSKAS, V. V., RAYL, M., SWIFT, J., ASPNES, D., and KANE, J. (1967) *Rev. Mod. Phys.* **39**, 395.

KAPLAN, T. A., STANLEY, H. E., DWIGHT, K., and MENYUK, N. (1965) *J. appl. Phys.* **36**, 1129.

KAUFMAN, B. and ONSAGER, L. (1949) *Phys. Rev.* **76**, 1244.

KAWASAKI, K. (1966a) *Phys. Rev.* **145**, 224.

—— (1966b) *Phys. Rev.* **148**, 375.

—— (1966c) *Phys. Rev.* **150**, 285.

—— (1966d) *Phys. Rev.* **150**, 291.

—— (1967) *J. Phys. Chem. Solids* **28**, 1277.

—— (1968a) *Prog. theor. Phys., Kyoto* **39**, 1133.

—— (1968b) *Prog. theor. Phys., Kyoto* **40**, 11.

—— (1968c) *Prog. theor. Phys., Kyoto* **40**, 706.

—— (1968d) *Prog. theor. Phys., Kyoto* **40**, 930.

—— (1968e) *Prog. theor. Phys., Kyoto* **39**, 285.

—— (1968f) *Phys. Lett.* **26A**, 543.

—— (1969) *Prog. theor. Phys., Kyoto* **41**, 1190.

—— (1970) *Ann. Phys., N.Y.* **61**, 1.

—— and TANAKA, M. (1967) *Proc. phys. Soc.* **90**, 791.

KELLY, D. G. and SHERMAN, S. (1968) *J. math. Phys.* **9**, 466.

KERRESK, J. and KELLER, W. E. (1967) *Bull. Am. phys. Soc.* **12**, 550.

KITTEL, C. (1958) *Elementary statistical physics.* Wiley, New York.

—— (1969) *Thermal physics.* Wiley, New York.

—— and SHORE, H. B. (1965) *Phys. Rev.* **138**, A1165.

KOMAROV, L. I. and FISHER, I. Z. (1962) *Zh. éksp. teor. Fiz.* **43**, 1927. English translation: *Soviet Phys. JETP* **16**, 1358 (1963).

KOUVEL, J. S. and COMLY, J. B. (1968) *Phys. Rev. Lett.* **20**, 1237.

—— and RODBELL, D. S. (1967) *Phys. Rev. Lett.* **18**, 215.

KRAMERS, H. A. (1936) In *Collected scientific papers of H. A. Kramers* (ed. H. B. G. Casimir). North-Holland, Amsterdam, 1965, p. 60.

—— and WANNIER, G. H. (1941) *Phys. Rev.* **60**, 252–263.

KUBO, R. (1943) *Busseiron–Kenkyu*, **1**, 1.

—— (1965) *Statistical mechanics.* North-Holland, Amsterdam.

—— (1968) *Thermodynamics.* North-Holland, Amsterdam.

KURAMOTO, Y. (1969) *Prog. theor. Phys., Kyoto* **41**, 604.

LANDAU, L. D. (1937a) *Phys. Z. Sowjun.* **11**, 26.

—— (1937b) *Zh. éksp. teor. Fiz.* **7**, 19.

—— (1937c) *Phys. Z. Sowjun.* **11**, 545.

—— (1937d) *Zh. éksp. teor. Fiz.* **7**, 627.

—— (1965) *Collected papers of L. D. Landau* (ed. D. ter Haar). Gordon and Breach, New York.

LANDAU, L. D., AKHIEZER, A. I., and LIFSHITZ, E. M. (1967) *General physics*. Pergamon Press, New York.

—— and LIFSHITZ, E. M. (1960) *Electrodynamics of continuous media*. Pergamon Press, Oxford.

—— —— (1969) *Statistical physics*, 2nd edn. Pergamon Press, Oxford.

—— and PLACZEK, G. (1934) *Phys. Z. Sowjun.* 5, 172.

LANGEVIN, P. (1905) *J. Phys. Radium, Paris* 4, 678.

LARAMORE, G. (1969) Ph.D. Thesis, University of Illinois.

LASTOVKA, J. B. (1967) Ph.D. Thesis, M.I.T.

—— and BENEDEK, G. B. (1966) *Phys. Rev. Lett.* 17, 1039.

LAU, H. Y., CORLISS, L. M., DELAPALME, A., HASTINGS, J. M., NATHANS, R., and TUCCIARONE, A. (1969) *Phys. Rev. Lett.* 23, 1225.

LEBOWITZ, J. (1968) *Ann Rev. Phys. Chem.* 19, 389.

LEE, M. H. and STANLEY, H. E. (1971) *J. Phys., Paris* 32S (in press).

LEE, T. D. and YANG, C. N. (1952). *Phys. Rev.* 87, 410.

LENZ, W. (1920) *Phys. Z.* 21, 613.

LIEB, E. H. (1967a) *Phys. Rev. Lett.* 18, 1046.

—— (1967b) *Phys. Rev. Lett.* 19, 108.

—— (1969) In *Lectures in theoretical physics*, Vol. 11D. University of Colorado Press, Boulder, Colorado.

—— and MATTIS, D. C. (1966) *Mathematical physics in one dimension*. Academic Press, New York.

LINES, M. E. (1971). *J. Phys., Paris* 32S (in press).

LUTHI, B. and POLLINA, R. J. (1968) *Phys. Rev.* 167, 488.

—— —— (1969) *J. Phys. Chem. Solids* 31, 1741.

MANDEL'SHTAM, L. I. (1926) *Zhurnal Russkogo Fiziko Khimicheskogo Obshchestva* 58, 381.

MARSHALL, W. and LOVESEY, S. W. (1971) *Theory of thermal neutron scattering*. Clarendon Press, Oxford (in Press).

—— and LOWDE, R. D. (1968) *Rept. Prog. Phys.* 31, 705.

MARTIN, D. H. (1967) *Magnetism in solids*. M.I.T. Press, Cambridge, Massachusetts.

MATSUNO, K. and STANLEY, H. E. (1970) *Phys. Lett.* 33A, 425.

MATTIS, D. C. (1965) *The theory of magnetism*. Harper and Row, New York.

MENDELSSOHN, K. (1966) *The quest for absolute zero*. McGraw-Hill, New York.

MERMIN, N. D. and WAGNER, H. (1966) *Phys. Rev. Lett.* 17, 1133.

MIGDAL, A. A. (1968) *Zh. éksp. teor. Fiz.* 55, 1964. English translation: *Soviet Phys. JETP* 28, 1036 (1969).

MILOŠEVIĆ, S., MATSUNO, K., and STANLEY, H. E. (1970). *Physica Status Solidi* 42 (in press).

—— and STANLEY, H. E. (1971) *J. Phys., Paris* 32S (in press).

MINKIEWICZ, V. J., COLLINS, M. F., NATHANS, R., and SHIRANE, G. (1969) *Phys. Rev.* **182**, 624.

MOLDOVER, M. R. (1969) *Phys. Rev.* **182**, 342.

MONTROLL, E. W. (1968) In *Statistical physics, phase transitions, and superfluidity*, Vol. 2 (eds. M. Chrétien, E. P. Gross, and S. Deser). Gordon and Breach, New York.

MOORE, M. A. (1969) *Phys. Rev. Lett.* **23**, 861.

—— JASNOW, D., and WORTIS, M. (1969) *Phys. Rev. Lett.* **22**, 940.

MORSE, P. (1969) *Thermal physics*, 2nd edn. Benjamin, New York.

MOUNTAIN, R. D. (1966) *Rev. mod. Phys.* **38**, 205.

—— and ZWANZIG, R. (1968) *J. chem. Phys.* **48**, 1451.

MUBAYI, V. and LANGE, R. V. (1969) *Phys. Rev.* **178**, 882.

MÜNSTER, A. (1966) In *Fluctuation phenomena in solids* (ed. R. E. Burgess). Academic Press, New York.

—— and SAGEL, K. (1958) *Molec. Phys.* **1**, 23.

NATHANS, R., MENZINGER, F., and PICKART, S. J. (1968) *J. appl. Phys.* **39**, 1237.

NJUS, D. L. and STANLEY, H. E. (1971) In *Proceedings of the Fordham conference on the dynamical aspects of critical phenomena* (eds. J. I. Budnick and M. P. Kawatra). Gordon and Breach, N.Y.

OGITA, N., UEDA, A., MATSUBARA, T., MATSUDA, H., and YONEZAWA, F. (1969) *J. phys. Soc. Japan* **26S**, 145.

ONSAGER, L. (1931a) *Phys. Rev.* **37**, 405.

—— (1931b) *Phys. Rev.* **38**, 2265.

—— (1944) *Phys. Rev.* **65**, 117.

OPECHOWSKI, W. (1937) *Physica* **4**, 181.

ORNSTEIN, L. S. and ZERNIKE, F. (1914) *Proc. Sect. Sci. K. med. Akad. Wet.* **17**, 793.

PATASHINSKII, A. Z. and POKROVSKII, V. L. (1966) *Zh. éksp. teor. Fiz.* **50**, 439. English translation: *Soviet Phys. JETP* **23**, 292 (1966).

PAUL, G. and STANLEY, H. E. (1971) *J. Phys., Paris* **32S** (in press).

PECORA, R. (1964) *J. chem. Phys.* **40**, 1604.

POKROVSKII, V. L. (1968) *Usp. Fiz. Nauk.* **94**, 127. English translation: *Soviet Phys. Uspekhi* **11**, 66 (1968).

POLYAKOV, A. M. (1968) *Zh. éksp. teor. Fiz.* **55**, 1026 English translation: *Soviet Phys. JETP* **28**, 533 (1969).

LORD RAYLEIGH (1871) *Phil. Mag.* **41**, 107, 274, 447.

—— (1899) *Phil. Mag.* **47**, 375.

REATTO, L. (1968) *Phys. Lett.* **26A**, 400.

RECHTIN, M. D., MOSS, S. C., and AVERBACH, B. L. (1970) *Phys. Rev. Lett.* **24**, 1485.

REIF, F. (1965) *Statistical and thermal physics*. McGraw–Hill, New York.

REITZ, J. R. and MILFORD, F. J. (1960) *Foundations of Electromagnetic theory*. Addison-Wesley, Reading, Massachusetts.

ROACH, P. R. (1968) *Phys. Rev.* **170**, 213.

—— and DOUGLASS, D. H., Jr. (1967) *Phys. Rev. Lett.* **19**, 287.

ROWLINSON, J. S. (1969) *Liquids and liquid mixtures*, 2nd edn. Butterworths, London.

RUSHBROOKE, G. S. (1963) *J. chem. Phys.* **39**, 842.

—— (1965) *J. chem. Phys.* **43**, 3439.

—— and WOOD, P. J (1958) *Molec. Phys.* **1**, 257.

RYTOV, S. M. (1957) *Zh. éksp. teor. Fiz.* **33**, 514. English translation *Soviet Phys. JETP* **6**, 401 (1958).

SCHOFIELD, P. (1969) *Phys. Rev. Lett.* **22**, 606.

SCHULHOF, M. P. (1970) *Ph. D. Thesis*, Brandeis University.

—— HELLER, P., NATHANS, R., and LINZ, A. (1970a) *Phys. Rev.* **B1**, 2304.

—— —— —— —— (1970b) *Phys. Rev. Lett.* **24**, 1184.

SCHULTZ, T. D., MATTIS, D. C., and LIEB, E. H. (1964). *Rev. mod. Phys.* **36**, 856.

SENGERS, J. V. (1966) In *Critical phenomena, Proceedings of a Conference* (eds. M. S. Green and J. V. Sengers), Natn. Bur. Stand. Misc. Publ. No. 273.

—— (1971) In *Proceedings of the Varenna summer school on critical phenomena* (ed. M. S. Green). Academic Press, New York.

—— and LEVELT-SENGERS, J. M. H. (1968) *Chem. Engng News* **46**, 104.

SIEGEL, L. and WILCOX, L. R. (1967) *Bull. Am. phys. Soc.* **12**, 525.

SKALYO, J., SHIRANE, G., FRIEDBURG, S. A., and KOBAYASHI, H. (1970). *Phys. Rev. B.* **2**, 1310.

SMART, J. S. (1966) *Effective field theories of magnetism*. Saunders, Philadelphia

STANLEY, H. E. (1967a) *Phys. Rev.* **158**, 537.

—— (1967b) *Phys. Rev.* **158**, 546.

—— (1967c) *Phys. Rev.* **164**, 709.

—— (1967d) Ph.D. Thesis, Harvard University.

—— (1968a) In *Solid state physics, nuclear physics and particle physics*, The ninth Latin–American School of Physics (ed. I. Saavedra). Benjamin, New York.

—— (1968b) *Phys. Rev. Lett.* **20**, 150.

—— (1968c) *Phys. Rev. Lett.* **20**, 589.

—— (1968d) *Phys. Rev.* **176**, 718.

—— (1969a) *J. phys. Soc. Japan* **26S**, 102.

—— (1969b) *J. appl. Phys.* **40**, 1272.

—— (1969c) *J. appl. Phys.* **40**, 1546.

—— (1969d) *Phys. Rev.* **179**, 570.

—— (1971a) *Scient. Am.*

—— (1971b) In *Critical phenomena, proceedings of the Battelle conference* (eds. R. E. Mills, E. Ascher, and R. I. Jaffee). McGraw Hill, New York.

—— (1971c) In *Proc. NATO summer institute on mathematical physics* (ed. A. O. Barut), Gordon and Breach, New York.

—— (1971d) In *Proc. NATO summer institute on magnetism* (ed. S. Foner). Gordon and Breach, New York.

—— (1971e) *Readings in phase transitions and critical phenomena* (to be published).

—— BLUME, M., MATSUNO, K., and MILOŠEVIĆ, S. (1970) *J. appl. Phys.* **41**, 1278.

—— HANKEY, A., and LEE, M. H. (1971) In *Proc. Varenna summer school on critical phenomena* (ed. M. S. Green). Academic Press, New York.

—— and KAPLAN, T. A. (1966a) *Phys. Rev. Lett.* **16**, 981.

—— —— (1966b) *Phys. Rev. Lett.* **17**, 913.

—— —— (1967a) *J. Appl. Phys.* **38**, 975.

—— —— (1967b) *J. appl. Phys.* **38**, 977.

—— and LEE, M. H. (1970) *Int. J. quantum Chem.* **4S**, 407.

—— PAUL, G. and MILOŠEVIĆ, S. (1971) In *Physical chemistry, an advanced treatise, vol. 8 (the liquid state)* (eds. H. Eyring, D. Henderson, and W. Jost). Academic Press, New York.

STELL, G. (1968a) *Phys. Rev. Lett.* **20**, 533.

—— (1968b) *Phys. Rev.* **173**, 314.

—— (1969) *Phys. Rev.* **184**, 135.

STEPHENSON, J. (1964) *J. math. Phys.* **5**, 1009.

STEPHENSON, R. L. and WOOD, P. J. (1968) *Phys. Rev.* **173**, 475.

—— —— (1970) *J. Phys. (C), Solid State Physics* **3**, 90.

SU, G. J. (1946) *Ind. Engng. Chem. analyt. Edn* **38**, 803.

SUZUKI, M. and KUBO, R. (1968) *J. phys. Soc. Japan* **24**, 51.

—— IKARI, H., and KUBO, R. (1969) *J. phys. Soc. Japan* **26S**, 153.

SWIFT, J. and KADANOFF, L. P. (1968) *Ann. Phys. (N.Y.)* **50**, 312.

SWINNEY, H. L. (1968) Ph.D. Thesis, Johns Hopkins University.

—— and CUMMINS, H. Z. (1968) *Phys. Rev.* **171**, 152.

SYKES, M. F. (1963) *J. chem. Phys.* **39**, 410.

—— ESSAM, J. W., and GAUNT, D. S. (1965) *J. math. Phys.* **6**, 283.

—— and FISHER, M. E. (1962) *Physica* **28**, 919.

—— MARTIN, J. L., and HUNTER, D. L. (1967) *Proc. phys. Soc.* **91**, 671.

—— and ZUCKER, I. J. (1961) *Phys. Rev.* **124**, 410.

TEMPERLEY, H. N. V. (1956) *Changes of state*. Cleaver–Hume and Interscience.

THOMAS, J. E. and SCHMIDT, P. W. (1963) *J. chem. Phys.* **39**, 2506.

THOMPSON, C. J. (1971) *Mathematical statistical mechanics*, Macmillan, New York.

TOWNES, C. H. (1961) In *Advances in quantum electronics* (ed. J. S. Singer). Columbia University Press, New York.

VAKS, V. G. and LARKIN, A. I. (1965) *Zh. éksp. teor. Fiz.* **49**, 975. English translation: *Soviet Phys. JETP* **22**, 678 (1966).

VAN DER WAALS, J. D. (1873) Ph.D. Thesis, University of Leiden.

VAN HOVE, L. (1954a) *Phys. Rev.* **93**, 268.

—— (1954b) *Phys. Rev.* **95**, 249.

—— (1954c) *Phys. Rev.* **95**, 1374.

VAN KAMPEN, N. G. (1969) In *Quantum optics, Varenna lecture notes* (ed. R. J. Glauber). Academic Press, New York.

VAN VLECK, J. H. (1937) *J. chem. Phys.* **5**, 320.

—— (1945) *Rev. mod. Phys.* **17**, 27.

VDOVICHENKO, N. V. (1964) *Zh. éksp. teor. Fiz.* **47**, 715. (English translation: *Soviet Phys. JETP* **20**, 477 (1965).)

—— (1965) *Zh. éksp. teor. Fiz.* **48**, 526. English translation: *Soviet Phys. JETP* **21**, 350 (1965).

VILLAIN, J. (1968a) *J. Phys., Paris* **29**, 321.

—— (1968b) *J. Phys., Paris* **29**, 687.

—— (1968c) *Phys. Stat. Sol.* **26**, 501.

VINCENTINI-MISSONI, M., LEVELT-SENGERS, J. M. H., and GREEN, M. S. (1969) *J. Res. Natn. Bur. Stnds. Sect. A* **73**, 563.

WANNIER, G. H. (1966) *Statistical physics.* Wiley, New York.

WATSON, R. E., BLUME, M., and VINEYARD, G. H. (1970) *Phys. Rev. B* **2**, 684.

WEGNER, F. (1967) *Z Phys.* **206**, 465.

WEISS, P. (1907) *J. Phys. Radium, Paris* **6**, 667.

—— and FORRER, R. (1926) *Annls Phys.* **5**, 153.

WIDOM, B. (1964) *J. chem. Phys.* **41**, 1633.

—— (1965a) *J. chem. Phys.* **43**, 3892.

—— (1965b) *J. chem. Phys.* **43**, 3898.

—— (1967) *Science, N.Y.* **157**, 375.

WOOD, P. J. (1958) Ph.D. Thesis, University of Newcastle.

—— and RUSHBROOKE, G. S. (1966) *Phys. Rev. Lett.* **17**, 307.

WU, T. T. (1966) *Phys. Rev.* **149**, 380.

YAHATA, H. and SUZUKI, M. (1969) *J. phys. Soc. Japan,* **27**, 1421.

YANG, C. N. (1952) *Phys. Rev.* **85**, 808.

—— and YANG, C. P. (1964) *Phys. Rev. Lett.* **13**, 303.

ZIMAN, J. M. (1964) *Principles of the theory of solids.* Cambridge University Press.

AUTHOR INDEX

SUBJECT INDEX